Economics and Management of Climate Change

Bernd Hansjürgens · Ralf Antes
Editors

Economics and Management of Climate Change

Risks, Mitigation and Adaptation

sponsored by
Stiftungsfonds Dresdner Bank
im Stifterverband für die Deutsche Wissenschaft

Editors
Bernd Hansjürgens
Helmholtz Center
for Environmental Research - UFZ
Leipzig, Germany
bernd.hansjuergens@ufz.de

Ralf Antes
Carl von Ossietzky University
Oldenburg, Germany
ralf.antes@uni-oldenburg.de

ISBN: 978-0-387-77352-0   e-ISBN: 978-0-387-77353-7

Library of Congress Control Number: 2008930110

© 2008 Springer Science+Business Media, LLC
All rights reserved. This work may not be translated or copied in whole or in part without the written permission of the publisher (Springer Science+Business Media, LLC, 233 Spring Street, New York, NY10013, USA), except for brief excerpts in connection with reviews or scholarly analysis. Use in connection with any form of information storage and retrieval, electronic adaptation, computer software, or by similar or dissimilar methodology now known or hereafter developed is forbidden.
The use in this publication of trade names, trademarks, service marks, and similar terms, even if they are not identified as such, is not to be taken as an expression of opinion as to whether or not they are subject to proprietary rights.

Printed on acid-free paper

9 8 7 6 5 4 3 2 1

springer.com

Preface and acknowledgements

This is the third book of our common project on Corporate Sustainability, funded by the Dresdner Bank Stiftungsfonds im Stifterverband für die Deutsche Wissenschaft. The first two books[1] focused on aspects of Emissions Trading at the interface of economics and business administration. This book takes a broader perspective and focuses on climate change, risks, and responses of society and business.

As is well-known, climate change is one of the biggest challenges for mankind. Although there is increasing evidence that climate change is already occurring, there is neither sufficient knowledge to what extent climate change poses risks to societies and companies, nor about adequate strategies to cope with these risks. Bringing together scholars from environmental economics, political science, and business management, this book describes, analyses and evaluates climate change risks and responses of societies and companies. The book contributes to the question of how climate change can be mitigated by discussing efficient and effective design of mitigation measures, in particular emissions trading and clean development mechanism. Placing special emphasis on the impact of climate change risks on business, the book investigates in what way selected sectors of the economy are affected and what measures they can undertake to adapt to climate change risks.

The book is based on a workshop held at the Leucorea, the former University of Halle-Wittenberg, Germany, from 28th to 30th November 2005. This workshop brought together young and established scholars from different disciplines and from different countries to discuss topics of climate change, risks, mitigation and adaptation. For the publication of the book the contributions of the participants were reviewed, revised and updated.

We would like to thank all those who have contributed to both the conference and the book. First of all, our thanks go to the participants and the authors of this book. David McKenzie and Paul Ronning did a great effort in improving the English style of those chapters which were not written by native speakers. Production of the book would not have been possible without the efforts of Sabine Linke and Susann Dietsch in assembling the text, tables, graphs, and reference from the various authors into a coherent manuscript. Kay Fiedler provided the

---

[1] Antes R, Hansjürgens B, Lethmate P (eds) Emissions Trading and Business, Heidelberg/Berlin, Springer-Physica 2006; Antes R, Hansjürgens B, Lethmate P (eds) Emissions Trading – Institutional Design, Decision Making and Corporate Strategies, Heidelberg/Berlin, Springer 2008.

layout, as he did for the two previous books. Last but not least, our special thanks go to Armin Sandhövel and Heike Heuberger from Allianz/Dresdner Bank for their strong support for the whole enterprise.

Leipzig / Oldenburg   February 2008

Bernd Hansjürgens and Ralf Antes

Contents

Preface and acknowledgements V

1 Introduction: Climate change risks, mitigation and adaptation 1
  Bernd Hansjürgens, Ralf Antes

Part I – Climate change risks and risk assessment

2 Rising natural catastrophe losses – what is the role of climate change? ... 13
  Peter Höppe, Tobias Grimm

3 Evaluation of risk in cost-benefit analysis of climate change 23
  Ingrid Nestle

4 A global climate change risk assessment of droughts and floods 37
  Wolfgang Knorr, Marko Scholze

5 Effects of low water levels on the river Rhine on the inland
  waterway transport sector 53
  Olaf Jonkeren, Jos van Ommeren, Piet Rietveld

6 Spillover impacts of climate policy on energy-intensive industry 65
  Vlasis Oikonomou, Martin Patel, Ernst Worrell

7 Risks, vulnerability, and participation: a layered management
  approach 79
  J Terry Rolfe

Part II – Mitigation: emissions trading and CDM

8   Intensity targets: implications for the economic uncertainties of emissions trading 97
Sonja Peterson

9   Three types of impact from the European Emission Trading Scheme: direct cost, indirect cost and uncertainty 111
Volker H Hoffmann, Thomas Trautmann

10 Participants' treatment of allowance price uncertainty: how are risk-aversion and real option values related to each other? 125
Frank Gagelmann

11 Climate change, sustainable development and risk: realizing a financial fund within the TEM model as an economic and business opportunity 145
Stefan Pickl

12 Current evaluation practice of the Clean Development Mechanism 157
Felicia Müller-Pelzer

13 Sustainable development, the Clean Development Mechanism, and business accounting 175
Pala Molisa, Bettina Wittneben

14 Risk management in the Clean Development Mechanism (CDM) – the potential of sustainability labels 193
Adrian Muller

15 Economic and social risks associated with implementing CDM projects among SME – a case study of foundry industry in India 209
Prosanto Pal, Girish Sethi

Part III – Adaptation: corporate responses to climate change

16 Developments in corporate responses to climate change within the past decade 221
Ans Kolk

17 Impacts of climate change on the electricity sector and possible adaptation measures 231
Benno Rothstein, Marion Schroedter-Homscheidt, Claudia Häfner, Susanne Bernhardt, Solveig Mimler

18 Modelling impacts of climate change policy uncertainty on power investment 243
Ming Yang, William Blyth

19 Reputational impact of businesses' compliance strategies under the EU emissions trading scheme 257
Corinne Faure, Arne Hildebrandt, Karoline Rogge, Joachim Schleich

20 Prudence, profit and the perfect storm: climate change risk and fiduciary duty of directors 271
Donna Lorenz

21 Business risks and opportunities from climate change in large developing countries – a case study focusing on China 293
Xianli Zhu, Xiangyang Wu

Introduction: Climate change risks, mitigation and adaptation

Bernd Hansjürgens[I,II], Ralf Antes[III]

[I] Helmholtz Center for Environmental Research – UFZ
Department of Economics
Permoserstraße 15, 04318 Leipzig, Germany
bernd.hansjuergens@ufz.de

[II] Martin Luther University Halle-Wittenberg
Faculty of Law and Economics
Professorship of Environmental Economics
Große Steinstraße 73, 06099 Halle, Germany

[III] Carl von Ossietzky University Oldenburg
Faculty II – Informatics, Economics and Law
CENTOS – Oldenburg Center for Sustainable Economics and Management
Uhlhornsweg 26, 26129 Oldenburg, Germany
ralf.antes@uni-oldenburg.de

Keywords: Climate change, risk, vulnerability, mitigation, adaptation, emissions trading, Clean Development Mechanism

1 Climate change, risks and risk assessment

Climate change is a fact. The latest report of the International Panel on Climate Change (IPCC), published in spring 2007, clearly indicates that climate change is already occurring and will definitely impact life on Earth. The 4th IPCC report does not change the messages of its predecessors. Instead it confirms the statements of the previous IPCC reports and underlines the fact that there is now more scientific evidence and a higher likelihood of the occurrence of climate change impacts. It is also stated that climate change is definitely caused by human influence. Major results put forward by the IPCC report are (IPCC 2007):

- Warming of the climate system is unequivocal, as is now evident from observations of increases in global average air and ocean temperature, widespread melting of snow and ice, and rising average sea level.
- Many natural systems are being affected by regional climate changes, particularly by temperature increases.
- Atmospheric concentrations of carbon dioxide (CO_2), methane (CH_4) and nitrous oxide (N_2O) have increased markedly as a result of human activities in the last 250 years and now exceed by far pre-industrialised values.
- Most of the observed increase in globally-averaged temperatures since the mid-20th century is very likely due to the observed increase in anthropogenic greenhouse gas concentrations. It is likely that there has been significant anthropogenic warming over the last 50 years averaged over each continent.
- Anthropogenic warming over the last three decades has likely had a discernable influence at the global scale on observed changes in many physical and biological systems.

It is important to note that it is not only the increase of the average temperature which causes problems to mankind, but in particular the expected increase in extreme weather events, such as floods, storms, long-term extreme droughts, hurricanes and so on. Figures and calculations of the Munich Re's Geo Risks Research reveal that an increase in natural catastrophes can already be observed and that the scientific evidence for human influence is increasing (Höppe and Grimm 2008). Although the strongest effects of climate change are expected in Africa and parts of Asia, Europe and the US will also be victims of these changes. For many authors major recent extreme events like Hurricane Katrina in the US or the winter storm Kyrill in Europe are seen as a forerunner of increasing extreme weather changes.

These developments cause *risks* to societies. Following the classical work by Frank Knight (1921), the definition of risk is based on two factors: (1) the expected damage a hazard can cause and (2) the probability of the occurrence of the damage. The problem is that in the field of climate change ex ante neither the damage nor the probability function is given. Due to high uncertainties in this field the assessment of risks is by nature characterised by incomplete knowledge or even a total lack of knowledge. In the field of climate change in most cases neither the damage function nor the probability of an extreme event are known. The problem of how to assess risks is thus of major concern.

In addition, it can be expected that climate change risks do not affect all members of the society equally. Instead, various societal groups as well as various sectors of the economy are affected differently. A decisive factor to determine the degree to which a person, a societal group or an economic sector is affected is its *vulnerability*. Here the working definition of vulnerability developed by Blaikie et al. (1994, p. 9) is helpful. In their definition vulnerability is "the characteristics of a person or a group in terms of their capability to anticipate, cope with, resist, and recover from the impact of a natural hazard." Instead of emphasising solely characteristics of the natural or technological hazard itself or the exposure (structure, building etc.) to the hazard, this definition focuses on the question of how communities and social groups are able to deal with the impact of a natural hazard. Hence, it is not so much the susceptibility of a community or a social group to a specific hazard that is of interest but the coping capacity, hence active behaviour, in a very general sense. It is obvious and has not to be repeated explicitly that vulnerability does not only relate to communities and social groups (although existing literature focuses on these actors), but also to companies.

Part I of this book focuses on climate change risks, risk assessment and vulnerability, with a special focus on business companies.

2 The Kyoto Protocol: instruments for mitigating climate change

Climate change research does not only analyse the changes in climate and their effects on society and companies but also mitigation and adaptation options for societies. The third part of the IPCC's Fourth Assessment Report explicitly deals with strategies and measures to combat and to adapt to global warming.

However, research on climate change had for a long time a one-sided focus on mitigation measures. It was (and in many cases still is) the reduction of greenhouse gases which was in the centre of climate change strategies. This perspective strongly influenced the international discussion on climate change as well as the design of institutions and organisations. It was in particular the 1997 Kyoto Conference which paved the way in the discussion of climate protection. Its final document, the Kyoto Protocol, established "quantified emission limitation and reduction commitments" to OECD countries and some economies in transition ("Annex I countries"). This heralded a completely new tack in climate policy: whereas the need to cut greenhouse gas emissions had already been acknowledged at the United Nations World Summit in Rio de Janeiro in 1992, it was at the Kyoto Conference that specific reduction targets for signatory countries were laid down for the first time.

Another direction brought about by the Kyoto Protocol was the introduction of new policy instruments for climate protection, namely the Clean Development Mechanism (CDM), Joint Implementation (JI) and emissions trading (ET). Since then ET in particular has become a widely discussed instrument for climate policy. One reason for the attention emissions trading has received is that it had already been the subject of intense debate in the United States owing to the introduction of several national US programmes in the late 1980s and 1990s (Ellerman et al. 2000; Ellerman et al. 2003; Hansjürgens 2005). Since 2004 also CDM has received increasing attention. This instrument contributes to both cost-effective environmental policy (by offering companies in industrialized countries opportunities to reduce emissions) and sustainable development (by defining adequate projects in developing countries). Although there are high transaction costs, this instrument plays an increasing role in climate change policy.

During recent years, intensive discussion has raged over these new policy instruments for climate protection. Recent developments in Europe and the United States have revealed that environmental economists' positions have started to converge regarding the choice of instrument, especially the usefulness of ET and CDM as instruments for climate policy. This is in particular apparent from the new EU proposal for a CO_2 emissions trading system in Europe, which was published in 2003 (European Union 2003). The European CO_2 trading scheme (ETS) is the largest emissions trading system worldwide and constitutes a "grand policy experiment".[1] Since then we observe a great number of further initiatives all over the world seeking to implement the instrument of emissions trading for climate protection (Antes, Hansjürgens, Lethmate 2008).

Thus ET and CDM form an integral part of the ongoing discussion of climate change policy – with companies being the major target of cost-efficient reduction of greenhouse gases. Therefore these instruments will be analysed in Part II of this book. As the instruments have been analysed in general, the contributions in this book focus on specific design options, and thus complement existing literature.[2]

3 The new challenge: adaptation as a response to climate change

In contrast to mitigation, the possibility for adaptation to climate change is seen as a topic which was for a long time strongly rejected by both climate change researchers and policy-makers. There was considerable concern that a far-reaching discussion on adaptation would reduce the pressure on mitigation – and thus foster (or at least accept) the ongoing anthropogenic emission of greenhouse gases into the atmosphere.

---

[1] It was Stavins (1998) who spoke of the US SO_2 trading scheme as a "grand policy experiment". As the EU CO_2 trading comprises 10,000–12,000 sources, it will exceed the SO_2 trading by a factor of five in terms of covered installations. See also Kruger and Pizer (2004); Hansjürgens (2005, p. 2).

[2] Thus Part II will in particular follow the lines of the two predecessors of this book: Antes, Hansjürgens, Lethmate (eds) (2006) and (2008). Both volumes deal with emissions trading, placing special emphasis on institutional design and corporate strategies.

However, this perspective has changed. Today, in addition to strategies and measures referring to mitigation, adaptation is seen as an essential and integral part of climate policy. Hence, the above mentioned Forth IPCC Report deals with both, mitigation and adaptation. In the current discussion on climate change, there are (at least) three arguments which are put forward in favour of adaptation (Pielke et al. 2007):

- Different time scales: It is quite obvious that reducing greenhouse gas emissions will take a long time – even if an international agreement would be reached in the near future. Historic emissions of the industrialised countries are responsible for the fact that climate change is already occurring. It will take a long time to change this path. In the words of IPCC (2007, p. 6): "There is high agreement and much evidence that with current climate change mitigation policies and related sustainable development practices, global GHG emissions will continue to grow over the next few decades." In the meantime, societies have no choice other than to adapt to the impacts of climate change.
- *Societies' vulnerability.* The vulnerability of many societies has increased during recent years. Responsible factors are, e.g., demographic change, access to water resources, social inequality, urbanisation, wars etc. Although these developments occur irrespective of climate change, they make societies more vulnerable to climate change. An increase in the coping capacities of societies (individuals, companies, regions) could improve their resilience and thus protect them from negative climate change impacts.
- *Stronger influence of those who are mainly affected by climate change.* At international summits the participating states, in particular states mainly affected by climate change impacts, increasingly demand strategies and measures for adaptation enabling them to cope with climate change impacts. This demand is put forward irrespective of the extent of mitigation which is still required in order to avoid the most severe impacts of global warming. This argument is particularly put forward by developing countries as affected states. These countries state that instead of spending money on potential threats to future generations one should spend the money on the improvement of the living conditions of the present generation.

Taking these arguments into account it becomes obvious that adaptation does not only refer to "passive" measures, i.e. measures that are taken as a reaction to an external and unavoidable trend, but also includes "active" strategies and measures which aim at making societies and companies more robust and less vulnerable. The final goal of such adaptation strategies and measures is thus to improve resilience of societies and companies

As companies form an integral part of the society and determine its capability to cope with climate change impacts, the contributions in Part III of this book deal with societal and in particular companies' responses to climate change

4 Overview of the book

The three parts of this book follow the line of argument presented in this introduction. The contributions in Part I focus on climate change risks, risk assessment and vulnerability. Part II includes contributions that deal with mitigation instruments, in particular design options for emissions trading and CDM. The contributions in Part III analyse the responses of societies and companies to climate change, placing special emphasis on companies' mitigation strategies and measures.

Following this introductory chapter 1, **Part I – Climate change risks and risk assessment** consists of six contributions:[3] (2) *Peter Höppe* and *Tobias Grimm* analyze to what extent the occurrence of rising natural catastrophes is caused by climate change. They show that natural catastrophes and damages have been increasing dramatically and causing more damages over time. As the upward trend in numbers of natural catastrophes is mainly due to weather-related events like windstorms and floods and is not apparent in the same way for events with geophysical causes like earthquakes, tsunamis, and volcanic eruptions, there is some justification for assuming that it is the result of changes in the atmosphere, most probably global warming. (3) *Ingrid Nestle* explores the evaluation of risk in cost-benefit analysis of climate change. In order to derive information about discounting in cost-benefit analysis, she develops an experiment on the de facto risk aversion of the sample population in the context of climate change. (4) *Wolfgang Knorr* and *Marko Scholze* analyse the risk assessment for droughts and floods. On the basis of an ensemble of climate model simulations they compute annual soil water run-off, and define as a risk a situation where the mean during 2071–2100 deviates from the 1961–1990 mean by more than one standard deviation of the current variability. They find that in general larger degrees of warming lead to larger areas and populations affected, with substantial risks for tropical countries even at low degrees of global warming. (5) *Olaf Jonkeren, Jos van Ommeren* and *Piet Rietveld* examine the welfare effect of water level variation on the river Rhine through its impact on freight prices in the inland waterway transport market. The authors demonstrate that freight price increases up to 100% at the lowest water levels compared to the situation of normal water levels. They estimate that the welfare loss for the economy in 2003 as a result of the increase in freight prices due to low water levels amounts to about € 172 million. (6) *Vlasis Oikonomou, Martin Patel* and *Ernst Worrell* focus on "spillover effects" of climate policy for energy-intensive industries. Energy-intensive industries are particularly vulnerable if climate policy would lead to higher energy costs, and if these industries would be unable to offset these increased costs. The authors find out that energy and carbon intensity of energy-intensive industries is rapidly declining in most developing countries, and reducing the "gap" between industrialized and developing countries. (7) Referring to the concept of Integrated Water resource Management (IWRM) in Canada, *J Terry Rolfe* develops a proposal to reduce vulnerability of watersheds by a layered management approach. This approach is building upon

---

[3] Note that the numbers in brackets represent the chapter numbers of this book.

consensus-building and creative problem resolution where communities themselves become leaders in adaptation, and moves away from traditional (technical) optimization models.

The eight contributions in **Part II – Mitigation: emissions trading and CDM** deal with mitigation measures, focusing in particular on intensity targets, emissions trading, and CDM. (8) *Sonja Peterson,* in a first paper, deals with intensity targets. Especially in the US such intensity targets are seen as an alternative to absolute reduction targets within a certain time horizon as has been prescribed by the Kyoto Protocol ('targets and timetables'). Peterson puts the existing theoretical and empirical results about intensity targets and uncertainty into perspective and augments them by additional data and findings. (9) *Volker H Hoffmann* and *Thomas Trautmann* identify three types of impact from the European Emission Trading Scheme on the affected companies: direct impact, indirect impact, and uncertainty. While direct impact refers to the cost for buying allowances, indirect impact refers to the increased input factor cost as suppliers price in their emission cost. Uncertainty refers to the limited planning reliability as many details of the regulation are still under negotiation or only last for a few years. Based on a Europe-wide survey the authors empirically show the relevance and industry dependence of these types of impact to a company's strategic decisions. (10) *Frank Gagelmann,* in his contribution on effects of different allocation methods of an Emissions Trading Scheme on participants' risk aversion, shows that under free allocation according to historic emissions a higher price risk is likely to lead to an aggregate reduction of abatement investment at any point in time. Innovative investment is also reduced, plausibly to an even stronger extent than investment in general. The following four contributions in Part II all deal with CDM. (11) The possibility of establishing a fund supporting the aims of the Kyoto Protocol is currently one of the central topics for discussion. Based on a mathematical approach which is based on the Technology Emissions Means (TEM) model, *Stefan Pickl* analyses how such a fund can be realized and designed and how it can be embedded in an optimal energy management. (12) *Felicia Müller-Pelzer* analyses the current evaluation practise of CDM project activities. She comes to the conclusion that the two main CDM goals – the achievement of additional emission reductions and the contribution to sustainable development in the host country – are not awarded equal attention. In her view an approach is needed that enables project developers to design and implement a sustainability strategy for their CDM project activities and to demonstrate their contribution to sustainable development. (13) Coming from the same starting point – that sustainable development is neglected in SDM projects –, *Pala Molisa* and *Bettina Wittneben* make proposals to improve the effectiveness of CDM by using alternative accounting procedures. Such accounting measures should foster not only the effectiveness of CDM, but also offer information about the degree to which CDM projects contribute to sustainable development. (14) In contrast, *Adrian Muller* sees sustainability labelling as a promising strategy to hedge against failures of CDM projects. Such labels could also serve as a business risk-hedging tool. (15) Finally, in a last paper, *Prosanto Pal* and *Girish Sethi* analyse economic and social risks associated with the implementation of CDM projects within a case study; they focus on the foundry industry in India.

In the six contributions of **Part III – Adaptation: corporate responses to climate change** the authors deal with various company responses and adaptation to climate change challenges. In a first paper, (16) *Ans Kolk* provides an overview of the research on business and climate change in the past decade as carried out by the author, with a particular focus on multinational companies. It covers a synopsis of policy developments, companies' political and market responses to climate change, as indicates areas for future research. (17) *Benno Rothstein et al.* analyse adaptation strategies to climate change in the energy sector. Within the framework of the CLIMAGY project (Climate & Energy: Adaptation Measures for Recent Climate Trends within the Energy Sector) the authors work out a ranking of vulnerabilities of electricity companies in the course of climate change and identify possible adaptation measures. The results show that, on the one hand, the electricity sector is being affected by climate change and weather risks in almost all business units (e.g. water availability for electricity generation). On the other hand, several options for actions are already being developed to help reduce the described impacts (e.g. load management). (18) *Ming Yang* and *William Blyth* present a computer model currently developed by the International Energy Agency (IEA) to quantify the impacts of climate change policy uncertainties on power investment. The methodologies used include the discounted cash flow approach to calculating project net present value, stochastic simulation to capture the characteristics of uncertain variables, and real option valuation to capture investors' flexibility to optimize the timing of their investment. The authors apply these modelling methodologies in a case study to evaluate the effects of the changing carbon prices on firms' decisions to invest in more energy efficiency power technologies. They conclude that the more uncertain the primary energy prices and carbon trading prices are, the more the economic case for lower emitting technologies deviates from the traditional discounted cash flow. (19) In order to assess the relevance of reputational effects on companies' climate strategies, *Corinne Faure et al.* conducted a survey of the 300 largest emitting companies in Germany. Results indicate that reputational effects matter, but are dominated by other factors such as compliance costs, technical and economic risks, or practicability. Long-run reputational risks are most relevant for the CDM and JI mechanisms. (20) *Donna Lorenz*, in her paper, analyses climate change risks and the fiduciary duty of company directors in Australia. She gives an overview of recent developments in climate change policy in Australia and describes the different risks which have to be managed by company director. (21) And, finally, *Zhu and Wu* conduct a case study on China about the climate change-related business risks and opportunities. They show that business risks and opportunities associated with climate change in large developing countries are different from those in the developed world.

References

Antes R, Hansjürgens B, Lethmate P (eds) (2006) Emissions Trading and Business. Berlin and Heidelberg, Springer/Physica Publishers

Antes R, Hansjürgens B, Lethmate P (eds) (2008) Emissions Trading – Institutional Design, Decision Making and Corporate Strategies. Berlin and Heidelberg, Springer Publishers

Antes R, Hansjürgens B, Lethmate P (2008) Introduction, In: Antes R, Hansjürgens B, Lethmate P (eds) Emissions Trading – Institutional Design, Decision Making and Corporate Strategies. Berlin and Heidelberg, Springer Publishers, pp 1–8

Blaikie P, Cannon T, Wisner B (1994) At Risk: Natural Hazards, People's Vulnerability, and Disaster. London and New York, Routledge

Ellerman AD, Joskow PL, Schmalensee R, Montero JP, Bailey E (2000) Markets for Clean Air. The US Acid Rain Program. Cambridge, Cambridge University Press

Ellerman AD, Joskow P, Harrison D (2003) Emissions Trading in the US: Experiences, Lessons, and Considerations for Greenhouse Gases. Washington, DC, PEW Center on Global Climate Change

European Union (2003) Directive 2003/87/EC of the European Parliament and of the Council of 13. October 2003 establishing a system for greenhouse gas emissions trading within the European Community and amending Council Directive 96/61/EC. ABL 275. Brussels, 25. October 2003

Hansjürgens B (ed) (2005) Emissions Trading for Climate Policy, Cambridge, Cambridge University Press

Hansjürgens B (2005) Introduction. In: Hansjürgens B (ed) Emissions Trading for Climate Policy. Cambridge, Cambridge University Press, pp 1–14

Höppe P, Grimm T (2008) Rising natural catastrophe losses – what is the role of climate change? In this volume: pp 13–22

International Panel on Climate Change (IPCC) (2007) Fourth Assessment Report. Climate Change 2007: Synthesis Report. Summary for Policymakers

Knight, F. (1921): Risk, Uncertainty, and Profit. New York

Kruger J, Pizer WA (2004) The EU Emissions Trading Directive: Opportunities and Potential Pitfalls. Discussion paper 04–24, Washington, DC, Resources for the Future

Pielke R, Prins G, Rayner S, Sarewitz D (2007) Lifting the taboo on adaptation. In: Nature, 8 February 2007, vol 445: 597–598

Stavins RN (1998) "What can we learn from the grand policy experiment? Lessons from SO_2 allowance trading". In: Journal of Economic Perspectives 12: 69–88

Part I

Climate change risks and risk assessment

Rising natural catastrophe losses – what is the role of climate change?

Peter Höppe[1], Tobias Grimm[1]

[1] Geo Risks Research Department, Munich Re
80971 Munich, Germany
phoeppe@munichre.com
TGrimm@munichre.com

Abstract

For more than 30 years Munich Re scientists have been analysing natural hazards throughout the world. Munich Re's NatCatSERVICE now has records of more than 22,000 single natural events having caused damages. Analyses of these data show very clearly that natural catastrophes have increased dramatically and are causing more and more damage. Inflation-adjusted economic and insured losses from these great natural catastrophes have risen to nearly US$ 180 bn in economic losses and around US$ 90 bn in insured losses in the record year of 2005. As the upward trend in numbers of natural catastrophes is mainly due to weather-related events such as windstorms and floods and is not apparent in the same way for events with geophysical causes such as earthquakes, tsunamis, and volcanic eruptions, there is some justification for assuming that this trend is the result of changes in the atmosphere, most probably global warming. Our analyses of the hurricane frequency in recent decades, taking into account the natural climate cycles (Multidecadal Atlantic Oscillation), indicates that current activity of strong hurricanes (SS 3 to 5) is higher than in any other previous period; this suggests an impact from global warming. Evidence for a significant impact of climate change in several sectors is undoubted. In the economy, climate change is no longer seen exclusively as a financial risk, but it also has been stated as offering a great opportunity.

Keywords: Natural catastrophes, climate change, rise in losses, risks and opportunities, insurance industry

1 Rising natural catastrophes and economic losses

In recent years, there have been increasing signs that the steady advance of global warming is progressively affecting the frequency and intensity of natural catastrophes. The following examples confirm that there has been a notable increase in such events over the past few years.

- The hundred-year flood in the Elbe region in the summer of 2002;
- The 450-year event of the hot summer of 2003, which caused more than 35,000 heat deaths in Europe;
- The record damages of the 2004 hurricane season;
- Japan's 2004 typhoon season, with an unprecedented ten landfalls;
- The first ever South Atlantic hurricane in March 2004, with damages in Brazil;
- India's highest 24-hour precipitation amount: 944 mm in Mumbai on 26 July 2005;
- 2005 the largest number of tropical cyclones (28) and hurricanes (15) in a single North Atlantic season since the beginning of records (1851);
- The 2005 hurricane season included the strongest (Wilma – core pressure: 882 hPa), fourth strongest (Rita), and sixth strongest (Katrina) hurricanes on record;
- Hurricane Katrina was the costliest single event of all times, with economic losses of over US$ 125 bn and insured losses of approximately US$ 60 bn;
- In October 2005, Hurricane Vince formed close to Madeira, subsequently reaching the northernmost and easternmost point of any tropical cyclone;
- In November 2005, tropical storm Delta became the first tropical storm ever to reach the Canary Isles;
- Larry, the strongest tropical storm (cyclone) recorded, reached the Australian coast in March 2006:
- Kyrill (January 2007) caused the second largest losses in Europe due to a winter storm;
- The flood series in the United Kingdom (June-July 2007) resulted in the largest flood loss ever in the UK.

Munich Re's Geo Risks Research unit has been researching loss events caused by natural hazards around the globe for over 30 years. These events are documented in the NatCatSERVICE database, which has been fed with data on all the major historic natural catastrophes. Munich Re's NatCatSERVICE now contains details of more than 23,000 individual events. The analyses undertaken by Geo Risks Research provide the most accurate estimate possible of the insured values exposed to natural hazards such as windstorm, flood and earthquake with a view to Munich Re's business.

The data analyses clearly show a dramatic increase in natural catastrophes around the globe, with ever-growing losses. The trend curve indicating the number of great natural catastrophes worldwide (involving thousands of fatalities, billion-dollar losses) reveals an increase from two per year at the beginning of the 1950s to around seven at the present time (Fig. 1).

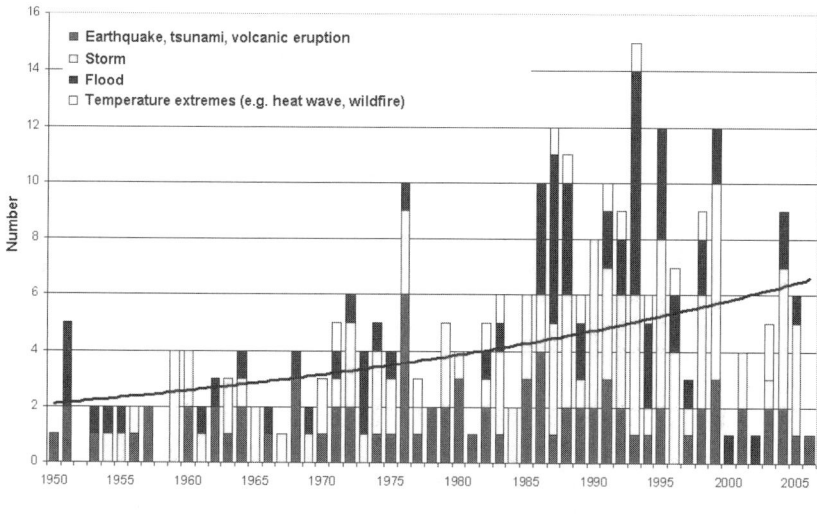

Fig. 1. Great Natural Disasters 1950–2006 (Number of events)

Economic and insured losses resulting from great weather disasters have risen even more sharply in real terms. In 2005, a record year, economic losses were as high as nearly US$ 180 bn and insured losses around US$ 90 bn (Fig. 2).

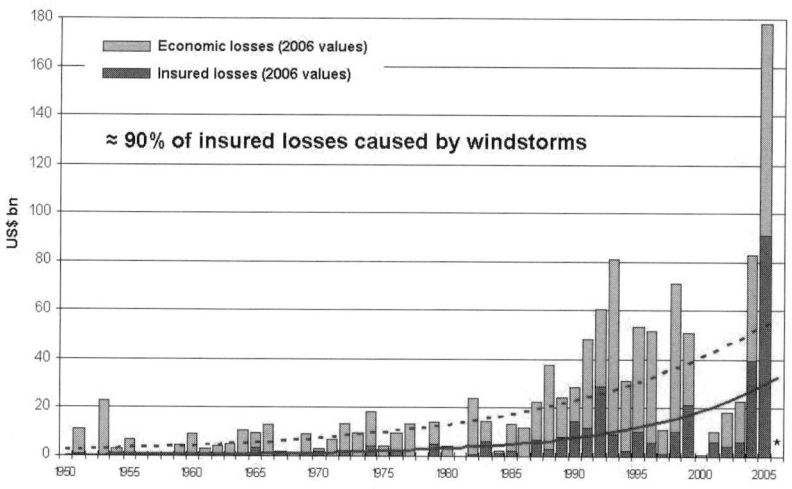

Fig. 2. Great Weather Disasters 1950–2006 (Overall and insured losses)

The main reasons for the sharp increase in losses from major, weather-related catastrophes are population growth, the settlement and industrialisation of regions with high exposure levels and the fact that modern technologies are more prone to loss. The state of Florida in the USA, which has always had a high hurricane exposure, is a good illustration of the way that socio-economic factors can act as natural catastrophe loss drivers. The population has grown from three million in 1950 to the current 18 million. The number of tourists visiting Florida each year recently passed the 80 million mark. It is clear, taking into account the increase in prosperity, that present-day hurricane losses in Florida are liable to be a multiple of those of a few decades ago.

Following 2005's record figures, the insurance industry reported relatively few large natural catastrophe losses in 2006. At the end of December 2006, economic losses from all loss occurrences amounted to US$ 45 bn and insured losses US$ 15 bn, less than one-sixth of the previous year's figure. The loss balance would have been higher had it not been for the fortuitous absence of severe North Atlantic hurricanes.

Only three Atlantic tropical cyclones caused losses in 2006, compared with 17 in the previous year. The lower level of hurricane activity was due to exceptional meteorological circumstances: dust particles carried from the Sahara to the hurricane breeding grounds absorbed solar radiation, thus warming the surrounding layer of air at medium altitude. The effect was to stabilise atmospheric stratification and hinder the formation of hurricanes, particularly during August. From October onwards, the El Niño phenomenon in the Pacific had a curbing effect. However, during September, in the absence of either El Niño or the Sahara dust factor, there were four hurricanes, which corresponds with expectations. A number of storms were steered away into the Atlantic by the dominant configuration of pressure systems without reaching the mainland, and so did not cause damage. This clearly shows that 2006 constitutes no more than a temporary respite in the general increase in weather-related natural catastrophes.

2 Increasing scientific evidence for anthropogenic causes

As the rise in the number of natural catastrophes is largely attributable to weather-related events like windstorms and floods (see Fig. 1), with no evidence of a similar increase in geophysical events such as earthquakes, tsunamis, and volcanic eruptions, there is some justification in assuming that anthropogenic changes in the atmosphere, and climate change in particular, play a decisive role. There has been more and more evidence to support this hypothesis in recent years:

- Analyses of air bubbles trapped in ice cores drawn from deep layers in the Antarctic ice suggest that the concentration of carbon dioxide, the principal greenhouse gas, over the past 650,000 years has never been even remotely close to the current 382 ppm (Siegenthaler et al. 2005).

- The ten warmest years on record since 1856, when systematic readings were first taken, have all been in the twelve-year period 1995–2006 (WMO 2007). The warmest year to date was 1998.

The fourth status report of the Intergovernmental Panel on Climate Change (IPCC 2007) regards the link between global warming and the greater frequency and intensity of extreme weather events as significant. The report finds, with more than 66% probability, that climate change already produces more heatwaves, heavy precipitation, drought and intense tropical storms and that the trend is rising. The expected rise in global average temperatures of up to 6.4°C by the end of the century, depending on emission and climate model, significantly increases the probability of record temperatures. Higher temperatures also enable air to hold more water vapour, thus increasing precipitation potential. Combined with more pronounced convection processes, in which warm air rises to form clouds, this results in more frequent and more extreme intense precipitation events. Even now, such events are responsible for a large proportion of flood losses. As a result of the milder winters now typical of central Europe, there has been a reduction in the snow cover over which stable, cold high-pressure systems used to form a barrier against low-pressure systems coming in from the Atlantic. This barrier now tends to be weak or to be pushed eastwards so that devastating winter storm series like those of 1990 and 1999 can no longer be considered exceptional, as also evidenced recently by Kyrill in January 2007. The wind readings of a number of representative German weather stations have shown a definite increase in number of storm days over the past three decades. At Düsseldorf Airport, for instance, the figure has risen from about 20 to 35 a year (Otte 2000).

In recent years, an increasing number of scientific publications have indicated that there is a causal link between climate change and the frequency and intensity of weather-related natural catastrophes:

- According to British scientists, it is more than 90% probable that the influence of human activity has at least doubled the risk of a heatwave like the one that hit Europe in 2003 (Stott et al. 2004).
- Hurricane models which take account of climate change show that, by 2050, maximum hurricane speeds will have increased by an average of 0.5 on the Saffir-Simpson Scale and the associated precipitation volume will have gone up by 18% (Knutson and Tuleya 2004).
- Publications by Emanuel (2005) and Webster et al. (2005) indicate a 50% increase in the duration and intensity of tropical storms in the North Atlantic and Northwest Pacific since 1970. This trend is expected to continue.
- The surface temperature of the world's oceans in the tropical cyclone breeding grounds has already increased by an average of 0.5°C as a result of climate change (Barnett et al. 2005; Santer et al. 2006).
- The only explanation for the increased intensity of tropical storms in the six ocean basins is the steady rise in sea surface temperatures over the last 35 years (Webster et al. 2006).

- Climate models show that winter storm losses will have more than doubled by 2085 in some European countries due to the effects of climate change (Schwierz et al. 2007).

Geo Risks Research has undertaken hurricane frequency analyses over the past few decades which take into account natural climate cycles (the Atlantic Multidecadal Oscillation, AMO). These indicate that the higher frequency and intensity of Atlantic tropical cyclones in recent years could be due to both the natural cycle (the current warm phase, which started in 1995) and global warming.

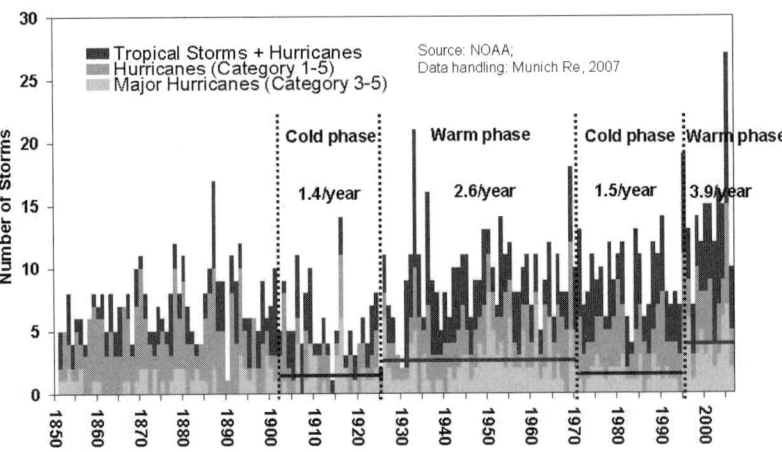

Fig. 3. Climate variability and hurricane activity

Figure 3 clearly shows that, on average, the number of destructive major hurricanes is significantly higher in the warm phases of the AMO than in the cold phases. This supports the theory that hurricanes form over very warm sea surfaces. However, it is also true that storm frequencies during the current warm phase (from 1995 onwards) have been much higher than in the previous warm phase in the middle of the last century. The difference can no longer be explained by natural fluctuation; it can only be due to global warming.

The analysis presented in Fig. 4 shows very clearly that sea surface temperatures which have already increased as a proven consequence of anthropogenic climate change have a considerable impact on hurricane losses. The graph shows the relationship between average annual USA hurricane losses and deviation in sea surface temperatures from the long-term average for the relevant season. The conclusion: the higher the temperature, the greater the loss.

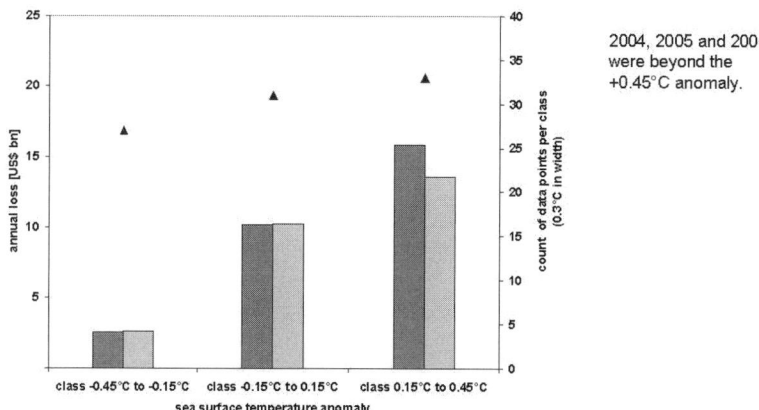

Yellow bars: mean annual losses according to R. Pielke's loss figures;
Orange bars: similar as above, but since 1954 Munich Re's annual loss figures were used
Blue triangles: number of data points per class (right-hand axis).
Source: Faust, Munich Re 2006.

Fig. 4. Mean annual normalized US hurricane losses in dependence on SST-anomalies

Even apparently anomalous events such as the unusually abundant snow in Europe during the winter of 2005 and the warm start to the winter of 2006 are in keeping with the scientific characteristics of climate change. As well as an increase in weather extremes and a general trend towards warmer winters, there is also likely to be greater variation in weather patterns.

Now that a number of changes have already happened and some of the predictions for the coming decades have already been seen, the key issue is no longer if and when there will be conclusive proof of anthropogenic climate change. The crux of the matter is whether the existing climate data and climate models can provide sufficient pointers for us to estimate future changes with reasonable accuracy and formulate adaptation and prevention strategies in good time. The insurance industry's natural catastrophe risk models have already been adjusted in the light of the latest findings. For instance, they now incorporate sea temperatures that remain above the long-term average due to the ongoing cyclical warm phase in the North Atlantic; the effects of this warm phase are reinforced by global warming. We can also expect the above-average water temperatures to increase further the intensities and probably also the number of cyclones.

3 Consequences for the insurance industry

Even before publication of the recent study by the well-known British economist Sir Nicholas Stern (2007) it was clear that climate change is not just an ecological problem; it is also an economic issue. If damage costs continue to rise, this also affects industry and primarily, of course, insurance companies.

Climate change affects the insurance industry in a number of ways:

- As extreme events increase in number and severity, loss frequencies and amounts grow correspondingly.
- Loss volatility increases.
- New exposures arise (e.g. hurricanes in the South and Northeast Atlantic).
- Unprecedented extremes are encountered (the strongest hurricane on record occurred in 2005).
- Premium adjustments have tended to lag behind rising claims, at least in the past.

Despite unfavourable loss trends, the insurance industry continues to offer a wide range of natural hazard covers whilst trying, at the same time, to encourage its clients to focus more on loss prevention. It is also making strenuous efforts to control its own loss potentials with the help of modern geoscientific methods. It is still difficult, however, to predict in quantitative terms the effects that future climate changes will have on the frequency and intensity of extreme weather events.

Munich Re in agreement with IPCC believes that the number of severe, weather-related natural catastrophes will increase in the long term as a result of continuing climate change. This, combined with the trend towards higher value concentrations in exposed areas, will increase loss potentials.

In order to at least slow down the rate of climate change – it is already too late to stop it – the emphasis needs above all to be on so-called no-regret or win-win strategies, such as reductions in energy consumption. Even if such strategies were to have less impact on the climate than expected, they would nevertheless help to conserve resources (including financial resources) and show that the industrial world was aware of its responsibility towards the Third World. To adopt such strategies, which are based on the precautionary principle, is to remain on the safe side and ensure winners all round.

Where the economy is concerned, climate change signifies opportunities as well as risks. It opens up many avenues for industry to develop low-emission, more climate-friendly technologies, or capture carbon dioxide released in the combustion process and store it underground (CO_2 sequestration), for example. It provides opportunities for insurers to develop new insurance products. One of Munich Re's new products is based on the Clean Development Mechanism introduced by the Kyoto Protocol. This mechanism enables investors from industrial countries to improve their climate balance sheet by investing in sustainable projects in the developing world. However, many would-be investors are deterred by the risks involved. In response, Munich Re has introduced the new Kyoto Multi Risk Policy.

The insurance industry has tremendous potential for promoting climate protection and climate change adaptation, and thus positively influencing future losses, by taking account of such issues in its products, investments, sponsoring activities, and communications. This has long been a Munich Re commitment. Munich Re's representatives share their knowledge at the annual world climate conferences (COP). The Munich Re initiated Munich Climate Insurance Initiative unites scientists, NGOs and the World Bank in an effort to find new insurance solutions designed to help above all poorer countries, which have no or limited access to the insurance market, to offset losses due to climate change. A number of Munich Re publications address the issue of climate change, for example "Weather catastrophes and climate change" (published by PG Verlag, Munich) and the Group has also produced "Winds of Change", a strategy game, in conjunction with the European Climate Forum. Munich Re is one of the signatories of the common statement of the Global Roundtable on Climate Change (GRoCC) on the need for climate protection as signed by 85 global companies, NGOs and scientific institutes on 20 February 2007 in New York City.

The objective of Munich Re's long-standing commitment is to help raise awareness of the risks posed by climate change and to prepare corresponding measures. Climate change, a global problem with decidedly adverse long-term consequences, clearly requires action based on international consensus. Regrettably, the results of last autumn's climate summit in Nairobi were disappointing.

There is every sign that the consequences of global warming are already evident, not least in Germany where this year's (2006/2007) warmest winter since records started is in line with climate model forecasts. Mild winters create ideal conditions for severe storms such as winter storm Kyrill, which swept across Europe in January causing losses running into billions of dollars, primarily in Germany and the United Kingdom. Kyrill also stood out because of its duration. It produced gale-force winds (over 63 km/h) that lasted for more than 24 hours in some places. Insured losses from Franz, another January winter storm which preceded Kyrill, amounted to several hundred million dollars.

In December 2007, Munich Re had already warned of the higher windstorm risk due to the unusually warm winter, and Kyrill confirmed this forecast. Although warm winters do not only result from climate change and warm weather does not necessarily produce severe winter storms, it is nonetheless true that the last winter has been a foretaste of the future climate and its extreme weather events.

References

Barnett TP et al. (2005) A Warning from Warmer Oceans. Science 309: 284–287
Emanuel K (2005) Increasing Destructiveness of Tropical Cyclones over the past 30 Years. Nature 436: 686–688
IPCC (2007) Climate Change 2007: The Physical Basis, Summary for Policymakers. Geneva
Knutson TR, Tuleya RE (2004) Impact of CO_2-Induced Warming on Simulated Hurricane Intensity and Precipitation: Sensitivity to the Choice of Climate Model and Convective Parameterization. Journal of Climate 17: 3477–3495
Otte U (2000) Häufigkeit von Sturmböen in den letzten Jahren. Deutscher Wetterdienst (ed) Klimastatusbericht 1999, Offenbach
Santer BD et al. (2006) Forced and Unforced Ocean Temperature Changes in Atlantic and Pacific Tropical Cyclogenesis Regions. PNAS 103: 13905–13910

Schwierz C et al. (2007) article submitted for publication in Climatic Change
Siegenthaler U et al. (2005) Stable Carbon Cycle-Climate Relationship During the Late Pleistocene. Science 310: 1313
Stern N (2007) The Economics of Climate Change: The Stern Review. University Press, Cambridge
Stott PA, Stone DA, Allen MR (2004) Human Contribution to the European Heat Wave of 2003. Nature 432: 610–614
Webster PJ et al. (2005) Changes in Tropical Cyclone Number, Duration, and Intensity in a Warming Environment. Science 309: 1844–1846
Webster PJ et al. (2006) Response to Comment on Climate in Tropical Cyclone Number, Duration, and Intensity in a Warming Environment. Science 311: 1713
WMO (2007) WMO Statement on the Status of the global Climate in 2006. Press release No 768

Evaluation of risk in cost-benefit analysis of climate change

Ingrid Nestle

Flensburg University
Internationales Institut für Management
Munketoft 3, 24937 Flensburg, Germany
nestle@uni-flensburg.de

Abstract

Assumptions on risk aversion have an important influence on the results of climate change cost estimates. However, it is highly controversial which values should be used. They can hardly be deduced from market behaviour, as the empirical risk aversion differs widely according to context. Therefore, an empirical survey has been conducted to enhance our understanding of empirical risk aversion in the context of climate change. First results give strong evidence that different values of risk aversion should be used for different damage categories. The risk aversion in the cases of human lives lost, serious health damages and irreversible damages in general is considerably higher than when moderate economic losses are at stake. On the contrary current climate cost models usually use one single value for risk aversion. The survey tends to support high values of risk aversion in the context of climate change.

Keywords: Climate change costs, risk aversion, empirical preferences

1 Introduction

Climate change poses serious risks for companies and society as a whole. An extensive overview of probable impacts is compiled in the Third Assessment Report of the Intergovernmental Panel on Climate Change (IPCC 2001). One method to support the decision on how much mitigation, adaptation and remaining damages our society should choose is cost-benefit analysis. In this paper I will look at the benefit-side of climate policies, especially at the possibilities to include risk assessment into cost-benefit analysis. Compared to the vast literature on the costs of climate change mitigation, there is little discourse on the direct benefits of climate policies (Morlot and Agrawala 2004, p. 11).

Empirical risk aversion differs widely depending on the context. For example, people tend to be risk-loving when participating in lotteries, moderately risk averse in some financial investment decisions and sometimes very strongly so when asking for insurance services. There is not one objectively right answer on which value of risk aversion to use in the valuation of climate change damages, but still the choice has a strong influence on the results. To overcome this problem one could try to deduce the right value from market observations – however, since climate change is a problem with different characteristics from day-to-day market decisions, this is very difficult. Another solution might be to ask professional philosophers who are experts in ethics – but this seems rather undemocratic. So the only option left is to ask people directly what their preference is in the context of climate change. (Dietz 2007)

This paper aims to make a contribution to exploring this new field. The first section delineates the importance and the limits of cost-benefit analysis for dealing with climate change. It concludes with a definition of what the role of CBA is assumed to be in the further elaboration of the paper. The second section attempts a literature review on risk treatment in cost-benefit analysis of climate change. This concludes the more theoretical part of the paper. The third and longest section of the paper describes an empirical experiment aiming to gain knowledge on the de facto risk aversion of the sample population in the context of climate change. Section 4 discusses the results and concludes.

2 Cost-benefit analysis and climate change

Cost-benefit analysis is a very important tool for support of rational decision making. But in the context of climate change it is faced with some difficulties:

Carlo Jäger points out that it might be an inferior approach compared to other risk management strategies because it identifies utility functions with monetary magnitudes – a very difficult task considering many non-market impacts of climate change. Furthermore, it does not deal with how decisions are reached and preferences are formed. He argues that either a portfolio analysis or the identification of core activities might be more helpful in this special context, although neither of these were capable of addressing cultural diversity. (Jäger et al. 1998) For more detailed discussion see IPCC 1996, chapter 5, and Portney 1998.

The following points give an overview of the most important problems linked to the quantification of external costs of climate change:

- The choice of the discount rate exerts huge influence on the result.
- It is difficult to attach a price to non-market goods.
- Risk has not been included satisfactorily so far (Schneider and Lane 1998, p. 161).
- The scientific basis still comprises huge uncertainties (Jacoby 2004, 3.1).
- The aggregation of utility between different individuals poses huge problems, a lot more so the aggregation between nations with widely varying income-levels (s.a. IPCC 1996, p. 206).
- The damage evaluations represent and attempt to predict the damages that global warming could cause in today's world, but it is very difficult to assess what could happen, far in the future, in a changed world under possibly very different circumstances (Tol 2002, p. 135).
- There is not enough knowledge about damage costs for higher greenhouse gas concentrations.

This considerable list of difficulties might cause doubt as to whether it is of any use to spend more time on the monetary assessment of benefits of climate change policies. Indeed, past studies come to widely varying results: monetary values of the same climate impact vary by as much as six orders of magnitude (Hohmeyer 2005, p. 164). Ethical assumptions play a major role in explaining the differences among various studies, especially the intertemporal discounting factor and international utility aggregation (Pittini and Rahman 2004, part 2; Hohmeyer 2005). This makes it next to impossible to considerably narrow the range of varying results by means of further scientific work.

Therefore Hohmeyer (2005) suggests that society should agree on a level of maximum greenhouse gas concentration in the atmosphere and assure with adequate climate policy instruments that this limit will not be surpassed. But even for the process of agreeing on a certain maximum level, an approximation of damage costs of different concentrations will be a very valuable source of information supporting a rational decision. Even in low concentration scenarios there will be damages due to climate change, and probably even irreversible damages, e.g. species extinctions. It is a question not only of preferences, but also of costs and benefits on which level societies should agree. Non-linear abrupt changes due to global warming are possible, but it is impossible to predict at which greenhouse gas concentration level this will occur (Schneider and Lane 2004). Again society is faced with a gradual decision: on how much risk to accept. Cost-benefit analysis certainly should not be the only instrument supporting the decision-making process, but it is too valuable a tool to give it up altogether.

Rather, it is necessary to carefully assess the aim of the damage evaluation before deciding on a method to deal with the inherent shortcoming. In a second step, ways to deal with these difficulties can be constructed in order to make the results as useful as possible.

From my point of view there are three main objectives of climate policy benefit evaluation:

1. **To give rational, scientific support for the decision-making process** regarding the maximum acceptable greenhouse gas concentration or the amount of mitigation chosen. The scientific input needs to be aggregated, because the results from natural sciences are often too complex and voluminous for decision makers to use. For example Net Ecosystem Productivity is a measure very few people will understand (Hitz and Smith 2004, p. 65).
2. **To facilitate a balanced discourse in society:** At the moment there exists more information on the costs of mitigation than on the benefits (e.g. IPCC 2001a). Furthermore, because of social inertia changes in behaviour are more difficult to introduce than persistence of the prevailing patterns (see Pittini and Rahman 2004, p. 201). Some more insight on the benefits of climate change mitigation can help to bridge the gap between expert insight about the necessity of action and public support for such a scheme.
3. **To set the right priorities in future climate change research:** In order to concentrate scarce research funds on the important fields, it is helpful to know which sectors are likely to experience the most comprehensive damages due to global warming.

In past years damage evaluation has advanced further. For example Kemfert (2004) has coupled a detailed economic model with a climate model – and presented interesting results from this WIAGEM model. But the results are still subject to the bias of normative assumptions that have to be taken. Tol (2002, 2002a) and Hitz and Smith (2004) have tried to achieve better results by combining the findings of many different studies. But as long as all of the underlying studies are based on more or less arbitrarily chosen normative assumptions, even a very sophisticated compilation cannot deliver value-free results. It is necessary to also advance the methodology of impact monetarization in order to avoid the situation in which wrong assumptions appear right simply because they are being used in a number of studies. Therefore, this paper will concentrate in the following on one of these normative issues: the valuation of risk.

3 Climate change and risk

As mentioned above the uncertainties connected with climate change are very considerable. And the level of uncertainty is greater in the benefit curve than in the cost curve of emission reduction (IPCC 1996, 5.4). To date most cost-benefit analyses have not sufficiently treated the risk of truly catastrophic outcomes (Schneider and Lane 2004, p. 161), and few studies have focused on increased climate variability and potential catastrophic changes (Hitz and Smith 2004, p. 65), although this is potentially a very considerable contribution to climate change damages (Callaway 2004, p. 156). However, some suggestions for how to deal with risk have been made:

- Portney suggests an approach to include risk aversion via a social insurance approach with a general referendum regarding how much the world population is willing to spend on an insurance scheme that would prevent a certain quantity of climate change risk for future generations (Portney 1998, p. 124ff).
- The authors of the ExternE study propose a risk premium on valuing damages (Markandya 1994, p. 30ff).
- Jones states that the best way to assess climate change damages in a probabilistic framework is to indicate the likelihood of a certain threshold being exceeded (Jones 2004, p. 249).
- Jäger et al. present the strategy of diversification (as many different policies as possible) or concentration on core activities as a way to deal with risk. They further elaborate that the market system provides an institutional mechanism to aggregate heterogeneous risk preferences into an outcome of considerable social legitimacy – but only where costs are internalised and the problem of missing future markets is solved (Jäger et al. 1998).
- Cline 85 proposes to include some form of weighting in order to take policy makers' risk aversion into account (Cline 1992, p. 85).
- In the private sector strategic hedging is the usual way to deal with low-probability high-impact risks; however, this is no viable strategy in the field of climate change, since no players exist who could satisfactorily insure world society against this risk.

So far the above mentioned suggestions have not led to quantitative inclusion of risk into calculations of climate policy benefits. Another possibility could be to integrate risk assessment into the discount rate. There is a very intensive discussion in the scientific community on what the right discount rate should be. However, within the scope of this paper it is impossible to present all the major arguments. For an overview read for example IPPC 1995, chapter 4; Lind and Schuler 1998; Markandya 1994, chapter 6; Maddison 1999. Here only the question whether risk assessment can be included into the discount rate is of interest. Friedrich and Bickel state that there are four possible reasons for discounting (Friedrich and Bickel, p. 124):

- Impatience
- Economic growth
- Changing relative prices
- Uncertainty (risk)

Thus, risk is one possible reason why future events might count less than present impacts: there occurrence is less certain. However, there is no reason why the quantity of uncertainty should be correlated directly to the time of the event. Friedrich and Bickel also conclude that changing relative prices and risk should rather be dealt with separately and not lead to a change in the discount rate. But if risk cannot be dealt with in the normal discount rate, there might still be the option to introduce a second percentage – adjusting various impacts for the probability of their occurrence. This is very similar to the proposal of the ExternE team.

The most explicit inclusion of risk into climate change cost analysis has probably been by Hope (2006) and Stern (2006). The model PAGE2002 developed by Chris Hope and also used by Nicholas Stern is a stochastic model. Uncertain input parameters do not enter the model as single best-guess values, but as probability distribution functions. Monte Carlo Analysis is used to assess the probability of different cost outcomes. Dietz et al. (2007) show that the limitation on best-guess values instead of probability distribution functions considerably underestimates the weight of possible high-cost outcomes.

With this explicit treatment of risk in their model, Hope and Stern have tried to include the objective probability of higher than expected future impacts. The consideration of risk aversion is still another issue. The Stern team use the negative of elasticity of the marginal utility of consumption η for the inclusion of risk aversion (Dietz in press, p. 12); for the Stern review a value of $\eta = 1$ was used in the base case (Stern 2006, p. 162). However, I argue that by the application of η only the transition from consumption to utility has been carried out. This procedure certainly captures the part of risk aversion explained by expected utility theory, but not necessarily the empirical risk aversion of real-life individuals (s.a. Rabin 2000).

Nordhaus and Boyer (2000, p. 88) use a relative risk aversion of 4 for the calculation of catastrophic risk in the DICE model. However, they explain no further the choice of this value for relative risk aversion. A study by the Swiss Institution for Population Safety (BABS without year) also states that, whereas the phenomenon of risk aversion is not controversial, its treatment remains unsystematic.

People value risk very differently under different circumstances. For example voluntary risk is usually rejected less than imposed risk (Markandya 1994, p. 32). In the case of climate change little of the risk perception is due to personal experience. Jäger et al. (1998) suggest that risk debates are not fundamentally about risk, but about trust, social and technological choice, credibility, power, legitimacy and control; perceived risk is the broader and more inclusive view from which objective risk (only the technical dimension) is abstracted. According to them the public is wary of conventional risk analysis: the objective science delivers very contradictory results and has not been able to prevent very serious environmental accidents. Calls for more science would not solve the problem.

4 Empirical survey

4.1 Aim of the empirical survey

As described above the estimation of the external costs of climate change is much disputed, because the results depend heavily on normative assumptions, which cannot be decided scientifically. Cost-benefit analysis of climate change is inherently faced with the problem that risk analysis and risk valuation cannot be clearly separated. One way to cope with the problem would be to make the calculations for various normative assumptions and let decision makers choose what they believe is adequate. But the result would be a huge range of numbers for different combinations of normative assumptions – thus destroying the charm of benefit evaluation and aggregation: namely to come up with numbers that are easy to handle and to communicate. Here another way to deal with this problem shall be ventured: An empirical survey to assess the actual preferences in the population regarding the risk aversion in the context of climate change. This is a first rough attempt to make it possible to calculate better founded estimates of direct climate policy benefits based on real risk preferences.

Of course, empirically founded preferences do not automatically need be the theoretically best possible preferences. This approach cannot make ethical and other theoretical considerations superfluous. Furthermore, subjective probabilities can be improved to be more objective with more information, although in the case of climate change no objective probabilities are available (Jäger et al. 1998). Still the empirically existing risk preferences are a very interesting clue towards the optimization of society's management of climate change. And as the whole field of impact monetarization is built upon preference theory, it is only consistent to resort to actual preferences for the valuation of risk as well.

This approach is in accordance with the postulation above, that the aim of the valuation of external costs should be kept in mind while setting the rules for the calculations: The first aim was to give scientific support to the decision-making process for a mitigation level. As it is a decision of society and not of science, it seems reasonable to base the treatment of risk at least partially on the existent preferences in society. One problem is that the interests of future generations can never be included empirically. This is a point where ethical theory remains important. The second aim was to facilitate a balanced discourse about climate change mitigation and adaptation in society. External cost figures that reflect the real preferences of people are a very good basis for supporting balanced discussion within society. As for the third point, prioritizing further research on climate change, empirical risk aversion can at least be a helpful indicator to acquire knowledge about the extent of damage in various sectors, taking into account public risk preferences.

Empirical risk aversion has been the focus of various research projects in the past; for example, Litai et al. (1983) have produced factors for the comparison of risks with different attributes. According to them, man-made risk is felt more severely than natural risk, involuntary risk more than voluntary risk, and new risks more than known risks. Plattner et al. (2005) have conducted a thorough literature review on risk valuation and tried to produce a model that could be used to value the risks of different natural catastrophes in Switzerland. They state that for each type of risk new aversion functions have to be found (Plattner et al. 2005, p. 37). Thus the results from the literature cannot be simply transferred to the case of climate change.

4.2 Construction of the empirical survey

In the risk section of the survey, valid answers were received from 137 1st semester students at Flensburg University in October 2005. 80 of the students study International Mangement (IM) and 57 Energy and Environmental Management (Energie- und Umweltmanagement – EUM). IM is a business career with international orientation. EUM is an industrial engineer dealing with different aspects of sustainable energy supply. The survey was done in the very first economics lecture of these students; they have therefore not yet been influenced by their respective careers. Still, they have chosen their respective careers for special interests and the EUM students are on average more likely to have spent time thinking about climate change and possible mitigation options.

The aim of this part of the survey was to assess the influence of risk and uncertainty on the evaluation of damage. The students were asked to compare two different theoretical cases: Case A, in which damages leading to a loss of 10% of the annual income would certainly occur in absence of climate change policies. And case B, in which the same loss would occur with a probability of 5% in the absence of climate change policies. The students compared these two cases and indicated which fraction of the climate change mitigation effort of case A they would consider to be reasonable in case B. The six possible answers were:

- The same effort
- Half the effort of case A in case B
- Exactly corresponding to the risk (5%)
- Less than the probability (1%)
- Nothing, although I would do something in case A
- In case A I wouldn't do anything either.

In the survey the question was not what the students were willing to do personally, but what they felt would be reasonable for society to do.

In a second step the students were asked to answer the same question not just for damages of 10% of the annual income, but for damages of 1% of the annual income, 10% of the annual income, 50% of the annual income, for irreversible damages, for serious health impacts and for loss of human lives. For each of these scenarios the same answers as above were available in a table. Basically the first part of the question was to make sure that the students would understand the somewhat complex task. They were able to get used to the scenario with just one possible damage estimate instead of a whole table. In the evaluation I will mainly rely on the results from the table, although it is interesting to see how many change their mind from the first question to the same issue later in the table. From a total of 137 students, 37 students indicated a lower level of effort, than they had in the previous question, as reasonable for the risk case in the table. 12 students switched to a higher level of effort in the table. It is possible that some students adapted their answers in the table so as not to appear inconsistent with their first answer. This problem especially afflicts the answers demanding a very high level of mitigation effort for the low-probability damage. Furthermore, no special budget limitation was given. Therefore, the absolute results may have a tendency to overstate the amount of reasonable effort in the 5%-case.

The respective distribution of values among the different damage categories most likely reflects the real preferences quite well. In the evaluation I have not aggregated the answers for each of the damage categories, but rather I have treated the answers of one student for the whole table as a meaningful pattern. In other words: because of the problems mentioned above, I do not use the absolute numbers as results, but the relative ordering of risk preferences for different damage categories. Table 1 gives a quick overview of the most important results:

4.3 Results from the empirical survey

A very clear result is that most of the respondents valued risk differently according to what kind of damage was at stake. 80.5% of the EUM-students and 71.3% of the IM-students were demonstrating a clearly ordered risk valuation (Table 1). Most of the students' risk aversion was growing with higher financial assets at stake and even more so when irreversible damages, serious health impacts or loss of lives were concerned. A few of the students showed a slightly different ranking giving more weight to the 50% income loss or to the irreversible damages than to the following categories. These findings are in line with other research (BABS without year, p. 9).

Table 1. Results from the student survey

	EUM	IM	Overall
Percentage who ranked the answers in such a way that, for each of the six damage categories, the reasonable effort is at least as high as for the previous category, but did not rank all equally.	77.2%	61.3%	67.9%
Ranked the answers in a similar way, but gave more weight to the 50% income loss or the irreversible damage	3.5%	10.0%	7.3%
Felt the same effort in case A and case B to be reasonable for all damage categories	15.8%	13.8%	14.6%
Saw 5% of the effort as reasonable in all damage categories	0.0%	3.8%	2.2%
Difficult to categorize	3.5%	11.3%	8.0%
Sum of the above	100.0%	100.2%	100.0%

This finding has important consequences for the treatment of risk. If, for example, the idea from the ExternE group was to be taken up and a risk premium attached to the uncertain impacts, this would clearly need to be a differentiated risk premium for different damage categories in order to be consistent with the existing risk preferences of the majority. The same is true if some form of weighting for risk aversion according to the suggestion by Cline were to be used. The factor may not only depend on the probability of the climate change impact, but also on the impact category. This adds difficulty to the estimation of external costs of climate change, because the value of the risk premium might well depend on the result of the calculation, i.e. the extent of climate change costs: 74.5% of EUM students and 71.4% of IM students had different risk preferences for different potential income losses.

Another finding of the survey is a very high risk aversion against climate change impacts. But as mentioned above there is probably some part of "social desirability" as a bias in the result, and actual decisions when faced with a constrained budget might be different. Still the risk aversion, particularly for irreversible damages, health damage and loss of human lives, is so pointed that it can be assumed to hold under more realistic assumptions. The problem of the missing budget constraint should not be too severe as the comparison is between two climate change mitigation scenarios and not compared to an option where no money is spent on climate policies. 56% of the students being risk neutral on at least one of the income damage categories can be taken as an indicator for not highly irrational answers. On the whole, 92.7% of the respondents were risk averse (a significant tendency to support efforts higher than 5% of those in case A), 1.5% were risk positive and 2.2% were risk neutral. The answers of 3.6% were difficult to put clearly into one of these categories. 78.8% felt that a lower damage probability would not justify lower mitigation efforts when irreversible damages, serious health impacts or human lives are at stake. Tables 2 and 3 show the corresponding figures. These results correspond well with the relevant literature: People feel they have very little personal control over large ecological risks, which strongly increases the perceived risk, and correspondingly, the risk aversion (Jäger et al. 1998).

Table 2. Risk Aversion

	EUM	IM	Overall
Risk averse	96.5%	90.0%	92.7%
Risk positive	0.0%	2.5%	1.5%
Risk neutral for all damage categories	0.0%	3.8%	2.2%
Difficult to place	3.5%	3.8%	3.6%
Sum of the above	100.0%	100.1%	100.0%
Risk neutral in at least one of the income damage categories	56.1%	56.3%	56.2%

Table 3. Number of damage categories with complete risk aversion

	EUM	IM	Overall
Ask for the same effort as in case A in all six categories	15.8%	13.8%	14.6%
Ask for the same effort as in case A in the last five damage categories	12.3%	16.3%	14.6%
Ask for the same effort as in case A in the last four damage categories	24.6%	21.3%	22.6%
Ask for the same effort as in case A in the last three damage categories	31.6%	23.8%	27.0%
Ask for the same effort as in case A at least in the last three damage categories (sum of the above)	84.2%	75.0%	78.8%

Some differences between the students from EUM and the students from IM can also be seen: The EUM students have a tendency to be more risk averse in the context of climate change: 84.2% ask for the same effort as in case A at least in the last three damage categories compared to 75% from IM. None of the EUM students were risk positive or risk neutral compared to 2.5% and 38%, respectively, from IM. The ranking of different damage categories was more clearly seen among EUM students: only 3.5% were difficult to place in the ranking schemes adopted for evaluation compared to 11.3% from IM. These differences can be explained by the fact that on average EUM students are more familiar with the issue of climate change than the IM students, even before they start their University education. A lot of them come to Flensburg because they are concerned about climate change and other problems connected to energy supply. In this respect they are not a typical sample from our society.

5 Discussion of the results and conclusions

These results have a clear consequence for the inclusion of high-impact low-probability events in the cost-benefit analysis of climate change. It does not seem to be appropriate to simply multiply the scale of possible damage by the probability of its occurrence. Furthermore, the results from this survey strongly indicate that external cost estimates should include some valuation of low-probability high-impact events. Scientific rigidity may demand cancellation of those damage categories for which the scientific foundation is extremely weak, but the corresponding results will be very far from the population's risk management preferences. Especially damage categories that are difficult to measure in monetary terms (irreversible damages, serious health impacts, loss of human lives) are those, for which risk aversion has proven to be strongest.

A possibility for making cost-benefit analysis useful for the future could be to leave a few categories that are especially difficult to value and at the same time connected to a large public risk aversion outside the monetarized value. For example an external cost estimate could provide monetarized costs for most damage categories, but indicate the expected number of human lives lost separately. The disadvantage is a more complicated result, but thus it is possible to avoid the problems of valuing human lives, aggregating these figures among nations and mixing this impact category with others that are subject to very different empirical risk aversion. A useful category might be "number of people at risk of hunger" (see Hitz and Smith 2004, p. 34) or "number of species extinct". Jacoby (2004, p. 302) also recommends a portfolio of benefit measures instead of one single measure. Thus some of the difficulties of evaluation of non-market impacts could be avoided. The trade off is always between losing clarity because of too many different categories and unities on the one hand, and making too many normative assumptions disappear in a high level of aggregation on the other hand. Wherever normative assumptions have to be made to provide clarity, the empirical evidence for factual preferences in the population should be taken into account. Further research is necessary to arrive at meaningful empirical values for normative issues in the context of climate change.

Of course the above survey is subject to several limitations: first of all the number of interviewees is rather small and can only give primary indications, but not final scientific evidence. The results have showed that risk preferences vary among different groups in society; a broader sample with less bias could lead to more general results. Furthermore, as calculations of global damages are the issue, people from different cultures would need to participate in a more sophisticated survey. As Jacoby (2004, p. 310) notes, different cultures have been found to have very different risk perceptions and preferences. The problem of not including people from future generations, on the other hand, can not be overcome. The issue regarding the budget constraint has equally been elaborated above. Finally, further research in this direction would be better served if it were rooted in the science of psychology.

References

Callaway J (2004) The benefits and costs of adapting to climate variability and climate change. In: The benefits of climate change policies. OECD-Publication: pp 11–158

Dietz S (2007) Presentation for the Electricity Policy Research group at Cambridge University on 19th November 2007. Most of his elaboration has been published in Dietz et al. 2007

Dietz S, Hope C, Patmore N (2007) Some economics of 'dangerous' climate change: reflections on the Stern Review. In: Global Environmental Change 17: 311–325

Hitz S, Smith J (2004) Estimating global impacts from climate change. In: The benefits of climate change policies. OECD-Publication, pp 31–82

Hohmeyer O (2005) Die Abschätzung der Kosten des anthropogenen Treibhauseffekts – dominieren normative Setzungen das Ergebnis? In: Vierteljahreshefte zur Wirtschaftsforschung 74 (2): 164–168

Hope C (2006) The marginal impact of CO_2 from PAGE2002: an integrated assessment model incorporating the IPCC's five reasons for concern. In: The Integrated Assessment Journal 6 (1): 19–56

IPCC (2001) Climate Change 2001: Impacts, Adaptation and vulnerability. A report of the Working Group II of the Intergovernmental Panel on Climate Change, ed By McCarthy JJ, Canziani OF, Leary NA, Dokken DJ, White KS. Cambridge University Press, Cambridge

IPCC (2001a) Climate Change 2001: mitigation. A report of the Working Group II of the Intergovernmental Panel on Climate Change, ed by Metz B and Davidson O. Cambridge University Press

IPCC (1996) Climate change 1995. Economic and social dimensions of climate change. Contribution of Working group III to the Second Assessment Report of the Intergovernmental Panel on Climate Change. Editors Bruce JP, Lee H, Haites EF. Cambridge University Press

Jacoby H (2004) toward a framework for climate benefit estimation. In: The benefits of climate change policies. OECD-Publication, pp 299–323

Jäger C, Renn O, Rosa E, Webler T (1998): Decision analysis and rational action. In: Rayner S and Malone EL (eds): Human choice and climate change, vol 3: The tools for policy analysis. Battelle Press, Columbus, pp 141–215

Jones R (2004) Managing climate change risks. In: The benefits of climate change policies. OECD-Publication: pp 249–297

BABS (Bundesamt für Bevölkerungsschutz der Schweiz) (without year): KATARISK – Katastrophen und Notlagen in der Schweiz. Eine Risikobeurteilung aus der Sicht des Bevölkerungsschutzes. Available at: http://www.bevoelkerungsschutz.admin.ch/internet/bs/de/home/themen/gefaehrdungen/katarisk.ContentPar.0005.DownloadFile.tmp/methode-monitor.pdf

Kemfert C (2004) Die ökonomischen Kosten des Klimawandels. Wochenbericht des DIW Berlin, 71 (42): 615–623

Lind R, Schuler R (1998): Equity and discounting in climate-change decisions. In: Nordhaus, Wiliam (ed): Economics and policy issues in climate change. Resources for the future, Washington DC, pp 59–96

Litai D, Lanning DD, Rasmussen NC (1983) The public perception of risk. In: V.T. Covello et al. (eds) The analysis of actual versus perceived risks. Plenum Press New York/London, pp 212–224. Cited after Plattner et al. (2005)

Maddison D (1999) The plausibility of the ExternE estimates of the external effects of electricity production. CSERGE Working Paper GEC 99–04 (Centre for social and economic research on the global environment at UEA and UCL)

Markandya A (1994) Externalities of Fuel Cycles. ExternE Project. Report number 9. Economic valuation – an impact pathway approach. European Commission, DG XII

Morlot JC, Agrawala S (2004) Overview. In: The benefits of climate change policies. OECD-Publication, pp 9–30

Nordhaus, William and Joseph Boyer (2000): Warming the world. Economic models of climate change. MIT press, Massachusetts

Pittini M, Rahman M (2004) The social cost of carbon; key issues arising from a UK review. In: The benefits of climate change policies. OECD-Publication, pp 189–220

Plattner T, Heinimann HR, Hollenstein K (2005) Risikobewertung bei Naturgefahren. Schlussbericht. Von der Plenarversammlung der Nationalen Plattform Naturgefahren (PLANAT) genehmigt

Portney P (1998) Applicability of cost-benefit analysis to climate change. In: Nordhaus W (ed): Economics and policy issues in climate change. Resources for the future, Washington DC, pp 111–127

Rabin M (2000) Diminishing marginal utility of wealth cannot explain risk aversion. Posted at the eScholarship Repository, University of California: http://repositories.edlib.org/iber/econ/E00-287

Schneider SH, Lane J (2004) Abrupt, non-linear climate change and climate policy. In: The benefits of climate change policies. OECD-Publication: pp 159–188

Stern N (2006) The economics of climate change. London, HM Treasury

Tol R (2002) Estimates of damage costs of climate change. Part I. Benchmark estimates. In: Environmental and Resource Economics 21: pp 47–73

Tol R (2002a) Estimates of the damage costs of climate change. Part II. Dynamic estimates. In: Environmental and Resource Economics 21: pp 135–160

A global climate change risk assessment of droughts and floods

Wolfgang Knorr[1], Marko Scholze[1]

[1] QUEST, Department of Earth Sciences, University of Bristol
Wills Memorial Building, Queen's Road, Bristol BS8 1RJ, UK
wolfgang.knorr@bristol.ac.uk
marko.scholze@bristol.ac.uk

Abstract

The use of information from climate model simulations by policy makers and other potential users poses great challenges. On the one hand, researchers are rightly cautious about prediction, as the uncertainties are large. On the other hand, policy makers may be used to dealing with great uncertainties, and may be quite prepared to use assessments of risk levels. A new way of assessing a comprehensive ensemble of climate model simulations is presented, using 16 climate models and various scenarios of future levels of greenhouse gas levels. We compute annual soil water runoff, and we define as a risk a situation where the mean during 2071–2100 deviates from the 1961–1990 mean by more than one standard deviation (1σ) of the current variability. We find that in general, larger degrees of warming lead to larger areas and populations affected, with substantial risks for tropical countries even at low degrees of global warming.

Keywords: Climate change, risk assessment, climate policy, drought, flood

Acknowledgements: We acknowledge the international modelling groups for providing their data for analysis, the Program for Climate Model Diagnosis and Intercomparison (PCMDI) for collecting and archiving the model data, the JRSC/CLIVAR Working Group on Coupled Modelling (WGCM) and their Coupled Model Intercomparison Project (CMIP) and the Climate Simulation Panel for organizing the model data analysis activity, and the IPCC WG 1 TSU for technical support. The IPCC Data Archive at Lawrence Livermore National Laboratory is supported by the Office of Science, U.S. Department of Energy. Further, we thank M. Drechsler, I.C. Prentice, N. Arnell, S. Cornell, J. House, W. Lucht and S. Schaphoff for helpful discussions and comments on previous versions. This work was support by the programme Quantifying and Understanding the Earth System of the Natural Environment Research Council, UK

1 Introduction

Any business or responsible policy maker essentially operates in a risky and uncertain environment. Modern risk management tries to minimize the negative impact of risky outcomes (through avoidance, adaptation, or burden sharing), and to maximize the opportunities from unforeseen developments (though preparedness and openness) (Head and Herman 2002). Risk management strategies always profit from better information and such data may be provided in terms of probabilities of given outcomes, even if those probabilities themselves are not very well known (Morgan and Henrion 1990).

The required probabilistic information may be provided by the research communities. However, there seems to be a widespread reluctance to quantify the degree of certainty of scientific discoveries and predictions, a reluctance that may be inherent in the scientific dialogue. Sometimes this discrepancy between policy-driven and the more traditional approaches to scientific discovery is described as the difference between 'normal' and 'post-normal' science (Funtowicz and Ravetz 1992). According to this characterization, post-normal science has a more extended peer-review community that includes policy stakeholders and characteristically applies a far more extensive description of uncertainty. It has been argued that the latter is also more relevant in the context of climate change, and that its prescriptions are followed to a large extent by the publication process of the Intergovernmental Panel on Climate Change (IPCC) (Saloranta 2001). Others have argued that there are still substantial challenges when it comes to incorporating the needs of policy makers for pragmatic approaches to tackle climate change starting from climate model information (Varis et al. 2004). There is also an important distinction to be made between normative and exploratory scenarios when future probabilities are assessed (Jones 2004). The IPCC, with its own climate change scenarios, largely follows the exploratory approach (Carter and La Rovere 2001), simply sketching out possible storylines without making claims as to which would more adequately address the normative challenges of the United Nations Framework Convention on Climate Change (UNFCCC).

Indeed, the case of anthropogenic climate change is an especially interesting one, because it is a field of science that from the start has been driven by an existing long-term policy problem that is increasingly being seen to have direct business implications (Nutter 1999; Dlugolecki 2004; Mills 2005; Dlugolecki and Lafeld 2005). Climate change research, which has grown both out of the more fundamental atmospheric sciences and the more practical weather forecasting, is currently moving in at least two separate directions that can be characterized again by a process-oriented traditional science approach, and a more application-driven approach (Jones 2004; Varis et al. 2004): one is the move towards including more and more processes in climate models, leading towards earth system models and earth system science (Brasseur et al. 2007); the other is the direct practical application of statistical climate predictions at the seasonal time scale (Dilley 2000; Meinke and Stone 2005). The attempt to bring both approaches together and provide century-scale climate forecasts complete with uncertainty assessments is a relatively new development (Hulme and Carter 1999; Murphy et al. 2004; Piani et al. 2005; Hare and Meinshausen 2006). There seems indeed to be a widespread consensus that the provision of uncertainty information has the potential of benefiting decision-making using both climate (Hobbs et al. 1997; O'Connor et al. 1999; Nutter 1999; Schlumpf et al. 2001; Jones 2004; Manning and Petit 2003) and seasonal forecasting (Suarez and Patt 2004; Meinke and Stone 2005; Linneroth-Bayer et al. 2005).

This consensus, however, still contrasts with a general system-analytical view of the climate change problem, one that is driven by the need for process analysis, and thus tends to see policy merely as a final application, rather than an essential part of the process that is driving the research forward. How deeply ingrained the system-analytical view is may be illustrated by the hierarchy that is inherent in the IPCC process: its three working groups (WG1: science; WG2: impacts and adaptation; WG3: mitigation) are not only numbered consecutively, information is essentially passed in the same direction, i.e. from process modelling to impacts to policy. As a consequence, a clear hierarchy emerges in which the more scientific parts of the IPCC process enjoy a freedom of not having to relate to the other, more relevant parts.

Given this situation, seeing climate change as a risk management rather than a system-analytical problem has the important advantage of forcing climate change research to incorporate its policy relevance right from its design, or at least forcing climate change researchers to consider ways of analysis and presentation in line with the needs of policy makers (e.g. Parry et al. 2001). Risks have to be quantified at least approximately in order to prepare for adaptation efficiently, allowing the most severe and unacceptable risks to be avoided. A first attempt in that direction has been presented by Scholze et al. (2006), who quantified the climate change-induced risks for three ecosystem properties (runoff, wildfires and habitat changes) for the end of the 21st century on a spatially explicit map. Risk management aims at providing messages that can be easily understood by the potential users of the information (Schlumpf et al. 2001). As a result, transparency and openness become generally more important (Dessai et al. 2004) as they vastly improve the practical value of forecasting (Suarez and Patt 2004).

The aim of the present paper is, therefore, to provide an illustrative example of climate research analysed and presented from a risk management perspective providing approximate, essential and quantitative information that is easy to understand, in this case aimed at the global policy makers responsible for setting climate stabilization targets. While using state-of-the-art modelling and scenarios, its aim is to provide a practical tool intended to help informed decision making. By virtue of simulating impacts in a geographically explicit fashion, the study is able to avoid controversial aggregation of positive and negative impacts. Instead of this, dual risks of deviations in opposite directions are explicitly quantified. It also does not attempt to reproduce the policy process itself or make any assumptions about optimal decisions (e.g. Mastrandrea and Schneider 2004), which would again mean following a system-analytical approach: these, or any decisions on what risk levels may be acceptable, are meant to be left to the user. The example risk levels given in this study are merely for the purpose of illustration.

Table 1. Climate model and emission scenarios used in the present study

Name of Climate Model	Scenario (SRES or committed)	Temperature change [°C] 1961–1990 to 2071–2100
Canadian Centre for Climate Model. & Analysis: CGCM3.1(T47)	B1	2.1
Météo-France/Centre National de Recherches Météorologiques: CNRM-CM314	A1B	3.1
	A2	3.8
	B1	2.0
	committed	0.6
CSIRO Atmospheric Research: CSIRO-Mk3.0	A1B	2.3
	A2	2.9
	B1	1.3
US Dept. of Commerce/NOAA/ Geophysical Fluid Dynamics Laboratory: GFDL-CM2.0	A1B	3.2
	A2	3.5
	B1	2.4
	committed	1.2
US Dept. of Commerce/NOAA/ Geophysical Fluid Dynamics Laboratory: GFDL-CM2.1	A1B	2.8
	A2	3.3
	B1	2.0
	committed	0.9
NASA/Goddard Institute for Space Studies: GISS-AOM	A1B	2.2
	B1	1.6
NASA/Goddard Institute for Space Studies: GISS-EH	A1B	2.2
NASA/Goddard Institute for Space Studies: GISS-ER	A1B	2.4
	A2	2.8
	B1	1.7
	committed	0.02
Institute for Numerical Mathematics: INM-CM3.0	A1B	3.1
	A2	3.8
	B1	2.5
Institut Pierre Simon Laplace: IPSL-CM4	A1B	3.4
	A2	3.9
	B1	2.6
	committed	1.1
Center for Climate System Research (University of Tokyo), NIES, and Frontier Research Center for Global Change: MIROC3.2(medres)	A1B	3.6
	A2	4.0
	B1	2.5
	committed	0.8
Max Planck Institute for Meteorology: ECHAM5/MPI-OM	A1B	3.6
	A2	3.9
	B1	2.6
	committed	0.6
Meteorological Research Institute: MRI-CGCM2.3.2	A1B	2.6
	A2	2.9
	B1	1.9
	committed	0.9
National Center for Atmospheric Research: CCSM3	A1B	3.3
	B1	2.1

Table 1. (cont.)

National Center for Atmospheric Research: PCM	A1B	2.5
	A2	2.7
	B1	1.7
	committed	0.8
Hadley Centre for Climate Prediction and Research/Met Office: UKMO-HadCM3	A1B	3.4
	A2	3.8
	B1	2.4
	committed	0.7

2 A case study of climate change risks

The UNFCCC requires its signatory members to avoid dangerous climate change by ensuring that climate impacts remain at a level that allows 'sustainable development' of societies and ecosystems to 'adapt naturally'. There has been an ensuing discussion about what level of global average temperature increase would constitute a practical threshold above which the risk of dangerous climate change would become unacceptable (Tirpak et al. 2005), with an often cited value of 2°C (Azar and Rhode 1997; Climate Action Network 2002; Grassl et al. 2003). In this study, we start from this debate and attempt to make a risk assessment based on the latest available climate model calculations of the IPCC (IPCC in press), which are themselves driven by a range of exploratory IPCC scenarios (Nakicenovic et al. 2000).

Instead of attempting to attach a specific probability to each combination of climate model and SRES scenario (which for the case of the scenarios many argue is impossible, see Jones 2004), we take the pragmatic approach of assuming that they are all equal. Because ecosystems and ecosystem services are central to climate impacts (Gitay et al. 2001), we combine the output from all climate model simulations with a state-of-the-art dynamic ecosystem model, LPJ (Sitch et al. 2003). We specifically consider risks to human systems by looking at rates of rainwater runoff. Runoff here serves as an indicator of the risk of either drought or flood, both being major threats to sustainable development. To analyse the outcome in a way that directly refers to the policy debate, we divide the available climate model simulations into three groups, depending on the simulated levels of global temperature increase: below 2°C, between 2°C and 3°C, and above 3°C. The analysis is based on the approach used presented by Scholze et al. (2006).

We further define a simple, globally uniform risk level based on the variability simulated within the current, observed climate. According to this definition, a dangerous level of climate change occurs locally at a given location if the mean runoff during the period 2071–2100 either exceeds the mean plus one standard deviation referring to the reference period 1961–1990 (flood risk), or if the mean during 2071–2100 is less than the mean minus one standard deviation of the reference period (drought risk).

The climate scenarios fed into LPJ were all constructed as a combination of the observed mean climate during the reference period (New et al. 1999) and anomalies of temperature, precipitation and radiation added from the climate model runs. As a consequence, all scenarios share the same mean observed climate during 1961–1990. Risks are then defined individually at each land grid cell of LPJ with a geographical resolution 1.5° by 1.5° globally. Risks for grid cells where annual runoff is less than 100 mm/year on average during 1961–1990 are not considered. (Such desert areas cannot sustain rain-fed agriculture so that drought is not normally a risk; they also tend to be sparsely populated so the risk of flooding, which certainly exists, can more easily be circumvented). The logic adopted here assumes that both ecosystems and societies are adapted to current climate variability and that a threat mainly comes from a change outside this range, which manifests itself by a major increase in extreme events. To illustrate this, consider that a shift of the mean by one standard deviation corresponds to a change in the return frequency of an original 100-year extreme event to approximately 10 years.

The study uses outputs from 16 coupled atmosphere-ocean general circulation models and four emission scenarios: a committed climate change with atmospheric levels of greenhouse gases held constant at 2000 levels, and SRES scenarios A1B, A2 and 1B (Nakicenovic et al. 2000). The first warming band (below 2°C) has a total of 16 model runs, the middle band (2 to 3°C) 20 runs, and the highest (above 3°C) 16 runs (see Table 1).

3 Hydrological risk and global average temperature increase

On the level of countries or political entities, a policy maker may be interested in the area under threat of a given risk, depending on a specific outcome as a result of a certain policy. The policy in our example is concerned with global climate change, and the different outcomes of the policy are the different levels of global average temperature increase. The information required by the policy maker is the area under threat within the political entity for which he or she carries responsibility. This type of information is given in Fig. 1 for four cases within the northern mid-latitudes. It shows the area under risk of increased runoff as an indicator of flood risk as a function of global mean temperature change. Each cross represents one of the 52 scenarios that enter the LPJ simulations. There is a large degree of variation between scenarios, with areas affected going up to 25% for the USA, and up to 65% for Russia. The area under risk, however, very clearly increases with increasing levels of warming. The statistical explained variance of the area affected against degree of warming (r^2) is 0.52 for Russia, 0.35 for EU25, 0.18 for USA, and 0.06 for China. This is consistent with the observation that an increase in the hydrological cycle accompanied by global warming affects the more northerly land areas particularly strongly (Raisanen 2005).

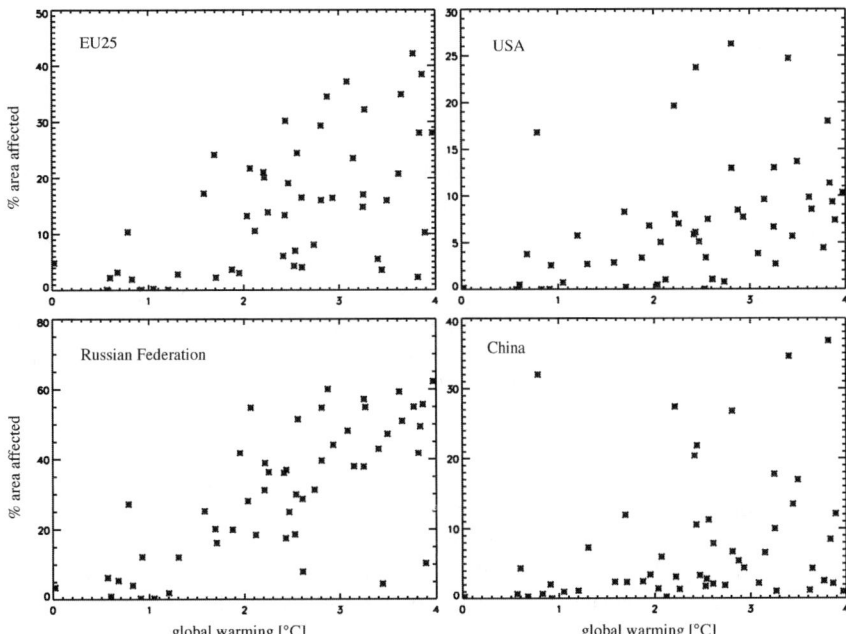

Fig. 1. Area fraction carrying a risk of *increased* runoff vs. global average warming 2071–2100 against 1961–1990; Each symbol represents one climate scenario; Note the variations in the percentage scales

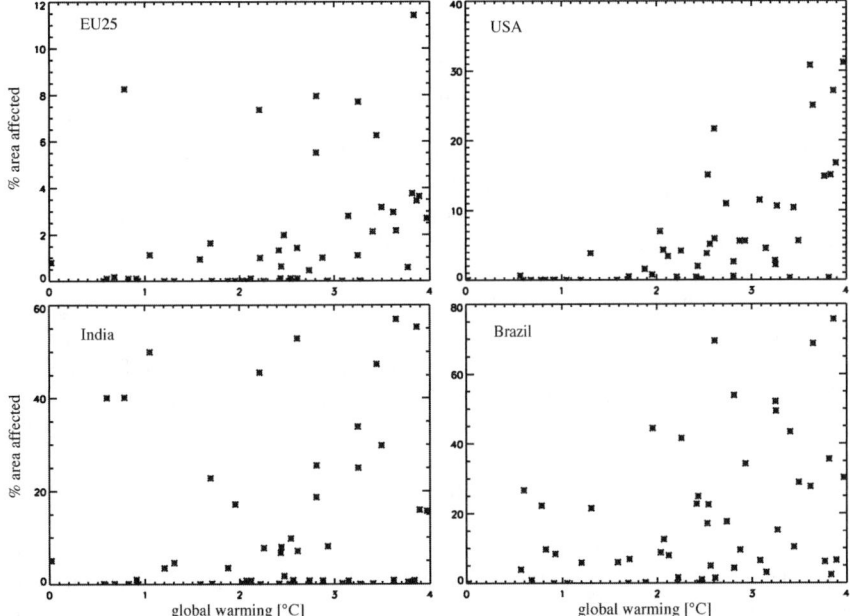

Fig. 2. Area fraction carrying a risk of *decreased* runoff vs. degree of global warming

Figure 2 shows the same type of analysis for the risk of decreasing runoff, again for the mid-latitude EU25 and USA, and for two large tropical countries, India and Brazil. For the mid-latitude cases, there is again an obvious link between warming and risk of drought, in this case it is the USA where the link is strongest ($r^2 = 0.39$), while for EU25 it is about the same as for flooding ($r^2 = 0.13$). The area affected for the USA is somewhat higher than for flood risk (up to 30%), while it is relatively low for EU25 (up to 12%). For India, we find less of a clear link with $r^2 = 0.02$, and large areas could be affected even at moderate warming levels. For the other tropical country considered, Brazil, there is still a strong link ($r^2 = 0.16$). In both cases, areas affected are very large, up to 60% for India, and 80% for Brazil. The correlation found between area under risk and global degree of warming was highly significant at the 99% confidence level in all cases, except for increased runoff in China, which was significant at a 95% confidence level, and decreased runoff in India, which was significant at an 80% level. In summary: as a first step in our analysis, we show that there is a strong indication that the outcome of a policy – more or less global warming – does indeed influence the risks that individual countries and ecosystems face.

Again from the policy-maker point of view, it should be of even more interest to assess not only the area, but also the population under risk. In keeping with our general approach of transparency and simplicity, we assess the amount of human population currently living within the area under risk. We do not attempt to use population scenarios which would introduce an additional level of uncertainty. Therefore, we re-weight the current analysis by multiplying risk levels at each grid cell by the population living in the area. We use the Gridded Population of the World, Version 3 (CIESIN 2004) unadjusted (for UN totals) for the year 2000.

For the EU25, as shown in Fig. 3, we find a population affected of up to 180 million, which incidentally is about the same proportion of the total as the area-based estimates shown in Fig. 1. Increasing risk with increasing warming is again obvious ($r^2 = 0.23$). The main difference is that now a greater number of scenarios show risks close to zero for populations. The same applies to the USA, where the explained variance drops to 0, but affected populations are up to 100 million, about a third of the total population, as against a quarter for area affected. The large number of cases where population affected is small suggests that there is a rather high probability of only a minor impact on a *per capita* basis, both for USA and EU25. Only for the highest impact simulated is relatively more population than area of the USA affected. For Russia, with its vast uninhabited stretches of land in high northern latitudes where the increase in rainfall is especially pronounced, we would expect less of an impact weighted as a proportion of population than of area. However, compared to EU25 and USA, we find fewer cases where the affected population is close to zero, and the maximum affected population, around 100 million, is similar in fraction of total to the maximum affected area (65%). There is also a stronger statistical link with level of warming (with $r^2 = 0.26$). Finally, for China we find a strong reinforcing effect of population distribution with up to 800 million or two thirds of the population being affected (area up to one third), while the number of cases with impact close to zero is about equal for population and area. However, as in the case of the USA, many scenarios predict close to zero risk, and the explained variance is low ($r^2 = 0.03$), which illustrates the high degree of uncertainty regarding hydrological impacts for China.

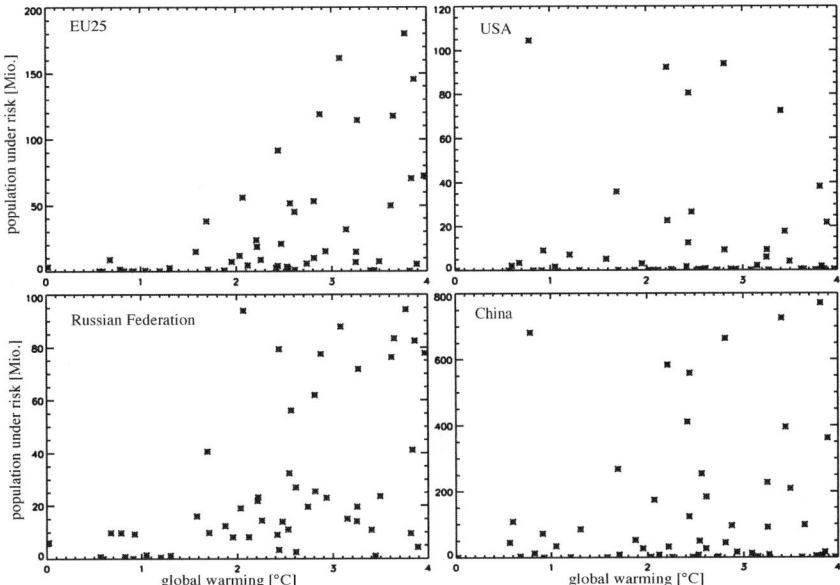

Fig. 3. Population of 2000 living in areas under risk of *increased* runoff vs. degree of temperature increase 2071–2100 against 1961–1990

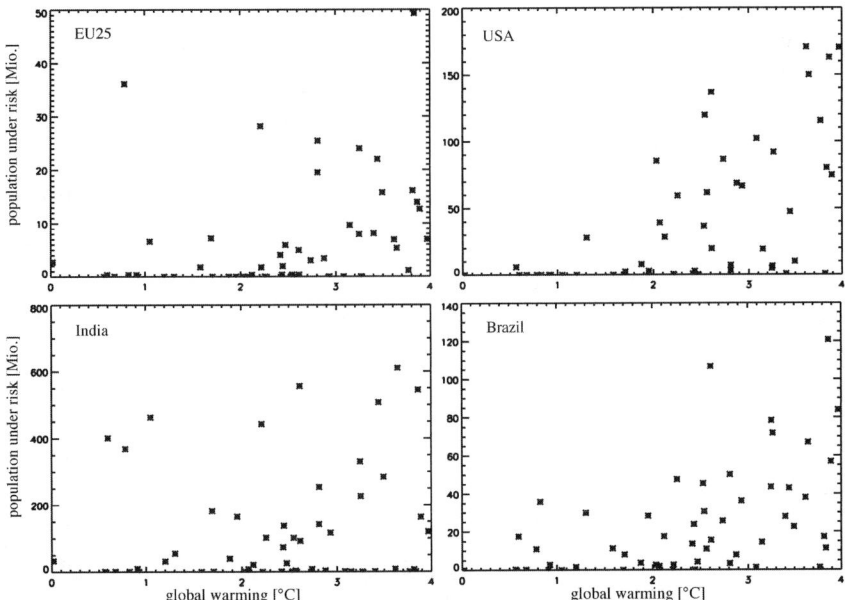

Fig. 4. Population under risk of *decreased* runoff vs. degree of global warming

For drought risk (see Fig. 4), the analysis again shows relatively low impacts for the EU25, but large impacts for the USA with up to 180 million affected, or 60% of the population, which is double the proportion compared to area affected. India shows up to 620 million of current population affected, or 60%, and Brazil with its lower population density still up to 120 million or 65%. Explained variances are 0.11 for EU25, 0.33 for USA, 0.02 for India, and 0.25 for Brazil. As for the area-based estimates, there are especially high risks for India even in the case of low degrees of warming. Thus, we find substantial risk exposure from both flooding and drought, which for many countries is reinforced by the current spatial distribution of the population.

Table 2. Probability in percent of exposing at least 5% population of the specified countries (or regions) to an increase or decrease in runoff that exceeds a specified risk threshold by 2071–2100

Risk	Decreasing runoff (drought)			Increasing runoff (flooding)		
Warming	<2°C	2–3°C	>3°C	<2°C	2–3°C	>3°C
Algeria	0	0	0	0	5	12
Angola	31	25	50	50	75	56
Congo	18	30	31	50	90	93
Ethiopia	18	40	43	31	40	68
Mali	12	20	56	0	20	6
Niger	25	25	43	25	50	81
South Africa	37	50	43	37	45	43
Sudan	37	50	50	43	50	62
Canada	25	55	62	12	25	37
Mexico	25	45	75	18	20	18
USA	6	60	68	12	25	25
Argentina	0	5	25	18	40	37
Brazil	37	60	87	31	40	50
Colombia	6	10	0	37	90	100
Peru	0	5	0	12	30	37
China	12	25	37	31	45	50
India	37	50	50	37	50	56
Indonesia	0	15	31	12	25	43
Iran	6	20	25	6	10	12
Kazakhstan	0	5	6	0	5	0
Saudi Arabia	0	0	0	6	15	12
Russia	6	20	6	50	90	87
Australia	0	10	37	6	25	31
EU25	6	10	18	6	35	56
Globe	**43**	**55**	**75**	**43**	**90**	**100**

4 Acceptable risk and probability of exceedance

As the previous analysis has shown, there are substantial risks associated with climate warming, and these risks generally increase with the degree of global average temperature increase. There is some indication that below a level of 2°C warming, risks become substantially lower. For example, the population of the USA under risk of drought for less than 2°C warming is less than 10 million in all scenarios but one (see Fig. 4). However, for other countries there are still substantial amounts of area and population under risk even below 2°C warming, and it is a moral question of high relevance as to what may be an acceptable degree of risk exposure. We therefore set an arbitrary threshold at 5% and 20% of population and compute the probability that this threshold will be exceeded separately for each warming bracket (<2°C, 2–3°C and >3°C). Again, we do this both for increased runoff (flood risk), and for decreased runoff (drought risk), where the risk is defined as a shift by at least one current standard deviation. We do this analysis for countries with at least 1 million sq. km of surface area (avoiding discrepancies between climate model resolution and country surface area) and at least 10 million inhabitants, as well as for the EU25 and the globe. We disregard the effects of migration and assume that population only changes proportionally within current geographical patterns.

If a potential policy-maker decides that coping with 5% of affected population is acceptable, he will want to calculate the probability that this policy will fail and that a larger number of people will be affected. Efficient risk management will usually demand that this probability stays low. As Table 2 shows, even for a warming of less than 2°C and in a scenario that contains several climate simulations where greenhouse gas levels were held constant at 2000 levels (see above), this risk is indeed very large, both for decreases and increases in runoff. Notable cases are Angola (drought: 31%; flooding: 50%), Sudan (37/43%), India and South Africa (37/37%), and Brazil (37/31%). The risks are thus especially high in tropical countries, as opposed to mostly richer countries of the more northern latitudes: 6% and 12% for the USA, and both 6% for EU25. Exceptions are again found at high northern latitudes: drought risk is still as high as 25% for Canada, and flood risk as high as 50% for Russia. On the global scale, the probability of exceeding 5% of population affected is considerably higher than for specific countries. We assume that the reason lies in the particular characteristic of the climate system: on a global scale, there are more possible realisations of changing climate patterns that leads to risks in some way.

Table 3. Probability in percent of exposing at least 20% population to the specified risks

Risk Warming	Decreasing runoff (drought)			Increasing runoff (flooding)		
	<2°C	2–3°C	>3°C	<2°C	2–3°C	>3°C
Algeria	0	0	0	0	0	0
Angola	6	10	31	50	55	50
Congo	6	0	0	37	70	68
Ethiopia	18	30	25	6	15	37
Mali	6	10	50	0	10	0
Niger	12	10	18	12	15	50
South Africa	12	40	18	31	35	37
Sudan	6	25	25	25	25	31
Canada	0	45	50	6	15	6
Mexico	18	45	50	6	5	6
USA	0	40	56	6	15	6
Argentina	0	0	6	18	20	31
Brazil	6	25	56	6	15	12
Colombia	0	5	0	25	50	93
Peru	0	0	0	6	0	6
China	0	5	18	12	25	25
India	18	15	37	6	25	31
Indonesia	0	10	25	6	25	18
Iran	6	10	6	0	0	0
Kazakhstan	0	0	0	0	0	0
Saudi Arabia	0	0	0	0	0	6
Russia	0	0	0	6	30	50
Australia	0	0	6	0	10	31
EU25	0	0	0	0	10	31
Globe	**0**	**5**	**12**	**6**	**15**	**18**

For a warming of more than 2°C, probabilities of not meeting the target of 5% of population affected generally increase. On a global scale, it can become as high as 100% for increased runoff and greater then 3°C of warming. We note in particular that the risk levels increase substantially for both the USA (68% for drought at >3°C) and the EU25 (56% for flooding at >3°C). It is also important that some countries show a clearly higher risk of decreases compared to increases (Iran, USA, Mali, Brazil, Mexico), while even more show the opposite (Argentina, Colombia, Peru, Russia, EU25, China, also global), which is consistent with the expectation of an increasing hydrological cycle due to climate warming (see above). A number of countries have similarly high risk levels in both directions (India, Australia, Indonesia, South Africa).

From a risk management point of view, it is evident that the probabilities calculated are mostly far beyond what is normally considered to be an acceptable level of risk of not meeting a policy goal. As a consequence, it appears more likely that a policy maker must be prepared for a fraction as high as 20% of population to be affected by climate. This is taken into account in Table 3, which lists the probability of exceeding this revised threshold. For a level of warming of less than 2°C, we find that for drought risk, the highest probabilities are still 18% for Ethiopia, Mexico and India, but 0% for the globe and a large number of other countries. For flood risk, the figure is still 50% for Angola, 37% for Congo and 31% for South Africa, while only 6% at the global scale. If global warming exceeds 2°C, the probability of policy failure again increases in most cases, e.g. for flooding for the EU25 (31% at >3°C, from 6%) and drought for USA (40% at 2–3°C, and 56% at >3°C, from 0%).

5 Conclusions

From a risk management point of view, the problem of climate change poses great challenges. Contrary to the widely used system-analytical approach that starts from scenarios and climate model simulations, we have tried to outline a use of such model calculations that starts from a normative, policy-driven question and applies methods of risk management to address it. The question here is what degree of warming constitutes 'risk manageable' climate change, and what is clearly 'dangerous'? In this case, we consider only changes in runoff, as both increases and decreases can lead some of the most common climate-related threats to both human and ecosystems.

Our first finding is that the degree of global average climate warming is indeed a reasonable predictor for the area affected by risky increases or decreases in runoff. The same is true if current population is used instead of area. Generally, risks of flooding are somewhat more pronounced than risks of drought. Secondly, we find a large degree of variability in the distribution of risks of runoff decreases (drought) and increases (flooding), with generally the highest risks in tropical countries, notably Africa, India, and a range of South American countries. Thirdly, independent of what acceptable amount of population at risk a given policy is prepared to tolerate, accept, or prepare for, we find that the probability of exceeding this target can increase substantially with the degree of global warming. Risk levels at even the lowest warming band below 2°C can still be substantial even if the accepted level is as much as 20% of the population. This notably applies to a number of especially exposed tropical countries, including India, China and a range of African and South American countries.

There is certainly room for improvement of the present analysis. For example, the analysis neglects that certain climate model simulations may be considered more likely, because those models agree better with observations than others. Also, population change was not considered in the present analysis, except indirectly through the use of emission scenarios. However, we believe that further refinement of model instruments designed specifically to guide policy-making should be kept as simple as possible. Instead, users should be provided with necessary background information enabling them to make an educated judgement of the outcome of the analysis. To this end, it will be more important to assess what changes in assumptions are likely to alter the general conclusions substantially. This makes the robustness and transparency of such an instrument the most important targets for further improvement.

References

Azar C, Rodhe H (1997) Targets for stabilization of atmospheric CO_2. Science 276: 1818–1819

Brasseur G, Kenneth Denman K, Chidthaisong A, Ciais P, Cox P, Dickinson R, Hauglustaine D, Heinze C, Holland E, Jacob D, Lohmann U, Ramachandran S (2007) Couplings between changes in the climate system and biogeochemistry. In: Contribution of Working Group I to the Fourth Assessment Report of the Intergovernmental Panel on Climate Change, Cambridge University Press, Cambridge

Carter RT, La Rovere EL (2001) Developing and Applying Scenarios, In: McCarthy JJ, Canziani OF, Leary NA, Dokken DJ, White KS (eds), Climate Change: Impacts, Adaptation, and Vulnerability, Contribution of Working Group II to the Third Assessment Report of the Intergovernmental Panel on Climate Change. Cambridge University Press, Cambridge, pp 145–190

CIESIN Center for International Earth Science Information Network, Columbia University; and Centro Internacional de Agricultura Tropical (2004) Gridded Population of the World (GPW), Version 3. Palisades, NY: CIESIN, Columbia University, http://sedac.ciesin.columbia.edu/gpw

Climate Action Network (2002) Preventing dangerous climate change. Climate Action Network, Bonn, 11 pp, http://www.climatenetwork.org/docs/CAN-adequacy30102002.pdf

Dessai S, Adger WN, Hulme M, Turnpenny J, Köhler J, Warren R (2004) Defining and experiencing dangerous climate change. Climatic Change 64: 11–25

Dilley M (2000) Reducing vulnerability to climate variability in southern Africa: the growing role of climate information. Climatic Change 45: 63–73

Dlugolecki A (2004) A Changing Climate for Insurance: A Summary Report for Chief Executives and Policymakers. Association of British Insurers, London, 24 pp

Dlugolecki A, Lafeld S (2005) Climate Change and the Financial Sector: an Agenda for Action. Alliance Group and WWF, München and Gland, 59 pp

Funtowicz SO, Ravetz JR (1992) The emergence of post-normal science. In: von Schomber R (ed) Science, Politic and Morality, Kluwever Academic Publishers, Dordrecht, pp 85–123

Gitay H, Brown S, Easterling W, Jallow B et al. (2001) Ecosystems and Their Goods and Services. In: Climate Change: Impacts, Adaptation and Vulnerability. Contribution of Working Group II to the Third Assessment Report of the Intergovernmental Panel on Climate Change. Cambridge University Press, Cambridge

Grassl H, Kokott J et al. (2003) Climate Protection Strategies for the 21st Century: Kyoto and Beyond. Berlin, German Advisory Council on Global Change (WBGU)

Hare B, Meinshausen M (2006) How much warming are we committed to and how much can be avoided? Climatic Change 75: 111–149

Head GL, Herman ML (2002) Enlightened Risk Taking: A Guide to Strategic Risk Management for Nonprofits. Nonprofit Risk Management Center, Washington, DC

Hobbs B, Chao PT, Venkatesh BN (1997) Using decision analysis to include climate change in water resources decision making. Climatic Change 37: 177–202

Hulme M, Carter T (1999) Representing uncertainty in climate change scenarios and impact studies'. In: Carter T, Hulme M, Viner D (eds) Representing uncertainty in climate change scenarios and impact studies. ECLAT-2 Workshop Report No I, Helsinki, Finland, 14–16 April 1999, Climatic Research Unit, Norwich, 11–37

IPCC Climate Change (2007) The Physical Science Basis. Working Group I Contribution to the IPCC Fourth Assessment Report, Intergovernmental Panel on Climate Change, in press

Jones R (2004) Incorporating agency into climate change risk assessment. Climatic Change 67: 13–36

Linneroth-Bayer J, Mechler R, Pflug G (2005) Refocusing disaster aid. Science 309: 1044–1046

Manning M, Petit M (2003) A concept paper for the AR4 cross cutting theme: uncertainty and risk. Intergovernmental Panel on Climate Change. http://www.ipcc.ch/activity/cct1 pdf

Mastrandrea DM, Schneider SH (2004) Probabilistic integrated assessment of dangerous climate change. Science 304: 571–575

Meinke H, Stone RC (2005) Seasonal and inter-annual climate forecasting: the new tool for increasing preparedness to climate variability and change in agricultural planning and operations. Climatic Change 70: 221–253

Mills E (2005) Insurance in a climate of change. Science 309: 1040–1044

Morgan MG, Henrion M (1990) A guide to dealing with uncertainty in quantitative risk and policy analysis. Cambridge University Press, Cambridge, 332 pp

Murphy JM, Sexton DMH, Barnett DN, Jones GS, Webb MJ, Collins M, Stainforth DA (2004) Quantification of modelling uncertainties in a large ensemble of climate change simulations. Nature 430: 768–772

Nakicenovic N et al. (2000) Special Report on Emission Scenarios. Cambridge University Press, New York

New M, Hulme M, Jones P (1999) Representing twentieth-century space-time climate variability. Part I: Development of a 1961–90 mean monthly terrestrial climatology. J Climate 12, 829–856

Nutter FW (1999) Global climate change: why U.S. insurers care. Climatic Change 42: 45–49

O'Connor RE, Bord RJ, Fisher A (1999) Risk perception, general environmental beliefs, and willingness to address climate change. Risk Analysis 19: 461–471

Parry M, Arnell N et al. (2001) Millions at risk: defining critical climate change threats and targets. Global Environmental Change-Human and Policy Dimensions 11: 181–183

Piani C, Frame DJ, Stainforth DA, Allen MR (2005) Constraints on climate change from a multi-thousand member ensemble of simulations. Geophys. Res. Let. 32: doi: 10.1029/2005GL024452

Raisanen J (2005) Impact of increasing CO_2 on monthly-to-annual precipitation extremes: analysis of CMIP2 experiments. Climate Dynamics 24: 309–323

Saloranta TM (2001) Post-normal science and the global climate change issue. Climatic Change 50: 395–404

Schlumpf C, Pahl-Wostl C, Schönborn A, Jaeger CC, Imboden D (2001) IMPACTS: an information tool for citizens to assess impacts of climate change from a regional perspective. Climatic Change 51: 199–241

Scholze M, Knorr W, Arnell NW, Prentice IC (2006) A climate-change risk analysis for world ecosystems. Proceedings of the National Academy of Sciences 103: 13116–13120

Sitch S, Smith B, Prentice IC, Arneth A, Bondeau A, Cramer W, Kaplan JO, Levis S, Lucht W, Sykes MT, Thonicke K, Venevsky S (2003) Evaluation of ecosystem dynamics, plant geography and terrestrial carbon cycling in the LPJ dynamic global vegetation model. Global Change Biology 9: 161–185

Suarez P, Patt AG (2004) Cognition, caution, and credibility: the risks of climate forecast application. Risk, Decision and Policy 9: 75–89

Tirpak D, Zhou D, Meira Filho LG, Metz B, Parry M, Schellnhuber J, Yap KS, Watson R, Wigley T (2005) Avoiding Dangerous Climate Change. International Symposium on the Stabilisation of greenhouse gas concentrations, Hadley Centre, Met Office, Exeter, UK 1–3 February 2005, Report of the International Scientific Steering Committee, May 2005, http://www.stabilisation2005.com

Varis O, Kajander T, Lemmelä R (2004) Climate and water: from climate models to water resources and management and vice versa. Climatic Change 66: 321–344

Effects of low water levels on the river Rhine on the inland waterway transport sector

Olaf Jonkeren[1], Jos van Ommeren[1], Piet Rietveld[1]

[1] Department of Spatial Economics, Vrije Universiteit
De Boelelaan 1105, 1081 HV Amsterdam, Netherlands
ojonkeren@feweb.vu.nl
jommeren@feweb.vu.nl
prietveld@feweb.vu.nl

Abstract

This paper examines the welfare effect of water level variation on the river Rhine through its impact on freight prices in the inland waterway transport market. We study this effect using a unique dataset including information about more than 3000 journeys of inland navigation vessels in 2003, of which 700 are useful for our analysis. We demonstrate that the freight price increases by up to 100% at the lowest water levels compared to the situation of normal water levels. The welfare loss for the economy in 2003 as a result of the increase in freight prices due to low water levels amounts to about € 189 million.

Keywords: Inland waterway transport, water level variation, freight price, welfare loss

Acknowledgements: We are very grateful to Dirk van der Meulen for provision of the Vaart!Vrachtindicator-data. Also, we would like to thank RIZA for provision of the data on water levels. This paper was written in the context of the Dutch ongoing Climate Changes Spatial Planning Research programme.

1 Introduction

In the year 2003, water levels on rivers in North-Western Europe were lower than usual. For months there was no precipitation and temperatures were high. Water levels in rivers dropped continuously and new low water level records were set. On some parts of the river Rhine, the main waterway in Western Europe, inland waterway transport was hampered due to the low water levels; there was simply not enough water for ships to navigate without the risk of grounding (CCS 2004).

The Rhine is a combined rain-snow river. As a result of climate change it is expected that the Rhine will be more rain-oriented in the future, i.e. that precipitation will increase in winter and that higher temperatures will cause a smaller proportion of precipitation to be stored in the form of snow in the Alps. Thus, more precipitation directly enters the river. As a result, average water levels will be higher in winter and the number of days with hindrance because of low water will decrease in winter.[1] In summer, besides the reduction in melt-water contribution, there will be less precipitation and more evaporation due to higher temperatures. As a consequence the Rhine will experience lower water levels and more days with low water levels in summer. For inland waterway transport, an increase in the number of days with hindrance because of low water levels in summer and autumn may be expected (Nomden 1997; de Ronde 1998; Middelkoop et al. 2000).

When water levels drop below a certain minimum level, ships cannot navigate without hindrance on certain parts of the Rhine river when they are fully loaded. In particular at one location, the town of Kaub, the effect is pronounced (see Fig. 1). Although for some of the trips that pass Kaub the maximum load factor may be determined by water levels in tributaries of the Rhine, the water depth at Kaub is the bottleneck for the large majority of the trips that pass Kaub. The estimated size of the welfare loss thus concerns cargo that is transported via Kaub during low water levels.

Water levels in the Rhine in 2003 were low from July to November. Such a long period of low water levels may arise more often in the future due to climate change.

In the current paper, we aim to examine the welfare effects of water level variation in the Rhine river through the inland waterway transport sector. Previous estimates suggest that the welfare consequences may be substantial. For example, RIZA et al. (2004) estimated the costs of low water levels for domestic inland waterway transport in the Netherlands based on assumptions about additional costs of low water levels. These extra costs concern the increase in the number of trips, in handling costs and costs as a result of longer waiting times at the locks; they amounted to € 111 million for the year 2003.

---

[1] Note however, that the number of days with hindrance because of high water levels in winter may increase. Nevertheless, this lies beyond the scope of this study as we only focus on low water levels.

Effects of low water levels on the river Rhine on the inland waterway transport sector 55

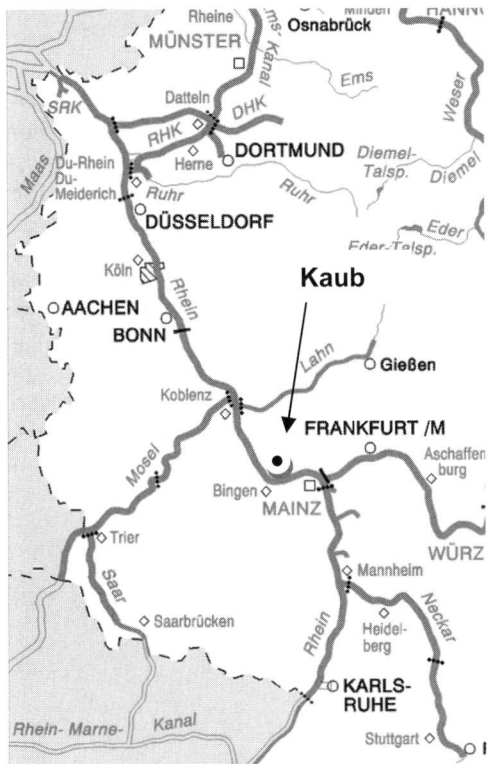

Fig. 1. Location of Kaub (Germany) on the Rhine river

Our calculation of the economic costs of low water levels is based on the relationship between water levels and freight prices in inland waterway transport on the Rhine river. First, we estimate the change in price in periods of low water levels compared to periods of normal water levels with an empirical model. Then, we use a theoretical model to estimate the welfare loss as a result of those higher freight prices.

The purpose of this paper is to contribute to the knowledge about the effects of climate change on the economy. Given the welfare loss due to low water levels, one is able to examine whether investment in projects that aim to make inland waterway transport more robust to low water levels might be economically sound. Furthermore, this paper contributes to the scarce literature on inland waterway transport compared to literature on other transport modes.

In the next section, the theory concerning welfare implications of water level variation will be addressed. In Section 3 we review the methodology and the dataset we use for our research. The results are presented in Sections 4 and 5. Section 6 offers some concluding remarks.

2 Theory

In order to determine the economic cost of water level variation we apply the concept of the economic surplus (e.g. O'Connell 1982; van den Doel and van Velthoven 1993; Perloff 2004). A surplus is a return or benefit above economic or opportunity cost (O'Connell 1982). We will first consider the consumer surplus. The consumer surplus arises when the utility or satisfaction derived from consumption of a commodity exceeds the price of the commodity. Figure 2 gives an example of an (inverse) demand function. For every possible quantity the inverse demand function (D) shows the marginal willingness to pay: the maximum amount of money that the consumer of the last unit is prepared to pay for consuming that unit. The area under the inverse demand function, bounded by the two axes and the vertical line at the equilibrium quantity consumed gives total benefits of consumption at that equilibrium level of consumption. This area represents the highest amount of money an individual would be prepared to pay for a specific quantity of a certain commodity.

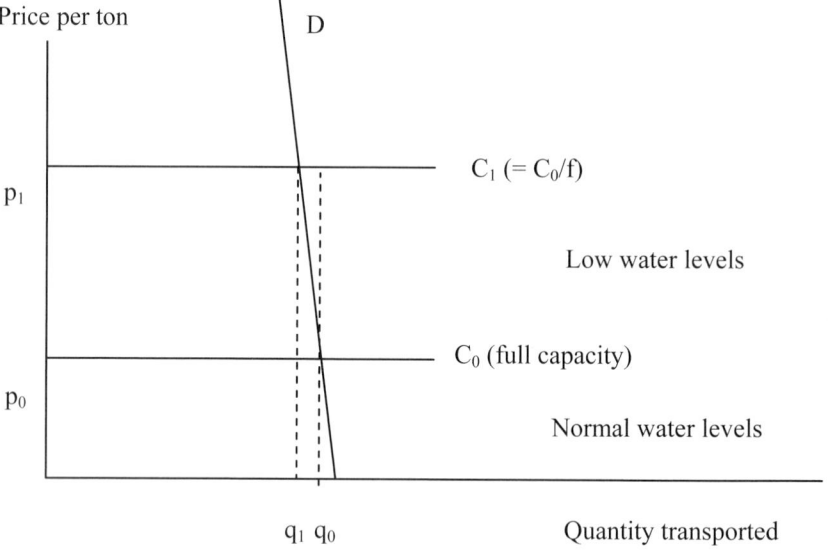

Fig. 2. Economic surplus and welfare loss in the inland waterway transport market

The marginal costs per ton (C_0) are assumed to be independent of the supply of the commodity. In equilibrium, demand equals supply, so $D = C_0$ and q_0 is determined. Thus, total costs are measured by the area under the horizontal line (C_0) for $q < q_0$. The area above the horizontal line C_0, bounded by the y-axis and the demand function represents the benefits less the costs. This area is called the consumer surplus. The consumer surplus is closely related to the economic surplus. The economic surplus is the total generalized surplus of consumers and firms and thus is the sum of the consumer surplus and the producer surplus. Given the absence of market imperfections[2], profits of producers are equal to zero[3], so the economic surplus is equal to the consumer surplus.

Although the inland waterway transport sector does not directly serve a consumer market, the assumption of 'no market imperfections' implies that the change in economic surplus on the inland waterway transport market, as shown in Fig. 2, is equal to the change in the consumer surplus on the market of the transported goods (Lakshmanan et al. 2001).

Now we investigate the effect of low water levels on the economic surplus. Suppose that the water level drops below a certain level, then inland navigation vessels have to reduce their load factor (f), to be able to navigate safely. At C_0, f is 1. At any cost level above C_0, $f < 1$. As a result, the marginal costs per ton at low water levels, $C_1 = C_0/f$, so transport costs per ton are higher. As a consequence inland waterway transport enterprises charge a higher price, p_1, so the economic surplus is reduced.

The above standard analysis is based on two main assumptions: demand is inelastic, and supply is perfectly elastic. The latter assumption seems reasonable, also in the short run, since there exists fierce competition between many inland waterway transport enterprises and entering the market is easy.

Demand for transport capacity of inland waterway transport is often thought to be inelastic due to several reasons. The price for transportation by inland navigation vessel for most bulk goods is substantially lower than transport by any other mode. Consequently, the price can rise substantially before other transport modes become competitive and modal shift effects are expected to be small. Besides, inland navigation vessels transport such large quantities that other modes of transport by far do not have enough capacity to transport all cargo originally transported by inland navigation vessels. Another even more fundamental reason for the inelasticity is that shippers[4] aim to prevent their production process from costly interruptions. Costs of inland waterway transport are only a small part of total production costs[5]. Thus, paying more for inland waterway transport in periods of low water levels is more cost-effective than having interruptions in the production process. As a result, demand for inland waterway transport is inelastic, however not totally inelastic. For example, shippers are able to postpone transport and make use of their stocks. Elasticity of demand for inland waterway transport is es-

---

[2] With no market imperfections we mean that there is perfect competition in which there are many players on the supply as well as on the demand side, each individual supplier does not have any market power, so net profit of all suppliers will be zero and the producer surplus is zero. This assumption is reasonable for the market under investigation.

[3] In reality it might be that inland waterway enterprises do generate some profits, however for reasons of clarity we neglect this possibility.

[4] Shippers must not be confused with inland waterway transport enterprises. Shippers are companies that want their products to be transported.

[5] Frederic Harris (1997) mentions that for most low value goods like coal and steel inland waterway transport is only a small part (about 2%) of total production costs.

timated to be in the order of about -0.2[6]. Due to substitution, a part of the cargo will be transported by other modes, so the elasticity for inland waterway transport demand will even be less sensitive.

Now, the economic costs (EC) of low water levels as a result of a price increase from p_0 to p_1 can be calculated. We have:

$$EC = (p_1 - p_0)q_0 - \tfrac{1}{2}(p_1 - p_0)(q_0 - q_1) \qquad (1)$$

and the definition of the price elasticity:

$$\varepsilon = (\Delta q / q_0)/(\Delta p / p_0) \qquad (2)$$

Then, we have for modest changes in p:

$$(q_1 - q_0) \approx \varepsilon q_0 (p_1 - p_0)/p_0 \qquad (3)$$

After substitution of the latter expression in (1) we have:

$$EC = (p_1 - p_0)q_0(1 + \tfrac{1}{2}\varepsilon(p_1 - p_0)/p_0) \qquad (4)$$

3 Dataset and methodology

3.1 Data

For our research, we employ a unique dataset, the Vaart!Vrachtindicator[7]. The Vaart!Vrachtindicator contains information about journeys made by inland waterway transport enterprises in Europe who report information about their journeys such as the price per ton, place and date of loading, place and date of unloading, capacity of the ship, weight of the cargo, type of cargo and more variables. The dataset only contains information on inland waterway transport enterprises that operate in the spot market[8] where the freight price per ton and low water surcharges are negotiated for each journey. Low water levels at Kaub are defined as 140 cm and lower (this will be justified later). Our dataset contains 3191 journeys which were reported during 2003. About 730 of these passed Kaub and did not include containers. We exclude container transport since its unit of measurement is

---

[6] Compared to the amount of tons transported in 2001, 2002 and 2004, in 2003 about 12% less was transported past Kaub (CCR 2002, 2004). Part of this percentage was transported by other modes. Using Eq. 2, ε is -0.19. Details of the calculations are available on request from the authors.
[7] More information can be found on the website www.vaart.nl.
[8] Inland waterway transport enterprises that operate in the long-term market (and work under contract) and receive a fixed price per ton throughout the year are not included in the dataset.

TEU (Twenty feet Equivalent Unit) whereas other products transported by inland navigation vessels are measured in tons.

3.2 Descriptives

To give an idea of the value of the most important variables we present the following descriptives in Table 1 and Fig. 3.

Table 1. Descriptives of most important variables

Variable	No of items	Minimum	Maximum	Mean	Std. Deviation
Water level ("Pegel"; in cm)	730	35.00	680.00	153.78	82.29
Price per ton (in €)	728	1.80	52.00	9.94	6.89
Logarithmic price per ton	728	0.59	3.95	2.11	0.58
Distance of transport (in kilometers)	714	127	997	681	154
Capacity of ship (in tons)	578	630	5391	1676	779

Note: Data originate from the Vaart!Vrachtindicator.

In Fig. 3 the water level is measured by the 'Pegelstand'[9] of the Rhine at Kaub and the freight price per ton is the price inland waterway transport enterprises that pass Kaub receive for transporting one ton of cargo. The figure shows a negative relationship between the freight price and the water level. For example, in September 2003 water levels were exceptionally low and prices were exceptionally high. As the water level drops, inland waterway transport enterprises have to reduce the load factor of their ships and will ask a higher freight price per ton in order to compensate for the tons of cargo they cannot transport. The freight price curve shows heavy variations. This is due to differences in the length of the trips. Every observation represents one trip made by an inland ship. For a long trip a bargeman receives a higher price per ton than for a short trip.

---

[9] 'Pegelstand' or 'Pegel' is the measurement for water levels in inland waterway transport in Germany. However, it is something different from actual water depth. There are several locations along the Rhine where the Pegelstand is measured. Each Pegel has its own 0-point. Thus, with Pegel Kaub it is only possible to determine navigation depth in the surroundings of Kaub. For other places other Pegels are valid. To give an idea of the maximum possible navigation depth at Kaub at certain Pegels: the water depth at Kaub is more or less one meter higher than the Pegelstand at Kaub. So, at Pegel Kaub 90 cm there is about 190 cm water between riverbed and surface. However, in this paper Pegels and water levels can be regarded as synonyms.

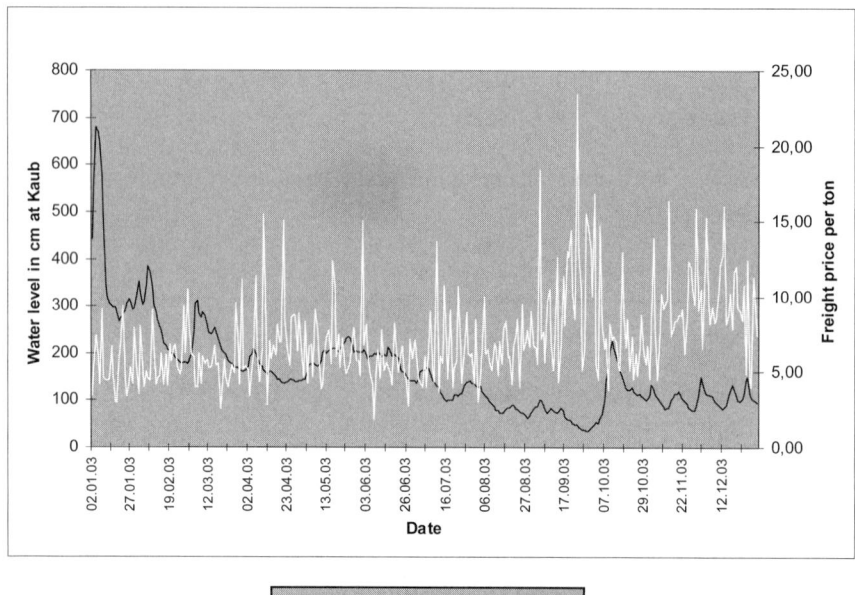

Fig. 3. Water levels on the Rhine and freight prices in inland waterway transport for 2003
Note: The data originate from RIZA and the Vaart!Vrachtindicator

With a scatter plot (Fig. 4) we can show a first rough estimate of the effect of water level variation on freight price. The plot also shows a negative relationship.

The observed average price per ton at low water levels[10] is € 12.47 while the observed average price per ton at normal water levels is € 7.45. Then, Δp is € 5.02 which means that on average at low water levels (≤ 140 cm Pegel Kaub) freight price per ton is 67% higher than at normal water levels (> 140 cm Pegel Kaub) in 2003 for all journeys without containers that passed Kaub. A more refined analysis is needed however, since in these figures we did not correct for several other variables that have an impact on freight prices.

---

[10] Low water levels start at 140 or 150 cm Pegel Kaub because that is the level below which low water surcharges are paid to the inland waterway transport enterprises.

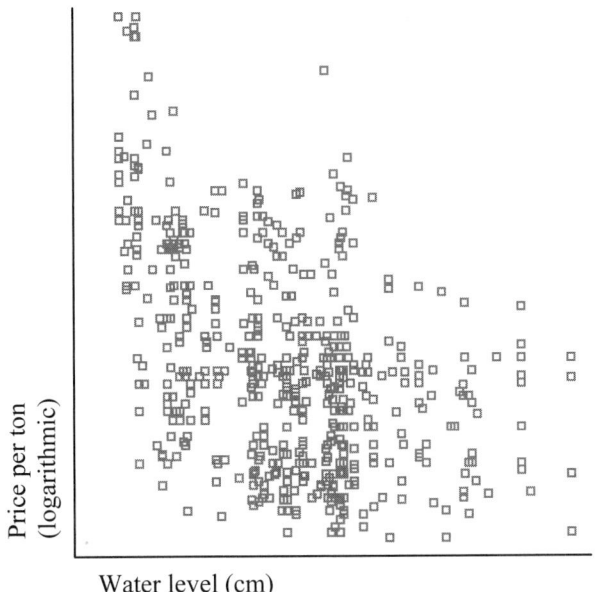

Fig. 4. Scatter plot of logarithmic price per ton and water level
Note: The data originate from RIZA and the Vaart!Vrachtindicator

4 Multiple regression analysis

We assess the impact of water level variation on freight price per ton using linear regression. We use the logarithm of the price per ton, and a set of independent variables including the water level at Kaub.

Among the independent variables are: (1) logarithm of journey distance because the larger the distance, the higher the price per ton is, (2) ship size (4 dummy variables), which allows for economies of scale, (3) cargo type (39 dummy variables), because due to differences in the density of each cargo type the price per ton will also depend on the cargo type, (4) upstream or downstream navigation because for upstream navigation more fuel is needed than for downstream navigation and (5) water level (7 dummy variables). Fuel price was not included as a variable in the regression because fuel prices stayed relatively constant throughout the year 2003.

As a result of the selection (and some missing values) the number of journeys was reduced to 728, which is about two journeys a day on average during the year considered. To examine the validity of our regression model we checked for multicollinearity and heteroskedasticity[11]. The standard logarithmic model is:

$$Y = \beta_0 + \beta_1 X_1 + \beta_2 X_2 + \ldots + \beta_{52} X_{52}, \qquad (5)$$

where Y is the logarithmic price per ton, ß$_0$ is a constant and ß$_1$ through ß$_{52}$ are independent variables. The R^2 of our model is 0.744. Further, the correlation between the variables water level and logarithmic price per ton is -0.335, which is significant at the 0.01 level[12]. Table 2 shows the estimated increase in freight price per ton as the water level drops one or more intervals taking into account the mentioned control variables.

Table 2. Increase in price at different water level intervals

Pegel Kaub (in cm)	Price per ton	Std. Error
140–131	+18%	5.4%
130–121	+39%	6.4%
120–111	+41%	4.7%
110–101	+40%	6.0%
100–91	+76%	5.4%
90–81	+53%	4.5%
< 81	+103%	3.6%

Note: The data originate from the Vaart!Vrachtindicator (2003)

We measure the water level variable by means of seven dummy variables. The reference-category is the group for water levels > 140 cm. We find that the results[13] are in line with the expectations derived from the Internationale Verlade und Transport Bedingungen[14] (VBW 1999).

Additional analyses show that for water levels between 141 cm and 150 cm there also exists a small effect of water level variation on freight price per ton. After correcting for the other explanatory variables we arrive at a pattern that is smoother than our initial estimate, although the drop for the interval 90–81 cm remains difficult to explain.

---

[11] We found that there hardly exists any correlation between the independent variables. Furthermore, the residuals of the regression follow a normal distribution and the variance of the residuals appears to be constant.
[12] Details on the regression results are available on request from the authors.
[13] Inconsistencies between the moment of reporting the freight price and the moment of passing Kaub may distort the results somewhat. However, the error is not systematic but random, so the estimate is biased towards zero.
[14] This document (the IVTB) determines rights and obligations of inland waterway transport enterprises and shippers in the European market and serves as a guideline for both parties when setting up short- and long-term contracts. The IVTB also gives guidelines for low water surcharges which can be used in negotiations. The IVTB states that low water surcharges can usually be charged at 140 or 150 cm Pegel Kaub.

5 Welfare analysis

The extra costs for inland waterway transport as a result of water levels below 140 cm imply a welfare loss. The size of the welfare loss can be estimated using Eq. 4 (see Section 2). For calculating the average price increase, first the average price of all journeys made at normal water levels is calculated. The results in Table 2 are used to calculate the average price increase at low water levels, taking into account the number of journeys in each water level interval.

Our estimate shows an average price increase of € 4.61 per ton (with $p_0 = $ € 7.45 and $p_1 = $ € 12.06), which is an increase of 62% compared to freight prices at normal water levels. Note that the p_1 of € 12.06 is the corrected price at low water levels while the p_1 of € 12.47 (see Section 3.2) is observed.[15] The value q_1 is the amount in tons transported past Kaub while water levels were lower than 140 cm at Kaub. Consequently, q_0 is q_1 plus the loss of demand due to low water levels; q_0 and q_1 turned out to be 43.6 and 38.4 million tons respectively in 2003. We already know ε is -0.19. Consequently, if we use Eq. 4 (see Section 2) the estimated welfare loss for all journeys that passed Kaub in 2003 is about € 189 million[16]. Translated to costs per unit of time this means that the average economic costs per week with low water were about € 6.7 million in 2003. Compared to the turnover in the Kaub related Rhine market of about € 680 million, the welfare loss in 2003 is about 28% of this turnover.[17]

One limitation to our research is the size of the standard errors (see Table 2). This leaves room for variation of the percentages mentioned in the table. To decrease the error terms in future research we will conduct analyses with a larger dataset.

Second, there could be periodic events that correlate with water level variation and that increase demand in certain periods. We cannot take into account such seasonal effects because we only observed one year.

Third, note that the welfare loss cannot be assigned to a certain geographical area, because the welfare loss is caused by all journeys that pass Kaub. These journeys have origins and destinations all over North-Western Europe.

Our assumption of no market imperfections implies that profits in the inland waterway transport market are zero. Furthermore, it is assumed that margins of each link in the chain remain unchanged and that the loss of welfare is finally paid for by the consumer (Lakshmanan et al. 2001). If this assumption does not hold, the welfare loss would be somewhat larger or smaller.

---

[15] So the price increase from € 7.45 to € 12.06 is only due to low water levels and is corrected for the effects of other (control) variables.
[16] Details on the calculations are available on request from the authors.
[17] The annual amount of cargo transported through the Kaub related Rhine market is about 80 million tons. The average price per ton for all journeys in the dataset that pass Kaub is about € 8.50.

6 Conclusions

In this paper we studied the effect of water level variation on freight prices per ton in inland waterway transport. For our estimation of the effect, several characteristics of inland waterway transport on the Rhine river were taken into account. A negative effect of water level variation on freight price per ton was found. With the results of our analysis we estimated a welfare loss of € 189 million due to water level variation on the river Rhine.

In future research, by also estimating the effect of water level variation on the load factor of ships and duration of journeys, we intend to analyse whether inland waterway transport enterprises are under- or overcompensated for the reduction in load factor.

References

CCR (2002) Economische ontwikkeling van de Rijnvaart, statistieken. Centrale Commissie voor de Rijnvaart
CCR (2004) Economische ontwikkeling van de Rijnvaart, statistieken. Centrale Commissie voor de Rijnvaart
CCS (2004) Kleinwasser. Combined Container Service GmbH&Co KG, Mannheim
de Ronde JG (redactie) (1998) Ministerie van Verkeer en Waterstaat, Instituut voor Marien en Atmosferisch Onderzoek (Universiteit Utrecht). De keerzijde van ons klimaat. Brochure RIKZ-RIZA-IMAU, Den Haag
Frederic Harris (1997) Invloed van klimaatverandering op vervoer over water. Frederic R. Harris BV, Den Haag
Lakshmanan TR, Nijkamp P, Rietveld P, Verhoef ET (2001) Benefits and costs of transport, classification, methodologies and policies. Papers in Regional Sciene, vol 80, 139–164
Middelkoop H, Rotmans J, van Asselt MBA, Kwadijk JCJ, van Deursen WPA (2000) Development of Perspective-based scenarios for global change assessment for water management in the lower Rhine delta. In: UNESCO-WOTRO International Working Conference "Water for Society"
Nomden E (1997) Een goed klimaat voor de binnenvaart? BERTRANC, Gemeentelijk havenbedrijf Rotterdam
O'Connell JF (1982) Welfare Economic Theory. Auburn House Publishing Company, Boston
Perloff JM (2004) Microeconomics, 3rd edn Pearson Addison Wesley, Boston
RIZA et al. (2004) Droogtestudie Nederland. Aard, ernst en omvang van de droogte in Nederland. Projectgroep droogtestudie Nederland
van den Doel H, van Velthoven B (1993) Democracy and welfare economics. University Press, Cambridge
VBW (1999) Internationale Verlade- und Transportbedingungen für die Binnenschiffahrt (IVTB). Verein für europäische Binnenschiffahrt und Wasserstraßen e.V., Duisburg

Spillover impacts of climate policy on energy-intensive industry

Vlasis Oikonomou[I], Martin Patel[II], Ernst Worrell[III]

[I] SOM, University of Groningen
PO Box 800, 9700 AV Groningen, Netherlands
v.oikonomou@rug.nl

[II] Copernicus Institute, Utrecht University
Heidelberglaan 2, PO Box 80.115, 3508 TC, Utrecht, Netherlands
M.k.patel@uu.nl

[III] Ecofys
PO Box 8408, 3503 RK, Utrecht, Netherlands
E.Worrell@ecofys.nl

Abstract

Energy-intensive industries play a special role in climate policy. These industries are particularly vulnerable if climate policy would lead to higher energy costs, and if they would be unable to offset these increased costs. The side-effects of climate policy on GHG emissions in foreign countries are typically referred to as "spillovers". This paper provides a review of the literature on the spillover effects of climate policy for carbon-intensive industries. Reviews of past trends in production location of energy-intensive industries show an increased share of non-Annex-I countries. This trend is primarily driven by demand growth, and there is no empirical evidence for a domestic cost increase as a result of environmental policy in these development patterns. Climate models nevertheless do show a strong fleeing of industries to developing countries as a result of such policies. Nonetheless, the energy and carbon intensity of energy-intensive industries is rapidly declining in most developing countries, thus reducing the "gap" between industrialized and developing countries. Despite the potential for positive spillovers in the energy-intensive industries, none of the models used in the analysis of spillovers of climate policies has an endogenous representation of technological change for the energy-intensive industries. This underlines the need for a better understanding of technological development in order to provide a better basis for the economic models used in the *ex-ante* assessment of climate policy.

Keywords: Spillover, technological change, carbon leakage, iron & steel industry, climate policy

1 Introduction

World-wide, industry is responsible for about 50% of greenhouse gas (GHG) emissions (Price et al. 2002). About three-quarters of these emissions are caused by energy-intensive industries (estimate based on IEA 2003a, 2003b) that produce iron and steel, aluminium, chemicals, fertilizers, cement, and pulp and paper. The emission intensity makes these industries an important target for climate policy. At the same time these industries are particularly vulnerable if climate policy would lead to higher energy costs and if they would be unable to offset these increased costs. A possible reaction of energy-intensive industry to climate policy can be the shift of production activities to countries with less stringent or no environmental policies, while simultaneously transferring production technologies to these countries. The side-effects of climate policy on GHG emissions in foreign countries are typically referred to as "spillovers".

Negative spillovers of climate policy, which are also referred to as *carbon leakage*, can be caused by:

- Relocation of energy-intensive industries to countries with a less stringent climate policy, which potentially could lead to lower production costs. However, energy-efficiency improvement due to climate policy may lower the energy costs and provide ancillary (productivity) benefits to energy-intensive industries.
- Increased net imports of energy-intensive goods from countries which have no or less stringent climate policies and more carbon-intensive production structures.
- Reduction in global energy prices due to reduced demand in climate-constrained countries, reducing the incentive for energy-efficiency improvement for energy-intensive industries in countries without climate targets.

Positive spillovers of climate policy can be caused by:

- Development of energy-efficient and low-GHG technologies in climate-restrained countries and implementation of these technologies around the world including countries that are not participating in climate stabilization regimes.

This introductory section is based on the project "Carbon leakages and induced technological change: the negative and positive spill-over impacts of stringent climate change policy" (Sijm et al. 2005) that contains a number of parallel assessment case studies on positive and negative spillovers of climate change policies. In this case study we focus on the effects of these spillovers on *energy-intensive industry*, based upon existing studies and empirical results. A more analytical work on the same issue has previously been conducted by Oikonomou et al. (2004). In the literature, there are many existing studies on the issue of carbon leakage, mainly from the perspective of economic theory with some others examining the application of models. This paper attempts to combine both approaches by comparing the theory with actual market behaviour and modelling results regarding the effects of climate change policy on energy-intensive industry.

In Section 2, we describe current production trends for energy-intensive products. Section 3 stresses the importance of production factors for relocation. This is followed by a discussion of the results of three detailed models for the steel sector in Section 4. Section 5 deals with potential positive spillovers of climate policy. Finally, we draw conclusions in Section 6.

2 Production trends

In our analysis of energy-intensive production we consider steel, aluminium, paper, nitrogenous fertilizers and cement, which together account for approximately 80% of the total energy use of the energy-intensive sectors in OECD countries (OECD 2002). Between 1971 and 2000 the OECD's overall production share of these energy-intensive products decreased on average from 88% to 43% (e.g. for steel from 89% to 57%). Oikonomou et al. (2004) make use of two indicators for these products in order to analyse whether changes in global production shares of energy-intensive products are predominantly competition-based (i.e. due to comparative advantage of production factors in developing countries) or predominantly demand-driven (i.e. due to development of new markets). The first indicator shows whether the production increase in non-Annex-I countries (of the Kyoto Protocol[1]) is demand- or rather export driven, while the second one shows whether or not non-Annex-I countries are gaining market shares in the rest of the world (developing markets included). The first indicator revealed that the production increase in non-Annex-I countries has been mainly demand-driven for steel, paper and cement, whereas it has been strongly export-driven for aluminium. The second indicator demonstrated that Annex-I countries lost some market share for aluminium in their own regions at the beginning of the 1990s but this development stopped quickly and the ratio of net imports into Annex-I countries to the consumption in Annex-I countries has been constant since then. All in all, a general outcome is that there has been no significant loss of market shares for Annex-I countries in their own regions and that the decrease of global production shares of Annex-I countries (leading to increase in non Annex-I ones) is predominantly demand-driven. While certain production factors may be more favourable in some developing countries, the developments to date do not seem to indicate that production in developing countries is clearly more competitive.

Climate policy has been relatively widely in force since the middle to late 1990s, while in a few countries first steps were already taken in the early 1990s. It could possibly be argued from the global production shares that so far, environmental policies are unlikely to have influenced the industrialized countries' market shares to a noticeable extent (since they were demand based). Furthermore, the

---

[1] Non Annex-I countries are developing countries under the United Nations Framework on Climate Change Convention that have no immediate restrictions to reduce their Greenhouse Gas emissions. Annex-I countries are industrialized and developed countries that have agreed to reduce their emissions to levels below their 1990 emission levels.

developments in other years and for other materials fit rather well with the trends in earlier periods when climate policy was not yet in force (Sijm et al. 2005).

3 Factors of production leading to relocation

In this section we provide an overview of production factors which are of key relevance for investment decisions in the energy-intensive sectors and we briefly discuss how these production factors differ between industrialized countries and developing countries. Furthermore, we summarize empirical analyses on the importance of environmental regulation for the decision about the investment location.

3.1 Overall factors

In relocation investment decisions, apart from the elements determined in a company's cost calculation for a given product there are also intangible or hidden costs (or benefits). In this section we take into account the most important investment criteria according to EBRD studies (Bevan and Estrin 2000) and the assessment method applied by the U.S. Country Assessment Service of Business International, as described by Wheeler and Mody (1991).

Many studies conclude that low wages and large market size (in million tonnes) and/or high market growth (in % GDP growth) that capture potential economies of scale are very decisive investment criteria (Bevan and Estrin 2000; Brainard 1993; Lankes and Venables 1996; Patibandla 2001; Singh and Jun 1995). However, proximity of product to market customers is also seen as a key criterion since it (partly) compensates for other cost factors. For energy-intensive industries, the Foreign Direct Investment (FDI) flows more to countries with high market growth, such as China (8%), India (5%) and Korea (5%), and less to the rest of Asia (UNCTAD 2001). These factors are also significant for relocation decisions.

Labour, energy (including taxes and environmental expenses) and – depending on the product and the country – raw materials and auxiliaries are often cheaper in developing countries than in industrialized countries. Other production factors tend to be more expensive. For example, transportation costs have been high due to monopolistic prices charged by shipping companies and/or the lack of critical mass and adequate infrastructure in the developing countries. Furthermore, depending on the product and the policy regime, import barriers may exist for the benefit of either the developing country or the industrialized country. With the increasing implementation of WTO agreements, trade-related barriers are, however, being reduced (Rumbaugh and Blancher 2004). Other important factors include export subsidies and public guarantees (for exports) as well as capital flow restrictions and price controls imposed by governments.

In total, these factors have led to higher production costs in developing countries than in industrialized countries, thus explaining why – until recently – there was only rather limited foreign investment in the developing world. However, the conditions are improving in developing countries. With globalisation and the advent of multinationals in developing countries, a substantial decrease of costs can be achieved for several production factors. A more detailed analysis of this trend is presented in Oikonomou et al. (2006).

Amongst the scarce quantitative information related to production factors is the data on cost categories by industrial sector as published by various sources. With regard to relocation induced by climate and energy policy, the most relevant indicator that can be extracted from this source is the share of energy cost as a fraction of total cost. For Dutch energy-intensive industry for instance, this fraction is around 15% for bricks and tiles, 12% for iron and steel, 8% for basic chemicals, 9.5% for pulp and paper, 7% for glass and 6% for cement (Ramirez et al. 2005). While the comparison of this type of information for developing countries and countries in transition could provide some insight, a meaningful cross-country analysis would need to correct for important differences between countries, especially regarding the product mix, the extent of further processing and the importance of non-productive activities such as trade and engineering services.

3.2 Environmental policy factors

In theory, in a competitive market, environmental regulations drive up fixed and variable costs, which can result in lower profitability and hence a reduction of competitiveness. These higher costs, as explained in the previous sections, may lead companies to relocate. However, available studies have shown that environmental policy in the *past* generally has *not* been a significant decision criterion for the location of the investment and hence does not represent a key explanatory factor for the investments in the developing world (relocation). An interesting point stated by Neumayer (2001) is that neither the World Bank nor the World Economic forum include environmental compliance costs in the competitiveness indicators (referred to as "attractiveness to invest in a country") which they publish (World Bank 1998; WEF 1999).

Most of the empirical analyses have dealt with the so-called "pollution haven" hypothesis. According to this hypothesis, industries relocate production facilities to countries with less stringent environmental requirements. Nevertheless, it is argued even in the early literature that the pollution haven hypothesis cannot be tested since it lacks empirical coverage for a number of reasons (Neumayer 2001). Firstly, pollution abatement costs, as calculated by OECD, are considered to account for less than 2% of the GDP for most countries.[2] The same expenditures as a percentage of total gross fixed capital formation amount (only in one case) to only 1.9%. Due to this low share we conclude that the extra expenditures of the industries have not been significant enough to justify relocation (OECD 2003). Indus-

---

[2] This study distinguishes between expenditure from the public and business sector. For the public sector, they range from 0.2–1.4% of the GDP, while for the business sector from 0.1–1.2% of the GDP.

tries with increasing returns according to scale tend not to relocate unless the pollution abatement costs rise to extremely high levels (Markusen et al. 1995). Another reasoning in the study of Neumayer (2001) is that even where environmental costs are high, international investors might not be deterred, as long as these standards provide clear market rules. In the developing countries the uncertainty with regard to policy changes is much higher. In general, the empirical literature does not confirm the pollution haven hypothesis unequivocally (Cole 2004). According to Bruvoll and Faehn (2006), econometric analyses have failed to find strong evidence that tighter environmental policies significantly affect investment patterns (Zarsky 1999; Smarzynska and Wei 2001; Eskeland and Harrison 2003) or trade patterns (Jaffe et al. 1995; Janicke et al. 1997). However, recent literature based on panel data at the industry level reveals that pollution havens could entail a moderate effect on industry relocation (Brunnermeier and Levinson 2004; Ederington et al. 2003; Ederington and Minier 2003).

In general, the dominant empirical studies show that the cost effects of the environmental regulation are very small, or even negligible.[3] Other cost factors seem to be more decisive investment criteria, with the most important ones being market size and growth (regional demand) and the wage level. There are two important limitations which must be borne in mind when drawing conclusions from the pollution haven literature regarding relocation due to environmental policy. Firstly, the pollution haven work has been mainly limited to the fixed-cost component of the abatement technologies because it refers, in particular, to end-of-pipe technologies. Environmental taxes that mostly affect variable costs are often not taken into consideration, because the focus of the pollution haven literature lies on direct regulations and cannot capture integrated solutions of environmental technology processes (Bouman 1998). Secondly, the competitiveness of plants in developing countries is nowadays higher than in the period to which most of the pollution haven analyses refer. In summary, we conclude that the existing studies cannot provide a clear picture about the effect of environmental policy on the relocation of energy intensive industries; however, they do indicate that if a relation between environmental policy and relocation does exist, then it is statistically weak.

4 Modelling results

In this section we compare the empirical findings with results from climate policy models that address energy-intensive industries. We have identified only three models that study the steel sector in detail: SIM (Steel Industry Model) (Mathiesen and Moestad 2002), STEAP (Steel Environmental strategy Assessment Program) (Gielen and Moriguchi 2001) and POLES (Prospective Outlook for the Long term Energy System) (Hidalgo et al. 2003). Each of the three models takes the various

---

[3] Such exhaustive literature is provided by Xing and Kolstad (1998)

steel production technologies into account.[4] According to these models even very moderate climate policies (tax or allowance levels of 10–25 $/t of CO_2) lead to severe leakage (and hence also to substantial relocation) as presented in Fig. 1. The leakage rate (shown in the X-axis) can be defined in this case as the ratio over a given time period of the gross climate-policy-induced change of GHG emissions in non-Annex-I countries to the gross climate-policy-induced change of GHG emissions in Annex-I countries. In the Y-axis, various levels of carbon tax (also reflecting the shadow price of marginal abatement costs) are presented.

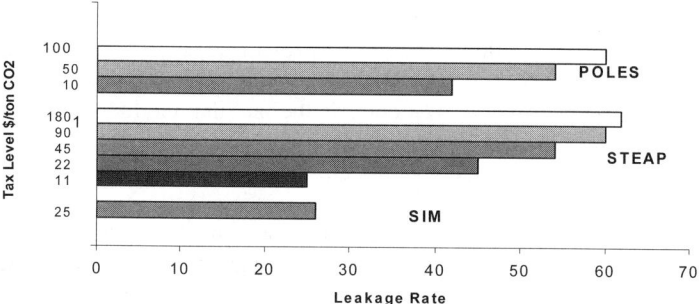

Fig. 1. Carbon leakage (%) in the steel sector for various policy scenarios in the three models

According to the model results shown in Fig. 1, climate policy would substantially affect steel production in various world regions and could lead to serious carbon leakage. As expected, stricter policy is predicted by the models to lead to higher leakage rates. The leakage rates at given tax levels differ substantially between the three models: at around 10 €/t CO_2, the leakage ranges between 25% (STEAP) and 40% (POLES). In contrast, at around 20–25 $/t CO_2 the leakage rates according to STEAP and POLES coincide well, while leakage according to SIM is nearly only half as high. Since it is not obvious how the differences in regional and temporal scope can explain these ranges, this diversity of results seems to indicate that the results are subject to major uncertainties. Together with the contrasting real experience made in countries which have introduced pollution control in the past, this raises questions about the reliability of the models. Since the climate policy models do not appear to account explicitly for differences in elasticities across countries/regions and time periods (past, present, future), we conclude that the modelling results are subject to major uncertainties. Another reason for doubting the reliability of the results is that none of the models reviewed seems to be have been calibrated for longer periods in the past (no publications are known on this subject matter). Moreover, if environmental policy has so far been a decision criterion of only subordinate importance (according to the pollution haven literature), one would expect this to be even more the case for climate policy: CO_2 emissions

---

[4] They encompass three ways of producing steel; the basic oxygen furnace (BOF), the electric arc furnace (EAF) and the open hearth furnace (OHF). The dominant form in the world production is the BOF, with 58% share, then EAF, 34% and OHF with other technologies (8%) (IISI 2001)

can be reduced by energy-efficient technologies, which often actually lower production costs. In contrast, emissions from traditional facilities can often only be reduced by add-on technologies, which generally increase production costs. At the same time it must be kept in mind that globalisation is gradually changing the business conditions. In addition, climate policy has been rather *soft* in the past; this also applies to most other environmental policies, which have led to the implementation of end-of-pipe technologies with inherently cleaner and more efficient processes. In contrast, climate policy, if undertaken seriously, could have a stronger impact on business decisions in future. However, this argument seems to explain only partly the contradictory results of empirical analyses and models because, according two of the three steel models reviewed, substantial leakage rates are to be expected even at (very) low CO_2 tax levels. One of the key factors that seems to be undervalued in current models is the fact that the location of production facilities is determined to a large extent by the demand for the products. Factoring this element into the models would clearly lead to much less drastic results for relocation.

The results obtained with energy and emission models when simulating the consequences of climate policy is often reported by means of a compact indicator called "*leakage rate*" or "*leakage*". Leakage quantifies how much of the policy-induced emission reduction in Annex-I countries is "eaten up" by emission increase in non-Annex-I countries. While the concept of this indicator seems plausible, its usefulness for policy-making is nevertheless limited (Oikonomou et al. 2006). This has to do with the fact that leakage is a derived indicator, which does not provide the full picture. Its application is limited for two reasons: firstly, the fact that it is a ratio of *gross* changes in emissions and does not address the *net* effect; and secondly, because it does not relate the changes to the absolute emission flow before implementation of the climate policy. It is therefore not advisable to make comparisons and to draw policy conclusions on the basis of the indicator "leakage" *only*.

5 Positive spillovers

Whereas up to here we have focussed primarily on the negative effects of climate policy for carbon emissions, this section deals with the positive spillover effects. The main positive spillover is the potential "technological spillover" from increased efforts in industrialized countries (with a GHG-emission reduction policy) to implement and develop technologies with a lower GHG-intensity. This spillover can take three forms (IPCC 2001):

1. R&D can refocus on low-GHG development paths for technology development. This enhanced focus can lead to reductions in all countries, whether or not they participate in policy regimes to reduce GHG emissions
2. Increased market share of low-GHG technologies can result in technology improvement and reduce the costs of these technologies

3. GHG policies in countries that focus on technology performance can send a strong signal to foreign competitors, which then need to adapt their R&D strategies in order to become competitive.

Some analysts have argued that these positive spillovers will counteract or even offset the negative spillovers or leakage (IPCC 2001). For example, if new production capacity in non-Annex-I countries were to use state-of-the-art technology to replace the reduced production from older, inefficient plants in Annex-I countries, the total emissions could be lower. Because modern, less energy- or carbon-intensive production technology[5] is mainly developed and produced in industrialized countries, policies in these countries will affect technology development and transfer paths worldwide.

Figure 2 presents the carbon intensity index (CII) for iron and steel production in selected countries. If a country were to produce steel at "best practice" efficiencies it would have a CII of one. A higher CII represents a higher potential for emission reduction. The energy and carbon intensity of energy-intensive industries is rapidly decreasing in most developing countries, reducing the "gap" between industrialized and developing countries (Kim and Worrell 2002), as shown in Fig. 2. This is due to the use of state-of-the-art technologies in the construction of new plants. In practice, very inefficient facilities may also be found in these countries, often next to modern state-of-the-art facilities. If climate change policies in Annex-I countries would affect the export of steel from developing countries, it remains unclear whether the marginal emissions of a new plant or the average emission intensity should be used to estimate the net emission leakage. In neither case is the answer straightforward, and it would be too simplistic to assume that production in non-Annex-I countries will be more energy and CO_2-intensive.

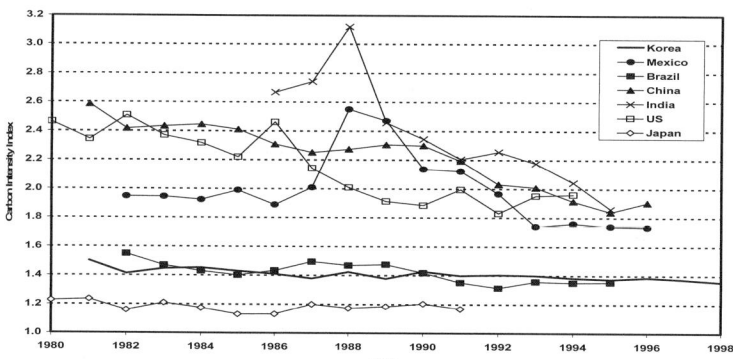

Fig. 2. Carbon intensity index for iron and steel production
Source: Kim and Worrell 2002

---

[5] The terms energy-intensive and carbon (or GHG) intensive industries are used as synonyms in this paper. Energy-intensive industries are by definition also GHG-intensive industries, except for those industries that use electricity from low-GHG technologies (e.g. hydropower) or fuels of renewable origin.

Considerable potential for emission reduction still exists, in both developing and industrialized countries. Technology development is likely to deliver further reductions in energy use and CO_2 emissions, when supported in a suitable manner. Although this development will mainly take place in industrialized countries, developing countries will be the most important markets for these technologies.

It is still unclear what will be the likely magnitude of the effect of climate policy on the rate of energy-efficiency improvement for advanced technologies in the energy-intensive industries. Research in the energy sector suggests that technological change in this sector is induced in response to market conditions (Grubb and Koehler 2001). Hence, it is likely that climate policy will affect technology development patterns. Some climate modellers have modelled endogenous technological change. Van der Zwaan et al. (2002) incorporated endogenous technological change as a function of cumulative capacity, predicting that this would reduce the costs for CO_2 emission reduction considerably. Lutz et al. (2005) modelled endogenous technological change in the German iron and steel industry, predicting a strong improvement in energy efficiency accompanied by a longer-term (beyond 2010) change towards less energy-intensive production processes in reaction to climate policy (modelled as a carbon tax).

Despite the potential for positive spillovers in the energy-intensive industries, none of the models used in the analysis of spillovers of climate policies has an endogenous representation of technological change for the energy-intensive industries. Recently, several groups have started to incorporate mechanisms for simulating changes in technology performance as a function of development and deployment, but none of these groups address demand-side technologies, especially with regard to the energy-intensive industries.

6 Conclusions

In order to assess the overall effect of climate policy, we distinguish between *negative* spillovers (e.g. relocation of energy-intensive industries to non-Annex-I countries, thus increasing energy intensity in those countries) and *positive* spillovers (e.g. increased development and deployment of energy-efficient technologies).

Based on the historical development of the production of energy-intensive products by regions, we conclude that the global production shares of Annex-I countries have been falling continuously in the last few decades. We reviewed the production factors that drive investment decisions to favour location in developing countries and tried to extract their significance. Furthermore, we examined the effect of the environmental regulations on relocation of the energy-intensive sectors. Most studies in the past have shown that environmental policies generally have *not* been a significant decision criterion for the location of the investment and hence do not represent a key explanatory factor for the investments in the developing world (relocation). However, new studies do reveal a slight effect of these policies on investment behaviour.

We compared these empirical findings with results from climate models. According to these models, even very moderate climate policies (tax or allowance levels of 10–25 $/t of CO_2) would lead to severe leakage (and hence also to substantial relocation). This is in contrast to the empirical studies on pollution control and raises questions about the reliability of the models. Since the models do not appear to account explicitly for differences in elasticities across countries/regions and time periods (past, present, future), we conclude that the modelling results are subject to major uncertainties.

The discrepancies of the results of the empirical studies on both positive and negative spillovers with the modelling results warrant further research in this field. Empirical research is needed in order to improve the understanding of technology development in industry, especially focussing on the role of policy and international technology transfer patterns (e.g. global suppliers, changing trade patterns, role of FDI, and potential spillovers to local firms). Further research is necessary with regard to the production factors and their importance for investment decisions. This could be carried out with interview-based and bottom-up analyses of the drivers, revealing the relevance of each of the production factors and evaluating the macro and microeconomic variables. This could help modellers to construct more realistic mechanisms for projecting carbon leakage and technological change in climate models.

References

Bevan A, Estrin S (2000) The determinants of foreign direct investment in transition economies. CEPR Discussion Paper No 2638, London, UK

Bouman M (1998) Environmental costs and capital flight. Tinbergen Institute Research Series No 177, University of Amsterdam, Netherlands

Brainard SL (1993) A simple theory of multinational corporations and trade with a trade-off between proximity and concentration. NBER Working Paper 4269, National Bureau of Economic Research

Brunnermeier SB, Levinson A (2004) Examining the evidence of environmental regulation and industry location. The Journal of Environment and Development 13 (1): 6–41

Bruvoll A, Faehn T (2006) Transboundary effects of environmental policy: Markets and emissions leakages. Ecological Economics 59: 499–510

Cole MA (2004) Trade, the pollution haven hypothesis and the environmental Kuznets curve: Examining the linkages. Ecological Economics 48: 71–81

Ederington J, Levinson A, Minier J (2003) Footloose and pollution-free. NBER Working Paper No W9718, National Bureau of Economic Research

Ederington J, Minier J (2003) Is environmental policy a secondary trade barrier? An empirical analysis. Canadian Journal of economics 36 (1): 137–154

Eskeland AE, Harrison GS (2003) Moving to greener pastures? Multinationals and the pollution haven hypothesis. Journal of Development Economics 70: 1–23

Gielen DJ, Moriguchi Y (2001) Environmental Strategy design for the Japanese Iron and Steel industry, a global perspective. Working document, NIES, Tsukuba

Grubb M, Koehler J (2001) Induced Technical Change: Evidence and Implications for Energy-Environmental Modelling and Policy. Working paper, Department of Economics, Cambridge University, Cambridge, UK

Hidalgo I, Szabo L, Calleja I, Ciscar JC, Russ P, Soria A (2003) Energy Consumption and CO_2 emissions from the world iron and steel industry. IPTS, EUR 20686EN, European Commission Joint Research Centre

IISI (2001) Steel Statistical Yearbook. International Iron and Steel Institute, Brussels

International Energy Agency (IEA) (2003a) Energy Balances of Non-OECD countries 2000–2001. IEA/OECD, Paris

International Energy Agency (IEA) (2003b) Energy Balances of OECD countries 2000–2001. IEA/OECD, Paris

Intergovernmental Panel on Climate Change (IPCC) (2001) Climate Change 2001, Mitigation (WG III). Cambridge University Press, Cambridge, UK

Jaffe AB, Peterson SR, Portney PR, Stavins RN (1995) Environmental regulation and the competitiveness of U.S. manufacturing: what does evidence tell us? Journal of Economic Literature 33: 132–163

Janicke M, Binder M, Monch H (1997) Dirty industries: patterns of change in industrial countries. Environmental and resource economics 9 (4): 467–491

Kim Y, Worrell E (2002) International Comparison of CO_2 Emission Trends in the Iron and Steel Industry. Energy Policy 30: 827–838

Lankes HP, Venables AG (1996) Foreign Direct Investment in Economic Transition: the changing pattern of investments. Economics of Transition 4: 331–347

Lutz C, Meyer B, Nathani C, Schleich J (2005) Endogenous Technological Change and Emissions: The Case of the German Steel Industry. Energy Policy 33 (9): 1143–1154

Markusen JR, Morey ER, Olewiler N (1995) Competition in regional environmental policies when plant locations are endogenous. Journal of Public economics 56: 55–77

Mathiesen L, Moestad O (2002) Climate policy and the steel industry: achieving global emission reductions by an incomplete climate agreement. Norwegian School of Economics and Business Administration, Discussion Paper 20–02, Bergen, Norway

Neumayer E (2001) Pollution havens: Why be afraid of international capital mobility? Presentation at Environmental Economics, org. by Smulders J and Bulte E, Tilburg University, Netherlands

OECD (2002) Energy Balances for OECD countries. Organisation for Economic Cooperation and Development, Paris

OECD (2003) Pollution abatement and control expenditure in OECD countries. Organisation for Economic Cooperation and Development, Paris

Oikonomou V, Patel M, Worrell E (2004) Does climate policy lead to relocation with adverse effects for GHG emissions or not? Report prepared within the project "Carbon leakages and induced technological change: The negative and positive spill-over impacts of stringent climate change policy". Project commissioned by the Dutch National Research Programme on Global Change (Scientific assessments and policy analyses, NRP-CC-WAB). Department of Science, Technology and Society/Copernicus Institute at Utrecht University, Utrecht, Netherlands

Oikonomou V, Patel M, Worrell E (2006) Climate Change: Bucket or Drainer. Energy Policy 24 (18): 3656–3668

Patibandla M (2001) Pattern of Foreign Direct Investment in Emerging Economies: an exploration. Working Papers from Copenhagen Business School, No 1, Department of International Economics and Management, Denmark

Price L, Sinton J, Worrell E, Phylipsen D, Hu X, Li J (2002) Energy Use and Carbon Dioxide Emissions from Steel Production in China. Energy 27: 429–446

Ramirez CA, Patel M, Blok K (2005) The non-energy intensive manufacturing sector. An energy analysis relating to the Netherlands. Energy 30 (5): 749–767

Rumbaugh T, Blancher N (2004) China: International Trade and WTO Accession. IMF Working Paper 04/36

Sijm J, Lako P, Kuik O, Patel M, Oikonomou V, Worrell E, Annevelink B, Nabuurs GJ, Elbersen W (2005) Spillovers of Climate Policy. ECN, Report C--05-014, Amsterdam, Netherlands

Smarzynska S, Wei SJ (2001) Pollution havens and foreign direct investment: dirty secret or population myth? CEPR Discussion Paper, vol 2966, CEPR, London

Singh H, Jun KW (1995) Some new evidence on determinants of foreign direct investment in developing countries. Policy Research Working Paper 1531, World Bank

United Nations Conference on Trade and Development (UNCTAD) (2003) World Investment Report. FDI policies for development: National and International Perspectives. United Nations, New York and Geneva

van der Zwaan BCC, Gerlagh R, Klaassen G, Schrattenholzer L (2002) Endogenous Technological Change in Climate Change Modeling. Energy Economics 24: 1–19

Wheeler D, Mody A (1991) International investment location decisions: the case of US firms. Journal of International Economics 33: 57–76

World Bank (1998) Competitiveness Indicators. World Bank, Washington DC

World Economic Forum (WEF) (1999) Global Competitiveness report. Geneva, Switzerland

Xing Y, Kolstad C (1998) Do lax environmental regulations attract foreign direct investment? Department of Economics, Departmental Working Papers, Paper wp28-98pt1, UCSB, California

Zarsky L (1999) Havens, halos, and spaghetti: untangling the evidence about foreign direct investment and the environment. Organization for Economic Cooperation and Development, Paris

Risks, vulnerability, and participation: a layered management approach

J Terry Rolfe

Integrity Research and Communications
3145 Semiahmoo Trail, Surrey, British Columbia, Canada
tsrolfe@aol.com

Abstract

Canada is evolving towards a layered water resource management approach which will empower Canadian communities and accelerate response to critical issues such as climate change and resource scarcity. Greater community involvement has occurred both spontaneously and been encouraged within the federal government's *transdisciplinary* approach, as reflected in the embrace of integrated water resource management (IWRM). This trend fits within the evolving global, democratic culture, with its increasing recognition of rights, interests and vulnerabilities. By using consensus-building and creative problem resolution, the layered management approach moves away from traditional optimization models and communities themselves become leaders in adaptation. This evolving culture will be set out in international context, explaining how Canada's goals have led to a greater awareness of individual, group, sector and community-based vulnerabilities. It will lay out the concept of *vulnerability* within a framework consistent with initiatives undertaken by the Intergovernmental Panel on Climate Change (IPCC).

Keywords: Risk, vulnerability, participation, layered management, integrated water resource management

1 Introduction

This paper flows from a team effort to link physical and socio-economic impacts due to water supply changes in the South Saskatchewan River Basin (SSRB). This CCIAP[1]-funded project provided "An Assessment of the Vulnerability of Key Water Use Sectors in the South Saskatchewan River Basin (Alberta and Saskatchewan) to Changes in Water Supply Resulting from Climate Change". These net impacts assessed for key sectors will invariably influence sub-regional social, economic and environmental characteristics, many of which can be monitored by way of climate change indicators. The primary task was to develop an end-to-end framework capable of meshing the more predictable dynamics with the many cascading uncertainties associated with climate change. This challenge called for pragmatism, recognizing the need to bridge scientific expertise with the front-line insights and commitment. The evolving layered management approach is thus of interest, as it creates synergy across scales and draws participation more broadly. This synergy is particularly relevant because the vulnerability of SSRB stakeholders rests not only upon activities carried out in the watershed proper, but also on the Basin's surface water contribution to other areas, peripherally and indirectly.

Water supply issues are known to contribute to socio-economic vulnerability. As stated by Kulshreshtha et al. (2002, p. 2), "water is fundamental to human life and its many socio-economic activities. Key water resources stresses now and over the next few decades relate to access to safe drinking water, water for growing food, environmental degradation due to overexploitation of water resources, and deterioration in water quality (Arnell et al. 2001). Socio-economic vulnerability is the joint product of environmental risk variables and social factors, where risks are related to the physical aspects of climate related hazards exogenous to the social (human) system". The modeling of risks posed by physical hazards and climate change, as well as related social program outcomes, has been advanced considerably as the UN Environmental Program (UNEP) Intergovernmental Panel on Climate Change (IPCC) has shifted its focus to the socio-economic study of adaptation. Given the diversity and variability of climate change impacts across regions, the IPCC has encouraged regional and sub-regional studies, many of which cover vast areas, such as the South Saskatchewan River Basin.

In this paper we analyze the implications of the IPCC's shifted focus to socio-economic adaption. In particular, we consider how human response should be viewed under an ecosystems approach which integrates physical and biotic systems and acknowledges the primacy of our shared, global life-support system. The intent here is to explain how multi-level participation can more effectively address risks and help reduce vulnerability under climate change.

We proceed as follows. In sections 2 and 3 we demonstrate how societal aspects can be taken into account under the ecosystems approach and how the IPCC integration of physical impacts and societal response has led to enriched decision-

---

[1] Natural Resources Canada, Climate Change Impacts and Adaptation Program.

making tools. Considering climate change, the advantages of integrating the dynamics of human social systems, especially the knowledge of engaged stakeholders, become apparent. The United Nations Framework Convention on Climate Change (UNFCCC) has evolved, now exploring critical concepts such as vulnerability and resilience, which are discussed in more detail in section 4. Section 5 explores the improved dynamics of participation and decision-making within a layered management approach, noting that comparative and multi-level analyses can be illuminating when shared resources such as water are at issue. Section 5.1 delves into the changing nature of public risk perceptions and how these influence layered management; Section 5.2 further considers how network perceptions, involvement and motivation also become key; Section 5.3 explains how a broader *transdisciplinary* approach helps foster community-based vested interests and creativity. In Section 6 we draw some conclusions.

2 Human response under an ecosystems approach

Under an ecosystems approach, the influence of climate change must be cast over multi-dimensional realms, including both terrestrial and aquatic physical and biotic systems, *including* human social systems. The human parameters of evolution and adaptation are part of this holistic puzzle. It is therefore imperative to understand both interests and motivations on the part of all resource stakeholders, and specifically, water interests both near and afar. The notion of property rights has been evolving under the umbrella of global common property regimes (Ostrom 1990) and transnational responses to local activities have intensified in particular for those with serious environmental implications.

For these reasons, it is both reasonable and practical for local community members and other key economic[2] stakeholders to be more fully engaged in water resource planning. Local residents are likely to be acutely and immediately aware of impacts, especially if they are tapped into scientific expertise. However, it is naïve to assume that local interests in the Basin can be isolated from interests elsewhere. A well-organized, inclusive process of local involvement might in fact be increasingly needed to respond, diffuse and counteract interests from "external" parties whose actions may at times be at odds with local interests and values[3]. This is one reason why a multi-dimensional, mixed method approach can be useful for examining resource issues and vulnerabilities.

Known risks and calculable socio-economic impacts can be estimated along with a range of lesser predictable quantitative and qualitative scenarios. This approach can not only satisfy the needs of local community planners, but also meet the compliance standards of government, provide input and feedback to policy makers, and appeal to other influential parties. It has been argued that exposure to

---

[2] These might be persons affiliated with any scale of business venture both directly through ownership and employment or indirectly through share-holding.
[3] See Reed (2003) on the multi-dimensionality of response in a remote resource-based community.

hazards[4] in particular sets the dimensions of vulnerability within the frame of resilience, with both vulnerability and resilience then identified as socially constructed "determinants" of adaptation (Smit et al. 2000). As Gallopin argues (2002), *resilience takes planning* and a mixed method approach which moves beyond clear determination may prove useful for taking into account scenarios and surprises, as well as predictable trajectories and branch points.

3 The shifting focus on impacts under the UNFCCC

There has been growing awareness under the leadership of the UNFCCC that an ecosystem approach is essential to our survival. The ecosystem approach integrates all atmospheric, terrestrial and oceanic physical and biotic systems. It highlights systems integrity – that the whole is more than the sum of the parts – and that system characteristics continually play out in terms of information flow, feedback loops, hierarchy, thresholds and equilibrium. A key acknowledgment under this approach is that the survival of human civilization rests upon its earthly life-support system. It is a daunting challenge to understand the complex linkages between human, ecological and physical systems, especially under a period of accelerated change. The full risk implications of habitat and biodiversity loss are inadequately understood and there is the likelihood of *cascading uncertainties* when physical and social dynamics are integrated. The ecosystem approach emphasizes that sub-system thresholds are important: that they may be breached alone, simultaneously or in sequence and that temporary chaos may occur but new equilibrium dynamics are likely to prevail which may or may not be as ideal for human life.

The UNFCCC requires its parties explore programs to facilitate adaptation to climate change. Until recently, there has been overwhelming emphasis on knowledge and technical solutions geared towards the mitigation of direct climate change impacts. This focus led to the development of methods and tools to systematically evaluate different adaptation strategies and resulted in the release of many useful outputs, including the 1999 *Compendium of Decision Tools to Evaluate Strategies for Adaptation to Climate Change*. Work is now underway to extend the range of tools to assess human adaptation, expanding the repertoire of decision-making under a socio-economic umbrella which considers the "entire process of vulnerability and adaptation assessment" (UN 1999, p. 2).

The UNFCCC focused earlier on the magnitude and urgency of climate change as driven by global general circulation models (GCMs). This focus was linked to scenarios driving impacts within regional biogeophysical sub-systems (such as the carbon and hydrological cycles) and resource sectors such as agriculture and forestry. However, there has been growing interest in how these climate change impacts might be mitigated or exacerbated by socio-economic systems and chal-

---

[4] This notion of "vulnerability" is at odds with interpretations in the fields of psychology and finance.

lenges associated with human security[5]. There has thus been a shift in focus from a more limited, market-based economic view of resource management to a broad socio-economic framework which casts human vulnerabilities in socio-political context. This increased emphasis on human geography (linking human situation to time and place) could provide an improved link between climate change and institutional development. Although as Folke et al. (1998) argued there is incongruency between conventional (formal) institutions and ecosystem properties, it is the case that informal institutions still exhibit systems characteristics with ingenuity derived from situational challenge and the synergy of heterogeneous participation.

The UNFCCC's shift in focus can also be related to the notion of vulnerability and its relationship to adaptive capacity, measured as resilience.

4 Vulnerability – the context of evolving interpretations

There have been important recent developments in the literature on vulnerability as it pertains to socio-economic systems and the management of water resources. Building on Kulshreshtha et al. (2002), the SSRB team has explored the evolving concept of "vulnerability" as enriched by interpretation across many disciplines and sectors. This *transdisciplinary* dynamic not only cuts across disciplines, forcing the examination of both assumptions and professional terminology, but opens the door to multi-dimensional participation which sites the notion of vulnerability in the realm of political economy.

The following interpretations of vulnerability are particularly useful for the SSRB end-to-end exercise. The SSRB project framework must not only dove-tail physical and socio-economic impacts but also allow for a wide range of policy extensions. The scope of policy considerations will likely benefit from a mixed method approach which uses both risk assessment and environmental indicators where practical (that is, where *known* outcomes can be anticipated) and scenario contingency planning where *unknown* outcomes flow more from cascading uncertainties driven by complex system dynamics.

As a concept, vulnerability continues to be of focus in IPCC work because of the word's inclusion in critical documents including the UNFCCC (Smit 2005). The IPCC defined vulnerability as: "the degree to which a system is susceptible to, or unable to cope with, adverse effects of climate change, including climate variability and extremes. Vulnerability is a function of the character, magnitude, and rate of climate change and variation to which a system is exposed, its sensitivity, and its adaptive capacity" (McCarthy et al. 2001, Box SPM-1). This definition refers specifically to the link between climate change exposures and resulting environmental and socio-economic impacts.

---

[5] This urgency is acknowledged now, but it was noted much earlier by Homer-Dixon (1991). Consideration of human security, as a dimension of climate change, is most prevalent in continuing work which views climate change more broadly within the continuity of sustainable development (see Munasinghe 2001).

The sequencing of exposures and associated sensitivities across human and ecological systems translates into a sector or community's resilience and susceptibility to harm. This susceptibility is a function of the characteristics of the exposures themselves (frequency, extent, magnitude and duration), as well as the very nature of the sector and community. These latter domains include physical and social capital, availability and liquidity of resources, and coping responses developed within the socio-political context. The extent to which sectors or communities can cope with – and perhaps even *benefit* from – either direct or indirect exposures derives a *net* vulnerability which in turn impacts on an organization's resilience.

However, there are many facets to sector and community vulnerability which make it difficult to separate physical from socio-political exposures. In the interests of further engaging stakeholders, it may not be that fruitful to dwell on fine-tuning measurement of vulnerabilities based on "isolated" physical exposures[6]. Berkes and Folke (1988) highlighted the inseparability of human and environmental systems, preparing the ground to explore linkages between ecological and social system resilience (Folke et al. 2002). Here "ecological resilience" can be viewed as both integrated and dynamic:

> "The amount of change a system can undergo and still remain within the same state or domain of attraction, is capable of self-organization, and can adapt to changing conditions (after Carpenter et al. 2001). Holling (e.g. 1986a) defined ecological resilience as the magnitude of disturbance that a system can experience before it moves into a different state (stability domain) with different controls on structure and function ..." (Folke et al. 2002, p. 36)

This broader picture of *vulnerability*, as a feature applicable to all biotic systems, including human sectors and communities, is essential if considering risks and uncertainty which are likely to accelerate and cascade under climate change. As sections 2 and 3 have shown, the societal aspects can be incorporated within the ecosystem approach, with aligned avenues of analysis and decision-making tools now pursued under the United Nations' initiatives on climate change. Vulnerability, as described above, is a useful focus for assessing risks and trade-offs under changing circumstances. As can be concluded to this point, any analysis of impacts and net vulnerability for particular human sectors have to consider the transient nature of resources such as water as well as the tendency of ecological actors and dynamics to breach human barriers. Section 5 will now argue in turn that a *transdisciplinary* approach is thus essential: it can draw insights across systems, scales, and academic disciplines, linking social scientists with ecologists and other physical scientists to more accurately assess risk in total. That risk as an outcome of impact exposure, sensitivities and resilience can be further understood with a layered management approach which broadly engages stakeholders and creates synergy and commitment across human and geographic scales. The success of identifying

---

[6] Prairie farmers, taking variable weather as a *given*, may not be so inclined to look separately at climatic and socio-economic change. For example, these stakeholders are drawn more so to discussions of how their global competitors will fair under climate change than how climate change may affect them here at home.

and mitigating these risks also rests, as section 5 suggests, on new global dynamics, including shifting human perceptions towards risks and how people define their allegiances to communities. For climate change adaptation, where impacts will often be highly localized or site-specific, the engagement of all residents and by extension, all interested parties, will be increasingly essential for assessing risks.

5 Layered management: participation and decision-making

In Canada, integrated water resource management (IWRM) has built upon the strong tradition of integrated assessment. Integrated assessment considers the "interactions of physical, biological, and human systems in order to assess long-term consequences of environmental and energy policies such as limits on greenhouse gas emissions, and other strategies to negate climate change" (Bell, Hobbs and Ellis 2003, p. 289). However, the IWRM approach openly pursues the principles of accountability, transparency and efficiency, as may take into account front-line perceptions and demonstrated preferences (Dupont 2005). Community-level satisfaction with respect to the achievement of these principles may nevertheless be subject to socio-political relativity.

As Chandler (1993, vii) argues, "it is impossible to understand fully any social system (in the absence of a comparative context)". Under a political economy lens, one compares state and local institutions, their structure and powers, intergovernmental relations, legal restraints, apportioned functions, and financial arrangements, as well as the role of political parties and pressure groups. All of these factors impact on both the real and perceived[7] influence of the electorate and local stakeholders[8] (Pocklington 1985; Chandler 1993), which in turn affect both government leadership and public motivations. Public motivations are clearly not limited to national fronts, and this rings true for Chandler's statement as well; this invariably casts IWRM analysis into a global trade and ideological frame even if the focus is on a sub-regional scale. Accordingly, there is much to be learned from sharing worldwide insights on efforts to develop local partnerships for better governance (OECD 2001). Many such initiatives have focused on problems of social exclusion: for example in Finland there has been local involvement in efforts to reduce unemployment (see Cinnéide 2001).

For water resource management, a regional and international trade perspective remains essential. Across the Prairies, co-operative local and inter-jurisdictional

---

[7] The government's assessment of public interest and concern about a particular event (i.e. a flood) will influence its decision to offer relief – an uncertainty which poses a deterrent to pro-active risk management.
[8] Stakeholders defined in light of financial interests in local enterprises (as consumers or suppliers) have yet another means of exercising influence. Those without financial means may be limited in their participation to public forums, pressure groups and political involvement (see Pocklington 1985).

agreements have evolved for over a century to deal with resource use and climate variability. Many of the local arrangements have evolved to offset geographic factors such as a dispersed population and perceived hinterland treatments (see Innis, in Drache 1995).

5.1 Public risk perceptions and layered management

Climate change is likely to enhance our understanding of public risk perceptions, both in immediate circumstances and in our place within the world's ecology. The foregoing political economy approach applies here as well, given Canadian communities are subject to the North American media's selective showcasing of events. While many Canadian citizens now regard themselves as global participants, aware of remote events and actively engaged in the resolution of disastrous events, there is now the possibility that the focus needed for important local issues is diffused. Such is evidently the case in the area of food security (Rolfe 2004) which leads many to overlook equity issues here at home (*Hunger Count*, Wilson and Tsoa 2001). Effective layered management may not only help overcome the diffusion on resource issues but also broader undercurrents of distrust and cynicism leveled at both the media and the government.

The potential of layered management might be limited by dichotomous, hierarchically-ordered assumptions. These assumptions are evident in communications gaps observed between "top down" versus "bottom up" information flows. Uncritical reference to these leveled distinctions might be best discarded altogether (Cohen 2005). It is nevertheless useful to observe differences between expert and public ways of talking about risks (Leiss and Powell 2004), as problematic gaps or vacuums may occur (ibid, p. 31)[9]. These gaps might be overcome by a participatory, circular story approach to reporting issues, drawing on both personalized and collective documentation of events and their impacts, linking the perception of risks with outcomes.

Such an approach underlies the work of heritage departments, which preserve local museums, archives and pictorial records. However, these community-based records do not always reflect the full circumstances that prompt large-scale transformation, such a mass migrations[10], and it must be noted that the legitimacy of public records impacts on community perceptions and both the group and individual sense of vulnerability. For this reason, the task of distinguishing between the severity of natural cycles (droughts, floods, famine, etc.) and human-induced climate change is not simply a challenge for scientific researchers, but also requires ongoing consultation with historians and communities.

---

[9] Leiss and Powell characterize the "expert" assessment of risk as: scientific, probabilistic, focused on population averages, and acceptable and comparative risks, given sensitivity to changing knowledge, but with a bottom-line assessment that: "A death is a death". In contrast, the "public" assessment of risk is intuitive and more focused on discrete events and personal consequences, with the bottom-line view that: "It matters how we die" (31).

[10] Essentially, people move away and their interests are no longer as prominent (see Thompson 1998).

This layered dialogue is essential if it is hoped that policy development will mesh with both individuals' and communities' predispositions to avoid certain risks and adapt in light of support systems developed to deal with adversity. Many of the challenges anticipated for adaptation relate to site-specific conditions and existing means of adaptation. Concerns remain that government may impose arbitrary solutions that do not provide the flexibility needed to negotiate workable local solutions. This finding was consistent with other studies, for example, with Shortt et al.'s (2004) observations of the Irrigation Advisory Committees in Southern Ontario. The Shortt et al. study also revealed the positioning of local satisfaction within a far broader spectrum, where the highest "buy-in" to local efforts was linked to the communities developing unique initiatives and thereafter usurped for communities modeling their efforts on the success of these innovators.

Voluntary stakeholder and public involvement in adaptation to or mitigation of climate change have often started at the local level. For example, the Bow River Basin Council[11] is a non-profit initiative bringing stakeholders and community members together to share experiences, assess impacts and uncertainties, and develop and recommend ideas for dealing with issues such as limited water supply. The use of formal, local organizations makes sense because, "global climate change does not necessarily imply that temperature or precipitation is increasing at specific locations" (Clark et al. 2000). The success of community-based initiatives such as these depends on the receptivity of government to layered management processes and policies flexible enough to allow particular locales to find solutions for their site-specific challenges. However, this is likely to raise resource debates. For example, one might question the applicability of uniform environmental standards (i.e. for point-source chemical discharge) versus more holistic, community-based arrangements with industrial stakeholders (where chemicals might be dealt with through wetland treatment).

5.2 Network perceptions, involvement and motivations

The challenge of adapting to climate change must not only take into account perceptions of both the problems and their solutions, but also opportunities to participate and adapt within a given network context. Risk perceptions have systematically shifted in light of the accelerated change, media coverage, and immediacy of a shrinking world (Douglas and Wildavsky 1982), desensitizing many to both physical and emotional response. This circumstance makes it even more imperative that the means of monitoring environmental change and adapting to impacts be meaningful to local stakeholders' interests and values.

A broader, *meta*-interpretation of transdisciplinarity is needed. It must not only recognize the benefits of knowledge shared across formal disciplines, but also the full spectrum of learning engaged to stir creativity and involvement of community participants of all ages. This locus of creativity may have been diffused by the growth of trans-continental networks which have "jumped scale" to a global de-

---

[11] See the Bow River Basin Council (BRBC) site at: http://www.brbc.ab.ca/.

mocratic culture. There is now the compensatory need to *re-ground* human interests, renewing linkages between local knowledge and ecosystem dynamics (Berkes and Folke 2002). While the use of creative problem-solving and consensus-building at a local level may on one hand reflect this broader democratic culture, it may also lead to a challenge of conventional methods used to organize exchange and assess impacts in economic terms.

For these reasons, vulnerability should not be assessed solely in economic terms. Net vulnerability remains dependent upon interacting exposures and sensitivities arising from inseparable human and environmental conditions. In turn, these *net* impacts are a result of both coping and opportunistic[12] strategies applied within short- and long-term socio-economic arrangements. These arrangements include all dimensions of human exchange – monetary, non-market, reciprocal and symbolic – which contribute to the momentum, predictability and resilience of human systems (Homans 2002; Blau 2002). Robust social systems retain stability-in-flux and the potential to be self-regulating (Holling's theory) if they are not over-regulated and can still draw upon heterogeneity for dialectic synergy (Murdoch 1997; Granovetter 1985; Alcouffe and Kuhn 2004).

Conventional economics and broader socio-economic analyses can be bridged under a transdisciplinary approach (see Rolfe 2004). Standard economic contributions are evidently useful, for example, for determining sector-based marginal sensitivities to price structures and overall impacts on regional economies (see Bruneau, 2004). However, these models typically assume rational decision-making without considering transaction reciprocity and associated costs (Williamson 1985). Community- and agent-based decision-making often relies on transaction histories, accepting sub-optimal but reliable and socially-acceptable solutions. Furthermore, these community-based solutions may exhibit greater sensitivity to the vulnerability of groups and individuals[13]. Many community-based reciprocal transactions (for example, farm family contributions) draw on open market, co-operative transactions and barter arrangements, where the element of trust is also key (Ostrom 1998). It is thus risky to simply assume the conventional economic approach is fully compatible with local adaptive responses.

5.3 Transdisciplinary bridges community-based creativity

Modern-day resource debates have challenged researchers and stakeholders to pursue transdisciplinarity to address environmental issues that span many disciplines and cross scientific, political and social realms. This pragmatic approach to problem-solving is compatible with the diverse, interdisciplinary needs of resource

---

[12] In the case of the Red River Flood and its impacts on residents of Grand Forks, North Dakota (Rolfe 2004), many residents looked at the deluge more so as an opportunity to reconfirm community networks and spiritual connections, often ignoring economic incentives to relocate elsewhere.

[13] Here, the parties may receive both reciprocal and non-market recognition for "generous" gestures.

managers, pursuing a holistic, ecological view that includes humans as an "integral part of ecosystems" (Berkes and Folke 1998). However, it remains reasonable to analyze demographics and behavioural systems[14] in concert with environmental data, to note the influence of geographic variation on adaptive systems (Endter-Wada et al. 1998). Much has already been accomplished under this expanded scope of "ecosystem management" which: ... focus(es) on ecological systems that may cross administrative and political boundaries, incorporating a "systems" perspective sensitive to issues of scale, and manag(es) for ecological integrity (e.g., conserving species and population diversity, dynamic processes, and representative systems). (ibid, p. 891).

At the heart of the transdisciplinary challenge is the corollary need to integrate models for how humans view their world(s) and create a social construction of reality (Berger and Luckmann 1967). This social construction is fundamental to how environmental values are perceived, defined, expressed, and linked to the broader context of global long-term sustainability (Rolfe 2004). The potential for further social science exploration of economic activity is enormous. For example, Frey (1992) argued the case for *homo oeconomicus* as a new social science paradigm which recognizes the links between economics and psychology, and provides a framework for exploring six areas commonly overlooked by conventional economics: the environment, politics, art, family, conflict and history. It is furthermore interesting that Frey's approach does not follow a strict psychological focus on individuals, but rather identifies psychology's compatibility with strict psychological focus on individuals, but rather identifies psychology's compatibility with economics and integrates human behaviour[15] more generally with institutions[16].

An integrated, layered resource management approach must thus be based on inclusive community and educational concepts. It must draw all community members into a cycle of creativity bridging formal research efforts from physical and social scientific disciplines with other information processes which include community history as well as the popular expression of thematic models and myths in humanities and arts[17]. These human expressions, like story-telling, have the potential to engage participants as active learners (Kowalski 1998), punctuating the acquisition of knowledge as meaningful within their personal and collective realities.

---

[14] These behavioural systems relate to motivation, valuation and participation.

[15] Frey's view of human behaviour also presupposes: 1) there are individuals agents, 2) incentives determine behaviour, 3) measurement is achieved by observing how preferences are demonstrated within a set of constraints, 4) individuals pursue self-interest (including benevolent action), and 5) institutional settings determine the constraints on human behaviour.

[16] Frey defines institutions as "agreements shaping repeated human interactions" (1992, p. 3), with three institutional dimensions given for: 1) decision-making systems, 2) norms, traditions and other behavioural rules, and 3) both formal and informal organizations.

[17] The European university legacy continues to mix science and humanities openly and addresses complex, transborder environmental issues.

The engagement of *all* learners is required for a Kuhnian paradigm shift[18], with profound insights into the limits to growth and the interconnectivity underlying equity problems worldwide. To this end, the power of popular literature and film media cannot be underestimated. One has only to reflect on the ground-breaking popularity of Carson's (1962) *Silent Spring,* which sold over 330,000 copies worldwide, or how Atwood's (1972) thematic interpretation, *Survival,* persists in Canada as an essential thematic work to this day. Both oral and written story-telling stirs the collective imagination by drawing on non-fictional romanticism (see Gayton's (1996) *Landscapes of the Interior: Re-Exploration of Nature and the Human Spirit*), gritty memoirs (Meyers' (1998) *The Witness of Combines*), and fiction itself (McMurtry's (1966) *The Last Picture Show*). The power of myth in science in particular becomes evident after reading Friedman and Donley's work, *Einstein As Myth and Muse* (1985), and Carpenter et al.'s (2002) analysis of collapse, learning and renewal in light of common myths, models and metaphors.

Often *what is not said* is just as important as *what is said* – something community members are acutely aware of when dealing with formal institutions. For this reason, it is not surprising that there is an increasing recognition of literature which draws upon less mainstream perspectives, for example Hogan's (1995) First Nations' account, *Solar Storms*. Often these accounts deal with vulnerability in the face of extreme events, such as human displacement due to the St. James Bay Project (ibid.) or the plight of migrants during the Dust Bowl.

6 Conclusion

A reliable water supply is essential for vibrant communities and resource sectors. Stakeholder interests within watersheds often rely upon the synergy of activity conducted on a wider basis and *visa versa*. Local involvement in resource decision-making, on issues such as water infrastructure use and pricing, is becoming increasingly important. These front-line efforts are critical for active monitoring and rapid response to both expected and unexpected change, as well as for gauging sensitivities and vulnerabilities. The inter-generational engagement of community learners might be an effective counter-balance to the diffusion associated with expanding, specialized global networks. The notion of property rights will continue to evolve under the umbrella of global common property regimes and the intensified focus on particular environmental issues from afar might be best tempered by the re-grounding of resource debates within local geographies, putting not only a human but a personal face on resource use.

Applied as a layered management strategy, the transdisciplinary approach can foster cross-disciplinary exchange, improve receptivity to mixed methods, and encourage both consultative exchange and inter-generational dialogue. All venues should be explored for stimulating creative energy and stirring both collective

---

[18] This notion of a "paradigm shift" is described in his seminal (1962) contribution, The Structure of Scientific Revolutions.

imagination and commitment. Use of both formal and informal venues is likely to improve the understanding of the interests and motivations of stakeholders. If one agrees that *resilience takes planning* and that history and momentum are significant factors underscoring socio-economic adaptation and vulnerability, then a truly broad approach to transdisciplinarity is justified. This approach must then draw upon the engagement and empowerment of the community-at-large through the creative process of the humanities and arts.

For issues related to climate change, it may not be enough for conventional disciplines to follow "simple prescriptions", to "replace inherent uncertainty with the spurious certainty of ideology, precise numbers, or action" (Gunderson and Holling 2002, XXII):

> The stereotypical economist might say "get the prices right" without recognizing that price systems require a stable context where social and ecosystem processes behave "nicely" in a mathematical sense (i.e., are continuous and convex). The stereotypical ecologist might say "get the indicators precise and right" without recognizing the surprises that nature and people inexorably and continuously generate. The stereotypical engineer might say "get the engineering controls right, and we can eliminate those surprises" without recognizing the inherent uncertainty and unpredictability of the evolving nature of the interaction between people and nature. (ibid. XXI-XXII).

References

Alcouffe A, Kuhn T (2004) Schumpeterian endogenous growth theory and evolutionary economics. J Evol Econ 14: 223–226

Armstrong R, Pietroniro E, Rolfe JT (2004) Water Use in the Saskatchewan River Basin. Technical Report, South Saskatchewan River Basin (SSRB) Project, Saskatoon, SK

Arnell N, Liu C, Compagnucci R, da Cunha A, Hanaki K, Howe C, Mailu G, Shiklamanov I, Stakhiv E (2001) Hydrology and Water Resources. In: McCarthy J, Canziani O, Leary N, Dokken D, White K Climate Change 2001: Impacts, Adaptation, and Vulnerability. Cambridge University Press, New York, pp 191–233

Bell ML, Hobbs BF, Ellis J (2003) The use of multi-criteria decision-making methods in the integrated assessment of climate change: implications for IA practitioners. Socio-Economic Plan Sci, 37 (4): 289–316

Berger PL, Luckmann T (1967) The Social Construction of Reality: a treatise in sociology of knowledge. Doubleday, Garden City, NY

Berkes F, Folke C (2002) Back to the Future: Ecosystem Dynamics and Local Knowledge. In: Gunderson LH, Holling CS (eds) Panarchy. Island Press, Washington, DC, pp 121–146

Berkes F, Folke C (eds) (1998) Linking Social and Ecological Systems: Management Practices and Social Mechanisms for Building Resilience. Cambridge University Press, Cambridge, UK

Blau PM (2002) Exchange and Power in Social Life. In: Calhoun C, Gerteis J, Moody J, Pfaff S, Schmidt K, Virk I (eds) Sociological Theory (edn 1964). Blackwell Publishing, Oxford, UK

Bruneau J (2004) Economic Value of Water in the SSRB. Technical Report, South Saskatchewan River Basin (SSRB) Project, Saskatoon, SK

Carpenter SR, Brock WA, Ludwig D (2002) Collapse, Learning, and Renewal. In: Gunderson LH, Holling CS (eds) Panarchy. Island Press, Washington, DC, pp 173–194

Carpenter SR, Cole JJ, Hodgson JR, Kitchell JF, Pace ML, Bade E, Cottingham KL, Essington TE, Houser JN, Schindler DE (2001) Trophic cascades, nutrients and lake productivity: whole-lake experiments. Ecol Mono 71: 163–186

Carson R (1962) Silent Spring. Fawcett Publications Inc., Greenwich, CO

Chandler JA (ed) (1993) Local Government in Liberal Democracies: An Introductory Survey. Routledge, London and New York

Clark JS, Yiridoe EK, Burns ND, Astatkie T (2000) Regional climate change: Trend analysis of temperature and precipitation series at selected Canadian sites. C J Ag Econ 48 (1): 27–38

Cinnéide MÓ (2001) Fighting Unemployment and Social Exclusion with Partnerships in Finland. In: OECD Local Partnerships for Better Governance. Paris, France, pp 175–211

Cohen S (2005) Comments made in Discussion Panel, Canadian Climate Change Impacts and Adaptation Annual Conference, Montreal, PQ, May 6

Douglas M, Wildavsky A (1982) Risk and Culture: An Essay on the Selection of Technical and Environmental Dangers. University of California Press, Berkeley, CA

Drache D (ed) (1995) Staples, Markets and Cultural Change: Selected Essays by Harold A. Innis. McGill-Queen's University Press, Montreal, PQ

Dupont DP (2005) Tapping into Consumers' Perceptions of Drinking Water Quality in Canada: Capturing Customer Demand to Assist in Better Management of Water Resources. CWRJ, 30 (1): 11–20

Endter-Wada J et al. (1988) A Framework for Understanding Social Science Contributions to Ecosystem Management. Ecol Appl 8 (3): 891–904

Folke C et al. (2002) Resilience and Sustainable Development: Science Background Paper commissioned by the Environmental Advisory Council of the Swedish Government in Preparation for the WSSD. International Council for Science (ICSU) Series on Science for Sustainable Development No 3, Paris

Folke C, Pritchard I, Berkes F, Colding J, Svendin U (1998) The Problem of Fit Between Ecosystems and Institutions. International Human Dimensions Programme (IHDP). IHDP Working Paper No 2. At: http://www.uni-bonn.de/IHDP/public.htm

Frey BS (1992) Economics as a Science of Human Behaviour: Towards a New Social Science Paradigm. Dordrecht, Boston and Kluwer Academic Publishers, London

Friedman AJ, Donley CC (1985) Einstein as Myth and Muse. Cambridge University Press, Cambridge

Gayton D (1996) Landscapes of the Interior: Re-Exploration of Nature and the Human Spirit. New Society Publishers, Gabriola Island, BC

Granovetter M (1985) Economic Action and Social Structure: The Problem of Embeddedness. Am J Soc 91 (3): 481–510

Hofmann N, Mortsch L, Donner S, Duncan K, Kreutzwiser R, Kulshreshtha S, Liggott A, Schellenberg S, Schertzer B, Slivitzky M (1998) Climate Change and Variability: Impacts on Canadian Water. In: Koshida G, Avis W (eds) National Sectoral Issues. vol VII of the Canada Country Study: Climate Impacts and Adaptation. Environment Canada, Ottawa

Hogan L (1995) Solar Storms. Scribner, New York

Homans GC (2002) Social Behavior as Exchange. In: Calhoun C, Gerteis J, Moody J, Pfaff S, Schmidt K, Virk I (eds) Contemporary Sociological Theory. Blackwell Publishing, Oxford, UK

Homer-Dixon TF (1991) On the threshold: Environmental changes as causes of acute conflict. International Security 162: 76–116

Holling CS (1986a) The resilience of terrestrial ecosystems: local surprise and global change. In: Clark WC, Munn RE (eds) Sustainable Development of the Biosphere. Cambridge University Press, Cambridge, pp 292–317

Holling CS (1986b) Myths of ecology and energy. In: Ruedisili LC, Firegaugh MW (eds) Perspectives on Energy: Issues, Ideas, and Environmental Dilemmas. Oxford University Press, Oxford, pp 36–49

Kowalski TJ (1998) The Organization and Planning of Adult Education. State University of New York, New York

Kuhn T (1962) The Structure of Scientific Revolutions. University of Chicago Press, Chicago

Kulshreshtha SN, Bruneau J, Armstrong R, Martz L (2002) Water Resources under Climate Change and Socio-economic Vulnerability in the South Saskatchewan River Basin: Some Methodological Issues. 57th Canadian Water Resources Association Annual Congress Proceedings. Montreal QC

Leiss W, Powell D (2004) Mad Cows and Mother's Milk: The Perils of Poor Risk Communication. 2nd edn McGill-Queen's University Press, Kingston, ON

Martz L, Rolfe JT, Armstrong R (2005) Climate Change and Water Resources in the South Saskatchewan River Basin: An Overview. 58th Canadian Water Resources Association Annual Congress Proceedings. Banff, AB

Martz L, Kulshreshtha S, Armstrong R, Pietroniro A, Horbulyk T, Bruneau J (2004) Climate Change and Water Resources in the South Saskatchewan River Basin. 57th Canadian Water Resources Association Annual Congress Proceedings. Montreal QC

McCarthy J, Canziani O, Leary N, Dokken D, White K (2001) Climate Change 2001: Impacts, Adaptation and Vulnerability. Cambridge University Press, New York

McMurtry L (1966) The Last Picture Show. Penguin Books, New York

Munasinghe M (2001) Interactions between climate change and sustainable development – an introduction. Inter J Global Envir Issues 1 (2): 123–129

Murdoch J (1997) Towards a Geography of Heterogeneous Associations, Prog Human Geog 21 (3): 321–337

Myers K (1998) The Witness of Combines. University of Minnesota Press, Minneapolis

Organisation for Economic Co-operation and Development (OECD) (2001) Local Partnerships for Better Governance. Paris, France

Ostrom E (1998) A Behavioral Approach to the Rational Choice Theory of Collective Action. Am Pol Sci Rev 92 (1): 1–22

Ostrom E (1990) Governing the Commons: the evolution of institutions for collective action. Cambridge University Press, Cambridge

Pocklington TC (ed) (1985) Liberal Democracy in Canada and the United States. Holt, Rinehart and Winston, Toronto

Reed MG (2003) Taking Stands: Gender and the Sustainability of Rural Communities. University of British Columbia Press, Vancouver, BC

Rolfe JT (2004) Using a Transdisciplinary Approach to Assess Food Security Issues in the North American Great Plains, PhD Thesis, Resource Management and Environmental Studies, University of British Columbia, Vancouver, BC

Shortt R, Caldwell WJ, Ball J, Agnew P (2004) A Participatory Approach to Water Management: Irrigation Advisory Committees in Southern Ontario. The 57th Canadian Water Resources Association Annual Congress, Montreal, QC

Smit B (2005) Comments made in plenary session, Canadian Climate Change Impacts and Adaptation Annual Conference, Montreal, PQ, May 3–7

Smit B, Burton B, Klein R, Wandel J (2000) An anatomy of adaptation to climate change and variability. Clim Change 45: 223–251

Thompson JH (1998) Forging the Prairie West. Oxford University Press, Oxford

Turner B, Kasperson R, Matson P, McCarthy J, Corell R, Christensen L, Eckley N, Kasperson J, Martello M, Polsky C, Pulsipher A, Schiller A (2003) Framework for vulnerability analysis in sustainability science. At: http://www.pnas.org/cgi/10.1073/pnas.1231335100

United Nations (UNFCCC) (1999) Compendium on methods and tools to evaluate impacts of, vulnerability and adaptation to climate change. Online UNFCCC since 2003: http://unfccc.int/adaptation/methodologies_for/vulnerability_and_adaptation/items/2674.php

Williamson OE (1985) The Economic Institutions of Capitalism. Free Press, New York

Wilson B, Tsoa E (2001) Hunger Count 2001: Canada's Annual Survey of Emergency Food Programs. Canadian Association of Food Banks, Toronto

Part II

Mitigation: emissions trading and CDM

Intensity targets: implications for the economic uncertainties of emissions trading

Sonja Peterson

Kiel Institute for World Economics
Department of Environmental and Resource Economics
Düsternbrooker Weg 120, 24105 Kiel, Germany
sonja.peterson@ifw-kiel.de

Abstract

Intensity targets that adjust to economic growth are discussed as one option to control greenhouse gas emissions without strongly affecting economic growth and with less uncertain economic cost than absolute targets. Intensity targets are especially promoted for developing countries but also increasingly for a general climate agreement for the Post-Kyoto period. The aim of this paper is to put the existing theoretical and empirical results about intensity targets and uncertainty into perspective and to augment them by additional data and findings. This allows initial conclusions to be derived and open research questions to be identified.

Keywords: Intensity targets; uncertainty, emissions trading, climate policy

1 Introduction

In order to establish emission targets without negatively affecting economic growth or the development process, dynamic emission targets are proposed that adjust in response to GDP. In most cases they imply a target for the emission intensity defined in terms of emissions relative to GDP. It is argued that such intensity targets may perform better than fixed targets especially for economies facing considerable uncertainty of economic development, which is typical for developing countries in particular. With the announcement of a US intensity target in 2001 this concept has also attracted new attention for developed countries. Currently, intensity targets are being discussed as one promising possibility for a Post-Kyoto agreement that integrates the USA and major developing countries.

However, both the reduction of economic uncertainty and the mitigating effects of intensity targets compared to absolute targets depend to a large degree on the relationship between GDP and emissions and the uncertainty associated with forecasting both variables. While it is not yet clear that intensity targets substantially reduce uncertainty of economic costs, they clearly lead to increased uncertainty about the emissions resulting from a reduction target. There are already a number of studies on intensity targets and on the relationship between emissions and GDP. The aim of this paper is to put the existing theoretical and empirical results into perspective and to augment them by additional data and findings. This allows initial conclusions about intensity targets to be derived and uncertainty and open research questions to be identified.

2 A brief history of intensity targets

Intensity targets for emissions are a special case of so called growth-indexed caps, which are in turn a special case of dynamic targets.

Dynamic Targets do not establish an absolute cap on a country's allowable emission level, but allow the level to fluctuate in response to some other measure (Kim and Baumert 2002). The target is in other words a function of this measure that could for example be population, previous emissions, cost of living or real wages. However, the most usual proposal is to let emission targets react to economic growth, measured as gross domestic product (GDP). A dynamic emission target that reacts in some way to economic growth is also denoted **growth-indexed cap**. If the emission target is adjusted to GDP growth in a linear fashion, the indexed cap becomes an **intensity target**. The target is in this case related to the emission intensity I of an economy, defined as the ratio of emissions to GDP.

Various forms of growth-indexed caps and intensity targets are discussed in the literature. An intensity target is usually understood as a target directly for I. A more general form of an intensity target is (Kim and Baumert 2002)

$$\text{Intensity target } IT = \text{Emissions}/GDP^\alpha \quad \text{or} \quad \text{Emissions} = IT \times GDP^\alpha \tag{1}$$

where α is fixed so that the actual emission target can be calculated from the actual GDP. For α = 1 this reduces to a direct target for *I*.

Ellerman and Sue Wing (2003) propose a convex combination of an absolute cap (AT) and an intensity target (IT) as a more general growth-indexed cap (GIC):

$$GIC = (1-\eta)*AT + \eta*IT*GDP \qquad (2)$$

For $\eta = 0$ this cap reduces to an absolute target; for $\eta = 1$, it reduces to a pure intensity target.

Dynamic targets were first proposed as a possibility to establish emission targets in developing countries that have not yet agreed to emission reductions in the Kyoto Protocol. It was argued that developing countries experience high, volatile or uncertain rates of economic growth so that absolute targets would be highly inappropriate (Baumert et al. 1999) while intensity targets "would provide the environmental benefit of reducing developing country emissions from business-as-usual levels while simultaneously accommodating developing country growth" (CCAP 1998). Other papers (Frankel 1999; Philibert and Pershing 2001) already developed this idea further shortly after the Kyoto Protocol was adopted. In parallel, Argentina committed itself as the first non-Annex-B country to an intensity target of the form defined in equation (1) with α = 0.5 (Bouille and Giradin 2002).

Soon, intensity targets also began to be seen as an option for industrialized countries (Philibert and Pershing 2001). They attracted new attention when, after the US withdrawal from the Kyoto Protocol in 2001, President Bush announced the target of reducing the US greenhouse gas intensity by 18% in the next ten years (Bush 2002). Even though it is controversial whether this target implies real effort or coincides with a business-as-usual scenario, intensity targets started to be discussed as a better alternative to absolute targets in general. A number of publications analyse the advantages and disadvantages of such an instrument (Ellerman and Sue Wing 2003; Kim and Baumert 2002; Jotzo and Pezzey 2005; Kolstad 2005, Müller et al. 2001; Pizer 2005; Quirion 2005; Sue Wing et al. 2005).

In the negotiations currently beginning regarding international climate commitments for the Post-Kyoto era after 2012, intensity targets will probably play a prominent role. In many existing proposals for Post-Kyoto regimes (see e.g. Bodansky 2004; Aldy et al. 2003) intensity targets are mentioned and proposed in various contexts.

3 The theory of intensity targets and uncertainty

Although there are several advantages and disadvantages of intensity targets compared to absolute targets, concerning e.g. equity issues or feasibility (see e.g. Dudek and Golub 2003; Kim and Baumert 2002; Kolstad 2005), the most important argument brought forward in favour of intensity targets is that they are able to reduce the uncertainty especially with respect to the economic costs of reaching a target. This section discusses the theory of the different uncertainties that are associated with intensity targets versus absolute targets.

3.1 Uncertainties associated with absolute and intensity targets

If the future GDP was known with certainty, absolute targets (AT) and intensity targets (IT) would be equivalent. Both are then linked via the equation

$$AT = IT*GDP \qquad (3)$$

and both would lead to exactly the same target emissions and abatement costs (see also Ellerman and Sue Wing 2003; Jotzo and Pezzey 2005). In reality though, GDP as well as emission growth and thus also the emission intensity are highly uncertain, as are the costs of reaching a given emission target. While an absolute target fixes future emissions, the future emission intensity remains uncertain. *Vice versa*, an intensity target fixes the future intensity level while future emissions are uncertain. Future GDP and the costs of reaching the emission target remain uncertain for both types of targets. It is thus not *per se* clear that intensity targets reduce the uncertainty of cost or the uncertainty in general compared to absolute targets.

The degrees of uncertainty associated with different instruments can e.g. be measured by two criteria that were introduced by Sue Wing et al. (2005).

Preservation of expectations: the degree to which each instrument preserves initial expectations e.g. about the level of emissions, abatement or abatement costs associated with the emission target.

Temporal stability: the degree to which each instrument minimizes the volatility of emissions, abatement and cost over time.

The analysis is complicated by the various linkages between emissions, GDP and abatement cost. Stricter emission targets can on the one hand be expected to reduce GDP to a smaller degree than the emissions, so that emission intensity decreases. On the other hand, although growth in GDP leads to more emissions, nevertheless it is also likely to reduce the emission intensity. Another possible linkage is that an economy that is able to use energy more efficiently may experience higher growth rates of GDP. Abatement costs are often assumed to depend only on the level of emission abatement. But since GDP growth may decrease the emission intensity of an economy, abatement costs may also depend on GDP.

The strength of the various effects depends on the structure of an economy, the direction of technical progress and its driving factors, the production processes and many other factors that are not yet well understood. There is some knowledge about the correlations between some of the variables but little knowledge about their causes. Conclusions about the uncertainty associated with intensity and absolute emission targets thus depend to a large degree on unproven assumptions in the underlying analytical framework. Furthermore, they also depend on the predictability of the variables. Before turning to the empirical relationships between the relevant variables and their predictability, Section 3.2 summarizes the existing analytical approaches for analysing the uncertainty of intensity targets.

3.2 Assumptions and findings of analytical models

Analytical models have focused on the question of whether intensity targets or more generally growth-indexed caps can reduce the uncertainties of economic costs, in terms of abatement costs or welfare costs, compared to absolute targets. Each of the models discussed below is based on different assumptions about the relevant relationships between the key variables and the associated uncertainties.

Kolstad (2005) decomposes the growth in emissions into the growth in GDP, the autonomous growth (or decline) of emission intensity and the growth (or decline) of intensity due to proactive effort. He thus assumes that there is an autonomous decline of the emission intensity in an economy. More importantly for his results on the uncertainty of abatement costs associated with either absolute or intensity targets, he assumes that marginal abatement costs are conditional on the level of economic output, or in other words, that they depend on the intensity. Concerning uncertainty, economic activity (GDP) and the costs associated with reducing GHG emissions while holding the level of economic activity constant are treated as random variables. Since in this framework abatement costs depend on GDP, his conclusion is that economic costs are less uncertain for an intensity target, since they are conditional on the level of the random variable GDP, so that the only remaining uncertainty stems from the uncertainty in the costs of abatement. In contrast, for an absolute target, the costs depend on two random variables: costs and GDP.

Sue Wing et al. (2005) make quite different assumptions about the relevant linkages and uncertainties. In their analytical model emissions and GDP are treated as random variables. There are no uncertainties associated with abatement costs, which are determined solely by the level of emission abatement relative to business-as-usual emissions. Thus, expectations of abatement costs under the absolute cap depend only on the expectation of counterfactual emissions, while under the intensity cap they depend on the expectation of both the counterfactual emissions and GDP. The authors then derive conditions under which, compared to absolute targets, intensity targets are associated with lower uncertainties of abatement costs in terms of both preservation of expectations and temporal stability (see Section 2). Whether intensity targets are able to reduce uncertainty now depends on the correlation between emissions and GDP and the variance in the predictions of these variables. Pure intensity targets are preferable if emissions and GDP are highly positively correlated and have similar degrees of variability. Using empirical data (see Section 4) they find that in terms of preservation of expectation the data for Japan, Mexico, Europe and the Former Soviet Union argue unequivocally for an intensity target. For the USA, Canada and China the results are less clear. Only for 45–55 percent (USA), 25–35 percent (Canada) and 10–45 percent (China) of abatement levels are intensity targets preferable. For the minimization of the volatility of abatement costs, the analysis shows that intensity caps are unequivocally preferable for the least developed countries and may be generally preferable for developed countries. Finally, the authors discuss the possibility of choosing a particular indexed cap as in Eq. (2). They show that it is always possible to choose an indexation parameter such that the partially indexed cap reduces uncertainty compared to an absolute target. Empirically, the optimal indexing parameter resulting from both uncertainty criteria is very similar, though it differs for the different countries. Nevertheless, in some countries the two criteria give conflicting recommendations.

Quirion (2005) adds another dimension to the discussion by considering marginal benefits from emission reductions. In an analytical model based on marginal abatement cost and marginal benefit curves, he determines the welfare deviations from the optimal policy that equalizes marginal costs and benefits for a given rate of abatement assuming that emissions and marginal abatement costs are uncertain. He finds that the ranking of an emission tax and absolute and intensity targets depends on the level of uncertainty about emissions and the relative slopes of the marginal benefit and costs curves. In most plausible cases intensity targets are either dominated by taxes or by absolute targets, but Quirion also concludes that in a case where e.g. a tax is not politically feasible in an international context, intensity targets may be preferable to absolute targets when marginal benefit curves are very flat. However, the expected welfare gaps are found to be very small.

Jotzo and Pezzey (2005) use yet another set of assumptions to design their analytical model that is then calibrated to empirical data. In this model, GDP, the emission intensity of output and the proportion of emissions that are linked to GDP are uncertain. In addition, various degrees of risk aversion can be implemented. Marginal abatement cost curves are linear in the emission level to be achieved and include an additional stochastic shift parameter. Thus, abatement costs are also uncertain. Furthermore, the model also includes deterministic benefit curves, which depend on global emission abatement and revenues from emissions trading. The multi-regional model is solved as a cooperative game, searching for an optimal level of emission abatement and an optimal differentiation of commitments between countries according to target type, parameters and simulation scenario chosen. It is also possible to calculate an optimal level of indexation to growth, comparable to the growth-indexed cap defined in Eq. (2). Again, this optimal intensity target always reduces uncertainty, while a standard intensity target may or may not reduce uncertainty. Standard intensity targets are expected to reduce uncertainty unless the degree of the GDP-emission linkage is very small compared to the stringency of the target. The main empirical result of the study is that standard intensity targets can reduce overall cost uncertainty and lead to better outcomes than absolute targets, but they can also systematically over- or undercompensate for GDP-related fluctuations. For all industrialized countries as well as Russia, China, India, Argentina, Mexico, Korea and South and North Africa an intensity target is preferable. In Brazil, Indonesia, South-East Asia and the rest of the world an absolute target is preferable to a one-to-one indexation.

Finally, Pizer (2005) also stresses in his argumentative analysis that the underlying premises for intensity targets being better able to accommodate unexpected growth (thus reducing uncertainty) are that emission fluctuations are tied to economic fluctuations, and that intensity behaves more predictably over time then emissions. If emissions and GDP were to be only weakly correlated, there would be no justification for an intensity target. Pizer also stresses the importance of the correlation between intensity and GDP. With a negative correlation, at least simple intensity targets flip the relationship between adverse economic shocks and the prospect of easier or harder targets. An intensity target becomes harder in the face of lower growth and easier in the face of higher growth. This is also noted by Dudek and Golub (2003).

Summarized, this discussion shows that:

- the assumed relationship between emissions, GDP, intensity and costs and the assumed uncertainties are all relevant for the effect of intensity targets on the uncertainty of costs and the reduction of uncertainty compared to absolute targets.
- in many models, the difference in cost uncertainty between standard intensity targets and absolute targets depends on the correlation between emissions, GDP and intensity and on the predictability of these parameters. Especially the correlation between GDP and emissions is important in several theoretical analyses.
- for an evaluation of "intensity targets and uncertainty" it is thus necessary to take a closer look at the empirical evidence for the relationships between the relevant variables and their predictability. This is done in Section 4.

4 Relationships and predictability of emissions, GDP, intensity and abatement cost

4.1 The relationship between GDP and emissions

Energy and emission intensity trends provide information on how effectively a country is using its energy and whether it is able to decouple economic growth and emissions. In 2002, the carbon intensity varied between ca. 0.2 kg CO_2/US$ GDP in Sweden, Japan, France and Austria, 0.6–0.8 kg CO_2/US$ GDP in the USA and Australia, up to more than 8 kg CO_2/US$ GDP in the Ukraine (International Energy Agency (IEA) 2004). Globally, the intensity has been falling constantly in the last 30 years and is now around 40% lower than in 1970. OECD countries, even if there are strong short-time fluctuations, are in line with this trend. However, the situation is different for developing and transition countries. In the former, the carbon intensity only started to decline in the early 1990s; in 2002 it reached about the level of 1970. In the latter, there is no stable trend. After rising from 1970 to 1980, the intensity started to decline until the early 1990's, then rose sharply and started to decline again in the mid 1990's. Today it is slightly higher then in 1970. The IEA projects that over the next 30 years the intensity will continue to fall further worldwide, with relatively constant levels in the OECD countries and a clear decrease within the transition and developing countries. The projected world intensity level in 2030 is 60% lower than the 1970 level. Nevertheless, the decline in intensities has been offset by the growth in GDP, so that actual emission grew world wide from 1970 to 2002 by more than 70% and in practically every country.

Höhne and Harnisch (2002) analyse historic emission and GDP data. They find that countries move through four basic stages of development, whereby the emission intensity first increases, reaches a maximum, stays constant and then decreases again. Assuming relationship (4) between emissions and GDP they perform a regression analysis for India, the Former Soviet Union (FSU) and the UK. For the USA they refine the formula as emissions and GDP show parallel fluctuations. The results are summarized in Table 1.

$$CO_2(t) = c*GDP(t)^{\alpha}; I = CO_2(t)/GDP(t) = c*GDP(t)^{\alpha-1} \qquad (4)$$

Table 1. Regression results

India	$\alpha = 1.23$ (1971–1991); $\alpha = 1.01$ (1991–1997)
USSR	$\alpha = 0.87$ (1971–1990); $\alpha = 0.78$ (1990–1999)
USA	$\alpha = 1.48$ (1985–1999); $\alpha = 0.45$ (1990–1999)
United Kingdom	No significant results

Source: Höhne and Harnisch 2002

Other authors have also analysed the relationship between GDP and emissions. Sue Wing et al. (2005) analysed emission and GDP forecasts and historical data for a number of countries. Though they find a strong positive correlation (> 0.85) between GDP and emissions in most countries and most time periods, in particular in developing countries, there are also some industrialized countries and some time periods where correlations are weak or even negative.

For data from 1971–2000, Jotzo and Pezzey (2005) calculate a significant positive correlation between the deviation of GDP from its trend and emissions from their trend for 23 out of the 30 largest emitting countries. However, the strength of the correlation varies. The authors also mention that so far all studies focus on CO_2-emissions from the burning of fossil fuels. From the scarce available data they found no evident correlation for methane, nitrous oxide and for CO_2 from land-use.

Based on data from Source OECD for CO_2 emissions and GDP from 1971 to 2000 I also undertook a correlation analysis for several countries. Some of the results are shown in the appendix. The results support the findings of the other authors that even though there is indeed a strong correlation between emissions and GDP in many countries, there are more then a few countries were it is quite weak. For the UK for example, the correlation is close to zero, supporting the findings of Höhne and Harnisch (2002). In some countries the correlation is even negative so that with a rise in GDP, emissions fall. However, the reasons are different in each case. While e.g. Sweden actively reduced its emissions and was obviously successful in decoupling them from GDP, emissions grew in the Eastern European countries and also Eastern Germany despite a dramatic fall in GDP in the early 1990s.

Summarized, there is evidence for a strong positive correlation between emissions and GDP, especially in developing countries. In some of the countries though, especially in the industrialized world, the correlation is weak or even negative. Since defining a suitable intensity target requires a correct picture of the relationship between emissions and GDP, this is thus not an easy task. Not only does the relationship change over time, it also is not always well defined.

4.2 The relationship between GDP and carbon intensity

As noted above, the relationship between GDP and the carbon intensity determines whether intensity targets flip the relationship between adverse economic shocks and the prospect of easier or harder targets. With a negative correlation between GDP and intensity, at least a simple intensity target would be easier to reach with more then expected growth and harder to reach with less then expected growth, which is not a desirable property of climate policy.

Pizer (2005) indeed finds a negative correlation for all seven industrialized countries in his sample. I also undertook a correlation analysis, also shown in the appendix. It turns out that the correlation between GDP and intensity is mostly very high. For most countries, it is highly negative, supporting the results of Pizer (2005). However, for some countries, including India and Bolivia, the correlation is positive. Thus even if a relationship like $CO_2 = b*GDP^\alpha$ or $CO_2 = a + b*GDP^\alpha$ holds, α is not equal to 1, i.e. the intensity changes with changing GDP.

4.3 Uncertainty of emissions, GDP and carbon intensity

Particularly relevant for the uncertainty of absolute and intensity targets are deviations from expected economic growth, emissions and intensity.

Analysing the range between lowest and highest projections of the US Energy Information Agency (EIA) for GDP, CO_2 and intensity as well as the change in the projections over time, Kim and Baumert (2002) find that future emission uncertainties are extreme in developing countries but intensities are much less uncertain. Future emissions in industrialized countries are less uncertain and the uncertainty in intensity is larger then in emissions. The authors conclude that CO_2 intensity targets are likely to be superior to absolute targets for developing countries, but that the latter may be preferable for some industrialized countries.

Jotzo and Pezzey (2005) examine forecasts for 2000 based on information available in 1985 for a large sample of countries. This yields a distribution of forecast errors, the standard deviation of which can serve as the magnitude of uncertainty. The study estimates uncertainty for GDP, energy sector emissions, energy sector emission intensity and non-energy sector emissions, each separately for OECD and non-OECD countries. The results indicate that uncertainty about future GDP is sizeable but significantly smaller than uncertainty about emissions and emission intensity. As the authors note, this is also in line with an analysis of forecast errors by the International Energy Agency and the US Energy Information Agency for a number of countries over the period 1995–2000.

Sue Wing et al. (2005) analyse forecasts for the years 2000 and 2010 and calculate both the correlation between GDP and emission and the variance in the forecasts of GDP and emissions. They find that emission and GDP projections are correlated positively except for 2012 projections in Japan and Mexico. The standard deviation of forecasts is in 60% of the cases higher for emissions and in 40% of the cases higher for GDP. There are examples where this is the case for both projection years and for industrialized and developing countries.

Pizer (2005) calculates the standard deviation of emission and intensity fluctuations for several industrial countries. He finds that, except for the USA where intensity fluctuates less then emissions, both variables vary considerably from year to year. He thus concludes that intensity targets fail to provide a more stable target.

Using the same data from 1971–2000 as before, I also calculated the standard deviation of these variables. It turns out that in industrial countries, and also for some developing countries, the standard deviation of emissions is an order of magnitude lower than that of GDP. In all transition economies that were analysed, and also in some developing countries such as India and China, the fluctuation of GDP is larger than that of emissions. Without exception, the carbon intensity shows the lowest variability. While the standard deviation varies between 5 (Sweden) and 785 (China) for CO_2 and between 6 (Hungary) and 1550 (USA) for GDP, it only varies between 0.03 (Brazil) and 1.6 (China) for intensity. However, what is more relevant is the predictability. A measure that can be calculated in this context is the deviation from a simple time trend, measured by the R^2 of an OLS regression of the variable against time. The resulting R^2 are shown in the table in the Appendix. With a few exceptions (e.g. India and China), GDP deviates less (shows a higher R^2) from the stable time trend than emissions. Comparing emissions and intensity, the former tends to deviate less from the trend in developing countries, and the latter deviates less in industrialized countries.

In summary, the results about the uncertainty of emissions, intensity and GDP are mixed and sometimes contradicting. Again, the results differ across countries and time periods. Overall, there is some weak evidence that emissions are indeed more uncertain than intensity, especially in developing countries, which would argue in favour of intensity targets if reducing uncertainties is the goal.

4.4 Determinants of abatement costs

The analytical analysis presented in Section 3.2 suggests that the determinants of abatement costs and the uncertainty of cost estimates also play a role for the evaluation of intensity targets. It is undisputed that marginal abatement costs depend on abatement levels. However, it is less clear, whether the costs are stable for certain abatement levels independent e.g. of business-as-usual emissions or choice of a policy instrument. Klepper and Peterson (2006) analyse analytically as well as empirically, based on a climate-economy model, whether marginal abatement costs are stable across different policy scenarios that lead to different business-as-usual emissions. While theoretically abatement costs for reaching a certain absolute as well as relative emission reduction target are not stable across different scenarios, this is empirically the case, at least for relative targets. That is, marginal abatement costs are stable for a certain percentage emission reduction relative to business-as–usual emissions across policy scenarios if all other factors remain unchanged. This is also the result of a paper by Ellerman and Decaux (1998) and would support e.g. the assumptions in the model of Sue Wing et al. (2005). Unfortunately there are no results about whether abatement costs are also influenced by e.g. carbon intensity, as assumed by Kolstad (2005), or GDP.

Along the lines of Ellerman and Sue Wing (2003) who use the EPPA model to simulate marginal abatement costs (MACs) for absolute and intensity targets under various assumptions about GDP growth, I use the climate-economy model DART (see Klepper et al. 2003) to compare the MACs for reaching a certain absolute or relative emission target or an intensity target under various assumptions about the future development of GDP. The "medium-GDP" scenario is the usual calibration of DART. For the "low-GDP" scenario I reduced the annual improvement in total factor productivity, that is mainly driving GDP growth in DART, by 1 percentage point. As a result, annual GDP growth is reduced by around 1 percentage point. In the "high-GDP" scenario I have *vice versa* increased annual total factor productivity growth and GDP growth by 1 percentage point per year. In the "medium-GDP" scenario I then simulate a 10% emission reduction in 2010 for all regions and calculate the equivalent absolute and intensity targets. The same relative, absolute and intensity targets are then simulated for the other two GDP scenarios. The resulting MACs for selected DART regions are shown in Figure 1.

Figure 1 shows that the MACs vary considerably with lower or higher GDP growth in all countries. For relative and absolute emission targets, MACs decrease with lower GDP growth and increase with higher GDP growth. For intensity targets – contrary to the findings of Ellerman and Sue Wing (2003) – the opposite is true. Comparing relative and absolute emission targets, the former are indeed more stable then the latter, and fluctuate less. Nevertheless, the deviations from the medium GDP scenario can reach 20%. Also contrary to the findings of Ellerman and Sue Wing (2003), intensity targets show the largest fluctuation in MACs. Even though these results depend on the model design and assumptions, they show that more research is needed in order to understand the determinants of abatement costs.

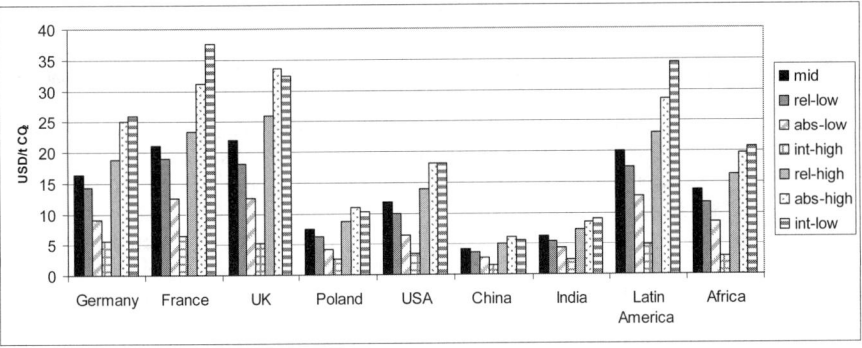

Fig. 1. Marginal abatement costs resulting under different assumptions on GDP growth

5 Conclusions

The discussion in this paper shows that intensity targets only reduce the uncertainty of economic cost compared to absolute targets under certain assumptions and conditions. Since the relationships between emissions, GDP, emission intensities and abatement costs are not yet fully understood it is too early for a final judgment. In particular, the determinants of abatement cost require further research. Nevertheless, it is possible to derive some preconditions for the reduction of economic uncertainty by means of intensity targets:

- Emissions and GDP must be highly correlated
- GDP and intensities must not have a greater degree of variability than emissions

The empirical results suggest that intensity targets generally tend to reduce cost uncertainty for developing countries rather than for industrialized countries, even though there are mixed results for both types of countries. Furthermore, simple intensity targets are likely to flip the relationship between adverse economic shocks and the prospect of easier or harder targets. In order to be able to design optimal intensity targets it is necessary to understand the relationship between emissions and GDP, which unfortunately not only differs across countries but also is not stable over time.

References

Aldy J, Ashton J, Baron R, Bodansky D, Charnovitz S, Diringer E, Heller TC, Pershing J, Shukla PR, Tubiana L, Tudela F, Wang X (2003) Beyond Kyoto – Advancing the international effort against climate change. Pew Center on Global Climate Change

Baumert K, Bhandari R, Kete K (1999) What might a developing country climate commitment look like? Climate Notes, World Resource Institute

Bodansky D (2004) International Climate Efforts Beyond 2012: A Survey of Approaches. Pew Center on Global Climate Change

Bouille D, Giradin O (2002) Learning from the Argentine voluntary commitment, Chapter 6. In: Baumert KA, Blanchard O, Llosa S, Perkhaus JF (eds) Building on the Kyoto Protocol: Options for Protecting the Climate. World Resource Institute, Washington, DC, pp 135–156

Bush Administration (2002) Global Change Policy Book – Executive summary of the global climate change initiative announced by Bush on February, 14, 2002

Center for Clean Air Policy (CCAP) (1998) Growth baseline: Reducing emissions and increasing investment in developing countries

Dudek D, Golub A (2003) "Intensity" targets: pathway or roadblock to preventing change while enhancing economic growth? Climate Policy 32: 21–28

Ellerman DA, Decaux A (1998) Analysis of Post-Kyoto CO_2 emission trading using marginal abatement cost curves. Report 40, Massachusetts Institute of Technology (MIT) – Joint Program on the Science and Policy of Climate change

Ellerman DA, Sue Wing I (2003) Absolute versus intensity-based emission caps. Climate Policy 32: 7–20

Frankel JA (1999) Greenhouse gas emissions. Brookings Institution. Policy Briefs No 52, Washington, DC

Höhne N, Harnisch J (2002) Greenhouse gas intensity targets vs. absolute emission targets. Paper presented at the sixth Conference on Greenhouse Gas Control Technologies, 1.–4. October 2002, ECOFYS energy and environment

International Energy Agency (IEA) (2004) CO_2 Emissions from fuel combustion 1971–2002. IEA, Paris

Jotzo F, Pezzey JCV (2005) Optimal intensity targets for emissions trading under uncertainty. Working Paper No 41, Program on Energy and Sustainable Development, Stanford University

Kim YG, Baumert K (2002) Reducing Uncertainty through Dual Intensity Targets, Chapter 5. In: Baumert KA, Blanchard O, Llosa S, Perkhaus JF (eds) Building on the Kyoto Protocol: Options for Protecting the Climate. World Resource Institute, Washington, DC, pp 109–133

Kolstad CD (2005) The simple analysis of greenhouse gas emission intensity reduction targets. Energy Policy 33: 2231–2236

Klepper G, Peterson S (2006) Marginal Abatement Cost Curves in General Equilibrium: the Influence of World Energy Prices. Resource and Energy Economics 28 (1): 1–23

Klepper G, Peterson S, Springer K (2003) DART97: A description of the multi-regional, multi-sectoral trade model for the analysis of climate policies. Kiel Working Paper No 1149, Kiel Institute for World Economics

Müller B, Michaelowa A, Vrolijk C (2001) Rejecting Kyoto – a study of proposed alternatives to the Kyoto Protocol. Technical Report, Climate Strategies – international Network for Climate Policy Analysis

Philibert C, Pershing J (2001) Considering the options: Climate targets for all countries. Climate Policy 1: 211–227

Pizer W (2005) The case for intensity targets. RFF Discussion Paper 05-02, Resources for the Future, Washington, DC

Quirion P (2005) Does uncertainty justify intensity emissions caps? Resource and Energy Economics 27 (4): 343–353

Sue Wing I, Ellerman AD, Song J (2005) Absolute vs. Intensity Limits for CO_2 Emission Control: Performance Under Uncertainty. Technical Report, University of Boston

Appendix

The correlation coefficient between $\ln(CO_2)$ and $\ln(GDP)$ measures the degree to which (*) $CO_2 = b*GDP^\alpha$ holds. If it holds, the correlation coefficient should be close to (-)1. In addition, if $0 < \alpha < 1$ then the correlation between the intensity and GDP is -1 if $\alpha \geq 1$ it is 1. For $\alpha < 0$ it is negative. Assuming a relationship (**) $CO_2 = a + b*GDP$, one can calculate the correlation coefficients directly. Here, if the relationship does not hold and/or additional variables are relevant to determine CO_2 emissions, the coefficients begin to vary from (-)1.

Using GDP and emissions for source OECD data from 1971 to 2000 for selected countries shows that the correlation coefficients are almost the same under (*) and (**). The direct correlation coefficients are reported in Table 2.

In order to assess the variability of CO_2, GDP and intensity I also regress these variables on a time trend using OLS. The resulting R^2 is also shown in Table 2 – the higher its value, the less the variable deviates from the time trend.

Table 2. Correlations of CO_2 emissions and CO_2 intensity (I) with GDP for selected countries

	Correlations				R^2 of an OLS regression on time		
	1971–2000		1990–2000				
	CO_2/GDP	I/GDP	CO_2/GDP	I/GDP	CO_2	GDP	I
USA	0.87	-0.94	0.99	-0.99	0.666	0.974	0.956
Japan	0.92	-0.92	0.95	-0.41	0.836	0.984	0.825
Germany	-0.84	-0.99	-0.90	-0.97	0.704	0.989	0.987
Great Britain	-0.68	-095	-0.77	-0.98	0.605	0.959	0.974
Sweden	-0.88	-0.92	-0.07	-0.91	0.830	0.965	0.896
Hungary	0.13	-0.73	-0.24	-0.86	0.215	0.575	0.927
Poland	-0.16	-0.79	-0.69	-0.99	0.057	0.618	0.504
Former Soviet Union	0.92	-0.88	0.97	-0.59	0.027	0.125	0.391
China	0.96	-0.94	0.94	-0.98	0.980	0.899	0.905
India	0.99	0.69	0.99	-0.82	0.958	0.931	0.694
Argentina	0.94	-0.73	0.91	-0.83	0.864	0.747	0.159
Brazil	0.96	-0.06	0.98	0.90	0.900	0.966	0.003
World	0.99	-0.99	0.98	-0.99	0.971	0.992	0.992

Source: own calculations based on data from Source OECD

Three types of impact from the European Emission Trading Scheme: direct cost, indirect cost and uncertainty

Volker H Hoffmann, Thomas Trautmann

Department Management, Technology, and Economics, ETH Zurich
Kreuzplatz 5, 8032 Zurich, Switzerland
vhoffmann@ethz.ch

Abstract

The authors identify three types of impact from the European Emission Trading Scheme on the affected companies: direct impact, indirect impact, and uncertainty. While direct impact refers to the cost of buying allowances, indirect impact refers to the increased input factor cost as suppliers price in their emission cost. Uncertainty refers to the limited planning reliability as many details of the regulation are still under negotiation or only last for a few years. Based on a Europe-wide survey, the authors empirically show the relevance and industry dependence of these types of impact to a company's strategic decisions.

Keywords: Emission trading, corporate strategy, uncertainty

1 Introduction

In January 2005 the European Union Greenhouse Gas Emission Trading Scheme (EU ETS) commenced operation. While the actual trading of emission allowances seems to be up and running, the question remains whether the EU ETS will induce technological innovation in order to reach the Kyoto targets (Gagelmann and Frondel 2005; Hoffmann and Trautmann 2006; Schleich and Betz 2005). There appears to be agreement that an incentive-based approach in environmental policy has a higher probability of inducing cost-effective technology innovation and diffusion than command and control approaches (Jaffe et al. 2004; Majumdar and Marcus 2001). However, for companies reliability in their planning remains crucial, as investment decisions in industries such as the power and the steel industry are currently committing a company for the next 20–30 years. Consequently, uncertainties resulting from implementation of the EU ETS may be important determinants for a company's technological reaction and influence whether proactive or reactive strategic decisions are taken (Aragón-Correa 1998; Aragón-Correa and Sharma 2003).

When investigating companies' decisions regarding their strategic response to the EU ETS, we propose that there are three types of impact from the EU ETS that need to be taken into account in order to understand the incremental effect of the EU ETS. The first type is the impact from *direct cost* resulting from a company's need to buy allowances[1]: The cost for covering its direct carbon emissions influences future investment decisions and technology choices. Most companies already reflect the allowance cost in both their daily operations and the evaluation of future investment decisions (PWC 2005).

Second, there can be an impact from *indirect cost* on strategic decisions. This effect results if a directly impacted company has the ability to integrate its direct carbon cost into its product prices: The company generates an indirect cost effect on other companies with positions in the value chain after the emitter. Thus the impact of indirect cost reflects the indirect emissions that a company generates with its demand for CO_2-intense products further up the value chain.

Third, we argue that the impact from *uncertainty* surrounding the implementation of the EU ETS is an important factor that needs to be considered when studying companies' strategic decisions (compare Hoffmann and Trautmann 2006). Indeed there is little understanding of the relevance of uncertainty in the context of implementing policy instruments. Marcus 1981 notes "it is difficult to know whether policy uncertainty is simply a rationalization for not innovating or

---

[1] Note that in most countries allowances were given for free to the installations. However, from an economic point of view the value of these allowances should be reflected in cost calculations first from an opportunity cost perspective (allowances could be sold) and second from a marginal production cost perspective (shortages of allowances need to be mitigated at current market prices). Furthermore, future allocations may have an increasing portion that is not allocated for free but via auctioning (Schleich and Betz 2005). Of course, it is a separate question if and how companies reflect the allowance cost in their product pricing.

whether there is a cause-and-effect relationship between policy uncertainty and technological change". Theoretical and empirical research presents a mixed picture: While findings from Paulsson and Malmborg 2004 indicate that in the context of the EU ETS policy uncertainty results in a "wait-and-see" strategy, some authors argue that environmental uncertainty in general increases the likelihood that a firm will develop a proactive environmental strategy (Aragón-Correa and Sharma 2003; Ettlie 1983).

Note that the third type of impact, *uncertainty*, cannot be viewed isolated from the impact from *direct* and *indirect cost*. Uncertainty can cause strategic reactions independent of cost effects (Aragón-Correa and Sharma 2003), but may also moderate the strategic response to cost effects (Marcus and Kaufman 1986). Hoffmann and Trautmann 2006 showed a correlation between the reported impact of the EU ETS on investment decisions and the level of perceived uncertainty, indicating interdependence between these types of impact.

Some authors have argued that the role of uncertainty can only be understood when taking a company's industry specific context into account (Hrebiniak and Snow 1980). We argue that the industry specific context is a particularly important parameter as it reflects the relative importance of direct or indirect emissions and the local or global character of competition.

In the following sections we will analyze the three types of impact, *direct cost, indirect cost,* and *uncertainty* in more detail and subsequently present empirical results that support both the presence of these three types of impact in the case of the EU ETS and the industry-dependence of these effects.

2 Three types of regulatory impact

We define *impact* as an influence on a company's decision making regarding its technology strategy: There is a higher impact if the EU ETS shows a higher influence on a company's technology strategy. The term *influence* refers to the incremental effect from the EU ETS. Thus we specifically analyze the effect that would not be there without the EU ETS. *Technology strategy* sets the focus on investment decisions rather than on decisions regarding the trading of actual allowances. *Type of impact* refers to the causal path through which the EU ETS has an impact. The causal path can be direct via emission allowance pricing (*direct cost*), indirect via passed-through allowance cost (*indirect cost*), or via *uncertainty*. Figure 1 shows all three types of impact. For the analysis of *indirect cost* we focus on the effects of electricity prices as this is of particular importance in the cost structure of power intensive industries such as steel, pulp and paper, and cement. Already in January 2004 the associations of power intensive industries pointed at possible windfall profits in the power industry from an increased electricity price and its impact on the competitiveness of power intensive industries (EUROFER et al. 2004).

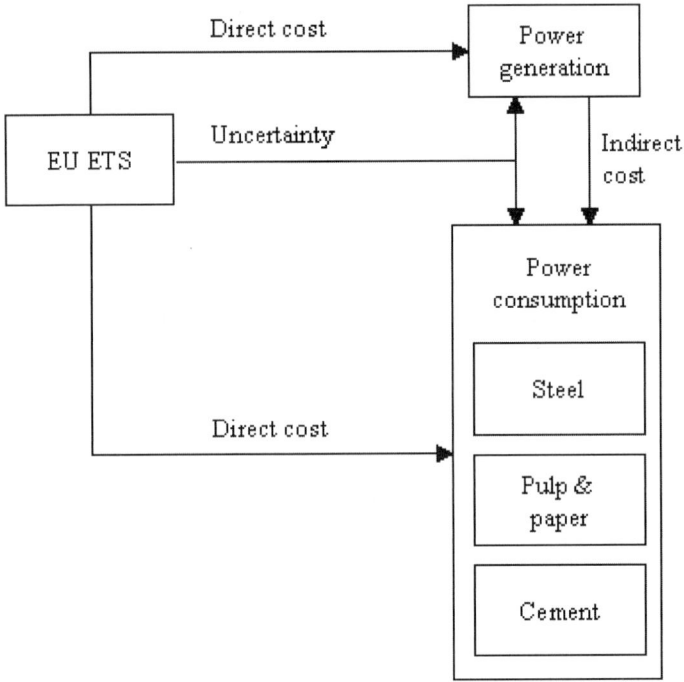

Fig. 1 Three types of impact from the European Emission Trading Scheme (main EU ETS industries)

2.1 Direct cost

The primary goal of the implementation of the EU ETS is to induce the introduction of low carbon technologies via the price for emission allowances that a company needs to compensate for its direct carbon emissions (Schleich and Betz 2005). We refer to this type of impact as impact of *direct cost*.

In the first half of 2005 the cost for allowances has ranged from 8 €/tCO$_2$ in January to 30 €/tCO$_2$ in June and stabilized at 20 €/tCO$_2$ in the subsequent months.[2] There have been many attempts to forecast companies' strategic decisions in response to the cost of their carbon emissions. An analysis of the International Energy Agency (IEA) for example showed that power plant operation would switch from coal to gas at an allowance price of 20 €/tCO$_2$ (Reinaud 2003). According to a study of Vattenfall also the investment in carbon capture and storage, discussed in both the power generation industry and power consuming industries, requires an allowance price beyond 20 €/tCO$_2$ to make this technology feasible (EUPD 2005).

---

[2] For current allowance prices see www.pointcarbon.com

Several authors have started to analyze companies' actual strategic responses. However, since the EU ETS has just started, the studies typically focus on the response to general climate policy and climate change. A cluster analysis of data gathered from the Carbon Disclosure Project[3] identified a set of typical strategies along two dimensions (Kolk and Pinkse 2005). Along the first dimension, companies can choose between innovation and compensation strategies, depending on an economic analysis of allowance price and mitigation cost. Following our definition of *impact*, we would only consider a company that chooses to innovate as being impacted, since a compensation strategy does not reflect technological change for this particular company. Along the second dimension these innovation strategies are observed on three different organizational levels: Companies may first innovate internally via process improvements or, second, vertically within the supply chain via product development. Third, Kolk and Pinkse argue that a company can even move horizontally beyond the supply chain to realize new product and market combinations, such as in the case of fuel cell development where oil and automobile companies are cooperating. According to our definition of *impact*, we would consider all three actions as responses to impact from *direct cost*, since all three actions reflect technological change in the affected company.

2.2 Indirect cost

The second type of impact from the EU ETS results from the ability of a company's supplier to pass through its cost from direct emissions. In the case of the power generation sector the cost from direct emissions may be reflected as opportunity cost in the electricity price. Consequently, the EU ETS has an indirect impact on power consuming industries such as the steel, the pulp and paper, or the cement industry. This effect may also impact industries such as the aluminum industry which, according to EU directive 2003/87/EC, is not covered by the EU ETS. As this impact reflects the emissions that a company causes indirectly by consuming power or other "non-clean" goods, we refer to this type of impact as impact from *indirect cost*.

The spot price for power at the European Energy Exchange (EEX) increased from 27.35 €/MWh in June 2004 to 29.74 €/MWh in December 2004 and 46.67 €/MWh in June 2005[4]. This corresponds to a 9% increase in the second half of 2004 and a 57% increase in the first half of 2005. Of course this increase is driven by multiple factors such as the oil and gas price and cannot be assigned entirely to effects from the introduction of the EU ETS. However, according to the Financial Times Deutschland, a German power utility was expecting 300–350 Mio. € additional profit in an internal estimate in September 2004 from pricing in the opportunity cost from emission allowances (Gammelin and Hecking 2005).

According to this same article, power intensive industries in Germany are expecting an additional cost of 1000 Mio € that cannot be explained without impact

---

[3] See www.cdproject.net
[4] See www.eex.de

from the EU ETS. This cost increase has an impact on strategic decisions of power consuming companies, which we call impact from *indirect cost*: large aluminum producers such as Norsk Hydro are currently assigning 10 €/MWh of their increase in power prices to indirect effects from the EU ETS. Since power accounts for 40% in the production cost of primary aluminum, the aluminum industry is expected to exit Europe and have its major production volumes outside Europe by 2015 (Gassmann 2005).

2.3 Uncertainty

The third type of impact from the EU ETS results from the limited predictability and planning reliability that the trading system creates in investment decisions (Trautmann et al. 2006). This limited predictability results from both price uncertainty regarding the market for emission allowances and from systemic uncertainties. The latter result from open questions regarding the objective, measures and rules, the implementation process of the EU ETS, as well as the interdependence with other regulations, and these are partly reflected in the allowance price. Companies are reacting quite differently to such uncertainties regarding the proactive vs. reactive dimension (Aragón-Correa and Sharma 2003; Paulsson and von Malmborg 2004) and we are referring to this type of impact as impact from *uncertainty*.

An analysis of the status of the EU ETS by the Pew Center (2005) concludes that key remaining uncertainties include not only the expectations of future targets and prices but also the readiness of all parties to trade, linkages to other trading programs, the availability and use of project-based allowances, the impact of Russian emission credits, strategies of new Central and Eastern European member states, the compliance role of governments, and progress in emission reductions from sectors outside the EU ETS. In one way or another, these uncertainties may be partly reflected in future allocations and ultimately in the allowance price. And, since the Kyoto Protocol covers the years 2008–2012, many of these questions remain open regarding the so-called post-Kyoto or post-2012 phase: According to a survey conducted in November-December 2004 companies' greatest concerns included uncertainties regarding allocation 2008–12 and policy post 2012 (PWC 2005).

Since the EU ETS has been implemented, there is only little research regarding companies' responses to the EU ETS and even less regarding companies' responses to uncertainty. However there are some indications that uncertainty is inducing a "wait-and-see"-strategy. An early investigation in Sweden revealed that, although Sweden and Swedish companies have a tradition of being proactive in environmental policy and management, companies are acting passively when it comes to emission trading (Paulsson and von Malmborg 2004). The researchers state that "ambiguous government policies are claimed to prevent the companies from making long term strategies on climate change mitigation in general and emission trading in particular." Furthermore, a cross-industry analysis of European companies affected by the EU ETS revealed a significant correlation between companies that indicated a strong impact of the EU ETS on their investment decisions and companies that perceived a high level of uncertainty (Hoffmann and Trautmann 2006). Thus, even if the impact from *direct* or *indirect cost* might induce a strategic investment from an economic point of view, the impact from *uncertainty* can alter the strategy, for example by simply postponing otherwise feasible investment decisions.

3 Empirical evidence

3.1 Method

In order to investigate the three types of impact from the EU ETS, we conducted a Europe wide survey six months after the launch of the trading scheme in January 2005. Since we wanted to understand the relative importance of direct and indirect cost impact, we chose to analyze the steel, pulp and paper, and cement industries. All three industries are affected directly as they are covered by the EU ETS according to EU directive 2003/87/EC[5]. Since electricity accounts for a significant portion of their production cost, all three industries are further impacted indirectly via the power industry. Finally, accounting for 40% of the emissions covered by the EU ETS, these industries represent a large share of the EU ETS and a solid basis for understanding the impact of uncertainty.

To investigate the relative importance of the impact of direct vs. indirect cost, we assumed that a company can either take action to reduce its direct carbon emissions or increase the efficiency of electricity consumption. Companies could answer on a three point scale stating either that their focus is more on reducing direct emissions (1), equal for both levers (2), or more on increasing energy efficiency (3). We would interpret a focus on reducing direct emissions as a focus on the impact from *direct cost*. A focus on increasing energy efficiency would be regarded as a focus on the impact from *indirect cost* resulting from indirect emissions further up the value chain.

Regarding the importance of the impact of uncertainty, we posed two questions distinguishing between two types of uncertainty: We assumed that a company can either respond to the EU ETS via the technology choice for its future production portfolio or via a geographic choice for its future production location. The latter reflects a company's option to move its production outside Europe as not being affected by the EU ETS. In each case we asked on a four point scale for the degree to which decisions were affected by increased uncertainty from the EU ETS. Respondents could choose between 1 (no increased uncertainty) and 4 (strongly increased uncertainty). We chose an even numbered scale in this case to force re-

---

[5] EU directive 2003/87/EC names installations from energy activities, production and processing of ferrous metals (mostly steel), mineral industry (mostly cement) and so called other activities (pulp and paper)

spondents to indicate a tendency and to avoid their choosing the middle of the scale.

3.2 Results

The survey resulted in a sample of 39 respondents with valid responses to all questions corresponding to a usable rate of 49%. 11 companies responded from the steel industry, 9 from the pulp and paper industry, and 19 from the cement industry. Tables 1 and 2 summarize the survey results.

Table 1. Descriptive statistics

		Total	Steel	Pulp & Paper	Cement
	N	39	11	9	19
Focus on indirect cost[a]	Mean	2.18	2.18	2.89	1.84
	(Std.)	(0.64)	(0.79)	(0.33)	(0.38)
Portfolio uncertainty[b]	Mean	2.51	2.73	2.11	2.58
	(Std.)	(0.91)	(0.79)	(0.93)	(0.96)
Geography uncertainty[b]	Mean	3.10	3.55	3.00	2.89
	(Std.)	(1.17)	(1.04)	(1.12)	(1.24)

[a]Three point scale: 1: Focus on reducing direct emissions (interpreted as a focus on direct cost), 3: Focus on increasing energy efficiency (interpreted as focus on indirect cost)
[b]Four point scale: 1: No increased uncertainty, 4: Strongly increased uncertainty

Table 1 shows the descriptive statistics. On a scale from 1 (focus on reducing direct emissions) to 3 (focus on increasing energy efficiency) a mean of 2.18 indicates that companies have a tendency to focus on increasing energy efficiency, which in this context is interpreted as a focus on the effects from indirect cost. However, the standard deviation of 0.64 indicates that there is a significant deviation between companies. While the pulp and paper industry shows a strong focus on indirect cost (2.89) the cement industry shows a stronger focus on direct cost (1.84).

On a scale from 1 (no increased uncertainty) to 4 (strongly increased uncertainty) companies indicated a medium increase in uncertainties in their portfolio decisions (2.51). Portfolio uncertainty hereby referred explicitly to uncertainty regarding the technology for future portfolio decisions resulting from the EU ETS. Again, the standard deviation 0.91 indicates that there are some differences between companies. Cement industry (2.58) and steel industry (2.73) indicate a bigger increase in portfolio uncertainty than the pulp and paper industry (2.11).

Compared to the portfolio uncertainty (2.51), geographical uncertainty appears to be even more important (3.10, same scale). The standard deviation (1.17) indicates a stronger variance between companies. Especially the steel industry indicates a strong increase in uncertainty regarding geographic location decisions (3.55). But also for pulp and paper (3.00) and cement (2.89), geographic uncertainty is rather high.

In order to test whether the analyzed questions are independent, we calculated Pearson correlations shown in Table 2. We considered that a high correlation between the focus on indirect cost, portfolio uncertainty and geography uncertainty would reveal a latent dependency making at least one analysis redundant. On the other hand, if there are no correlations, we would expect to analyze independent effects. Focus on indirect cost appears to be independent of both portfolio uncertainty ($R = -0.07$) and geographic uncertainty ($R = -0.10$) at very low significance of the observed effect ($p > 0.1$ for both cases). Portfolio uncertainty and geographic uncertainty show a positive correlation ($R = 0.27$) with moderate significance ($p = 0.09$) and thus are still considered worthy of independent discussion.

Table 2. Correlations

		Focus on indirect emissions	Portfolio uncertainty	Geographic Uncertainty
	N	39	39	39
Focus on indirect cost	Pearson	1	-0.07	-0.10
	(Sig.[a])		(0.67)	(0.56)
Portfolio uncertainty	Pearson		1	0.27
	(Sig.[a])			(0.10)
Geographic uncertainty	Pearson			1

[a]Significance of Pearson Correlation (2-tailed)

3.3 Discussion

The results suggest that all three types of impact from the EU ETS, *direct cost, indirect cost,* and *uncertainty,* play an important role in companies' decision making under the EU ETS. The focus on either reducing direct emissions or increasing energy efficiency is not exclusively on one lever but there is a tendency towards increasing energy efficiency with significant differences between industries. The fact that the pulp and paper industry shows a relatively strong reaction to the impact from *indirect cost* can be understood when looking at the cost structure specific to this industry: While costs from direct emissions mainly show up in paper making and chemical pulping, all other processes for producing pulp (wastepaper recovery, mechanical or thermo-mechanical pulping) are mainly affected by indirect cost from indirect emissions: In total 65% of the total electricity consumption is purchased from the grid mainly by mechanical pulp mills and some stand-alone paper mills causing a stronger focus on increasing energy efficiency for purchased power than on reducing direct emissions (CEPI 2004).

The cost structure of the cement industry shows a different picture: Direct emissions from clinker production and fuel use may increase total production cost by 30% at current allowance prices (Demján 2005). Consequently, the cement industry shows a stronger reaction to the impact from *direct cost*.

An analysis of variance has shown that, in general, industry dependence is more important for the impact of the EU ETS on investment decisions than for the level of perceived uncertainty (Hoffmann and Trautmann 2006). This can be understood when considering that investment decisions take place in an industry-specific context, whereas all industries ultimately face the same kinds of objective uncertainties.

The variance of the level of uncertainty within an industry is relatively high, compared to the variance of reported focus on indirect versus direct cost.[6] At the same time industries show smaller differences in impact from *uncertainty* than in impact from direct and indirect cost. This indicates that industry alone does not give a sufficient explanation for lower or higher impact of uncertainty. Given the variation within industries, the reason for different perceptions appears to be partly driven by the individual company context. Still, the relatively low impact of uncertainty on portfolio decisions in the pulp and paper industry may partly be driven by the relatively small overall cost impact of 3% that the EU ETS has on the pulp and paper industry (Hübner 2006). Furthermore, the relatively strong impact of uncertainty on geographic decisions in the steel industry reflects the fact that transport cost for steel plays a much smaller role than for pulp and paper or cement, thus the steel industry faces more international competition. Plant relocations, acquisition into non-Kyoto countries and low labor cost countries such as China are observed trends in recent years, also driven by the high demand of steel in China (Burgert 2003).

Summing up, all three types of impact are perceived by companies affected by the EU ETS. Differences in the impact of direct or indirect cost and uncertainty from the EU ETS can partly be explained as stemming from industry specific factors. However, variances within industries indicate that company specific context factors that are independent of the industry need to be taken into account further.

4 Conclusion and outlook

This study investigated the presence of three types of impact from the EU ETS on companies' technology strategy: *direct cost*, *indirect cost* and *uncertainty*. Furthermore, the relative importance of these types of impact was to be understood as including industry specific effects.

The presented results show how companies perceive these three types of impact. While some companies see a stronger impact on their strategic decisions from direct cost, other companies see a stronger indirect cost impact. Companies generally perceive an increased level of uncertainty on their strategic decisions.

The variations between companies in the perception of the relative importance of the three types of impact could partly be interpreted by analyzing industry specific results. However, there remain some variations that cannot be simply explained by industry-dependence. Thus, further research should take company specific factors into account such as country, company size, or the investment cycle.

---

[6] This holds true after taking the effect from the different scales (3 vs. 4 points) in Table 1 into account

In order to understand why even within a given industry some companies act similarly and others show varying strategies, future research might draw from the complementary perspectives of institutional theory (Scott 2001) and the resource-based view (Barney 1991): While institutional theory states that, particularly in an uncertain environment, isomorphic pressures lead to similarity of organizations (Milstein et al. 2002; Oliver 1997), the resource-based view underlines the importance of firm-specific resources and complementary assets for competitive advantage (Christmann 2000).

Furthermore, the analysis thus far focused on power consuming industries directly covered by the EU ETS. Further investigation could look at the power sector for impact from *direct cost* and *uncertainty* and non EU ETS industries such as the aluminium industry in order to investigate impact from *indirect cost* and *uncertainty*.

In particular, the impact of uncertainty needs to be understood in more detail. So far we have only showed that companies do perceive uncertainty and that companies assign high significance to the impact of uncertainty on their decisions. However, it remains unanswered what the resulting effect is: Are companies driven towards a "wait-and-see-strategy" as proposed by Paulsson and van Malmborg 2004? Or are companies moving towards a proactive environmental strategy as proposed by Aragón-Correa and Sharma 2003?

Finally, given the young age of the EU ETS, the presented analysis was dependant upon respondents' perceptions regarding the impact of the EU ETS on their decisions. Future research would have the benefit of analyzing the actual impact ex post by looking at quantitative indicators such as investment volume or the reduction of specific emissions.

References

Aragón-Correa JA (1998) Strategic proactivity and firm approach to the natural environment. Academy of Management Journal 41 (5): 556–567

Aragón-Correa JA, Sharma S (2003) A Contingent Resource-Based View of Pro-active Corporate Environmental Strategy. Academy of Management Review 28 (1): 71–88

Barney JB (1991) Firm Resources and Sustained Competitive Advantage. Journal of Management 17 (1): 99–120

Burgert P (2003) Steel funds flow east as investors bet on hottest card. American Metal Market 111 (24): 24A–25A

CEPI (2004) European pulp and paper industry: Annual statistics 2004: Confederation of European paper industries

Christmann P (2000) Effects of "Best Practices" of Environmental Management on Cost Advantage: The Role of Complementary Assets. Academy of Management Journal 43 (4): 663–680

Demján Z (2005) The possible impact of CO_2 trading on competitiveness of cement industry. Linking the Kyoto project-based mechanisms with the EU ETS, Vienna

Ettlie JE (1983) Policy implications of the innovation process in the U.S. food sector. Research Policy 12 (5): 239–267

EUPD (2005) EU pushes to cut CO_2 capture cost, European Power Daily, 7
EUROFER, EULA, CEMBUREAU, CEPI, EUROMETAUX, CPIV (2004) Energy intensive industries call upon EU decision-makers to pay more attention to the impact of emission trading upon their competitiveness: Associations of power intensive industries
Gagelmann F, Frondel M (2005) The impact of emission trading on innovation - science fiction or reality? European Environment 15 (4): 203–211
Gammelin C, Hecking C (2005) Trittin fordert Rechenschaft von RWE, Financial Times Deutschland, 19/Aug/2005. Hamburg
Gassmann M (2005) Norsk Hydro erwartet Massenverlagerung, Financial Times Deutschland, 31/Oct/2005. Hamburg
Hoffmann VH, Trautmann T (2006) The role of industry and uncertainty in regulatory pressure and environmental strategy. Academy of Management Best Conference Paper 2006 ONE
Hrebiniak LG, Snow CC (1980) Research Notes: Industry differences in environ-mental uncertainty and organizational characteristics related to uncertainty. Academy of Management Journal 23 (4): 750–759
Hübner JH (2006) The effects of emission trading on the pulp and paper industry. Unpublished Doctoral thesis (to be published), University Hamburg
Jaffe AB, Newell RG, Stavins RN (2004) Technology Policy for Energy and the Environment. NBER Innovation Policy & the Economy 4 (1): 35–68
Kolk A, Pinkse JM (2005) Business Responses to Climate Change: Identifying Emergent Strategies. California Management Review 47 (3): 6–20
Majumdar SK, Marcus AA (2001) Rules versus discretion: The productivity con-sequences of flexible regulation. Academy of Management Journal 44 (1): 170–179
Marcus AA (1981) Policy uncertainty and technological innovation. Academy of Management Review 6 (3): 443–448
Marcus AA, Kaufman AM (1986) Why it is so difficult to implement industrial policies: Lessons from the synfuels experience. California Management Review 28 (4): 98–114
Milstein MB, Hart SL, York AS (2002) Coercion Breeds Variation: The Differential Impact of Isomorphic Pressure on Environmental Strategies. In Hoffman AJ, Ventresca MJ (eds) Organizations, Policy, and the Natural Environment: 151–172: Stanford University Press
Oliver C (1997) Sustainable competitive advantage: Combining institutional and resource-based views. Strategic Management Journal 18 (9): 697–713
Paulsson F, von Malmborg F (2004) Carbon Dioxide Emission Trading, or not? An Institutional Analysis of Company Behaviour in Sweden. Corporate Social Responsibility and Environmental Management 11 (4): 211–221
PewCenter (2005) The European Union Emissions Trading Scheme (EU-ETS) In-sights and Opportunities: Pew Center
PWC (2005) Responding to a changing environment: Applying emissions trading strategy to industrial companies: PricewaterhouseCoopers
Reinaud J (2003) Emissions trading and its possible impacts on investment decisions in the power sector: International Energy Agency
Schleich J, Betz R (2005) Incentives for energy efficiency and innovation in the European Emission Trading System. ECEE 2005 Summer Study -what works & who delivers? 1495–1506
Scott WR (2001) Institutions and Organizations (Second Edition ed) Sage Publications, London

Trautmann T, Hoffmann VH, Schneider M (2006) A Taxonomy for Regulatory Uncertainty – Application to Flexible Mechanisms of the Kyoto Protocol. Proceedings of the International Federation of Scholarly Associations of Management VIIIth World Congress 2006, Berlin

Participants' treatment of allowance price uncertainty: how are risk-aversion and real option values related to each other?

Frank Gagelmann

German Emission Allowance Trading Authority (DEHSt)
Bismarckplatz 1, 14193 Berlin, Germany
frank.gagelmann@uba.de

Abstract

The way in which participants of a tradable allowance system treat the allowance price uncertainty has so far been analysed either as reductions in "market exposure" resulting from risk-aversion, or as "wait-and-see" strategies in the sense of real option theory. This paper analyses how these two reactions could interact. The following conclusions can be drawn from this integration under the assumptions that participants are at least on aggregate risk-averse, that relative market positions (seller or buyer) are a result of differences in abatement cost and not of systematic differences in risk-attitude, that investments are to a significant degree irreversible, and that no dominant "flipping" of buyers becoming sellers occurs: Under free allocation according to historic emissions, a higher price risk is likely to lead to an aggregate reduction of abatement investment at any point in time. *Innovative* investment (i.e, investment employing new technologies) is also reduced, plausibly to an even stronger extent than investment in general. Auctioning generally leads to higher abatement investment than free allocation under risky allowance prices, since under auctioning all agents are "buyers" in the market. The overall effect of price risk on abatement investment is ambiguous under auctioning and depends on the relative importance of the revenues lost while "waiting-in-line", compared to the option value of waiting, which depends on the relative importance of the "random" price factors.

Keywords: Innovation, investment, risk-attitude, risk-aversion, real option theory, real option value, tradable permits, volatility.

Acknowledgements: I would like to thank Thomas Burkhardt, Thomas Krause, Sonja Peterson and Thomas Trautmann for valuable comments. All remaining errors are those of the author.

1 Introduction

"Market-based" environmental policy instruments are believed to be effective and efficient since they make use of "market signals" regarding the price of pollution (be it a pollution tax rate or the market price of tradable allowances). Economic actors decide on which abatement measures to perform by comparing their abatement costs with this market signal. Usually, it is assumed that economic actors know the prospective market price and/or act risk-neutral. However, since the price of an emission allowance fluctuates stochastically (and does so independently of the market form due to external "shocks" resulting from fuel prices, economic growth, weather periods, technical change, etc.), this is a strong assumption and high volatility can affect actors' reactions.

Two reactions are possible, in principle, and have been described in the literature: reduction of "market exposure" because of risk-aversion, and "wait-and-see" strategies in the sense of real option theory. Models have incorporated either of these, but no systematic look has been taken as to how the two might interact in the context of emissions trading.

Ben-David et al. (2000) put forward that risk-averse "sellers" (firms holding spare allowances that they can sell on the market) reduce their abatement level relative to a situation of certainty, in order to reduce their "exposure" to the impact of allowance price uncertainties, which increases with the number of spare allowances they own (or generate). Risk-averse buyers, on the other hand, increase abatement to lessen their purchase exposure. However, in their empirical results from an experiment, they find that price uncertainty does not lead to the expected exposure reductions. They suggest an explanation that potential buyers are using wait-and-see strategies. This explanation does not seem sufficient since as a result, allowance prices would have to increase unless sellers abate more – in spite of their own possible wait-and-see wishes. Since Ben-David et al. (2000) also report no significant price difference in their experiment between the situation with certainty and the situation with uncertainty, their explanation of wait-and-see strategies would mean that buyers must have a much higher real-option valuation than sellers, which is a proposition yet to be investigated.

Zhao (2003) suggests that real option valuation is irrespective of risk-attitude, as it only affects the discount rate, and real option theory can be applied irrespective of the discount rate. This proposition seems too short-sighted since it only covers investments which increase dependency on the stochastic variable (here: the allowance price), while in fact there may also be investments which reduce this dependency (reduce the investor's "exposure" to the allowance market). There are more concepts, which are discussed in this paper, for integrating risk-aversion other than just discount rate.

For judging potential investment and innovation effects of fluctuating allowance prices, at least some structure of the different models seems necessary. This is done in the following paper. It builds on the author's paper presented at last year's Wittenberg workshop, in which the interplay of risk-aversion and real options theory had been presented as an open question.

The remainder of this paper is organised as follows: Chapter 2 describes the different possible concepts of integrating risk-aversion in investment decisions under tradable emission allowances, and derives a proposed overall effect of risk-aversion on aggregate investment and on (investment related) innovation. Chapter 3 first describes the basic idea of real option theory and then analyses how it could be intergrated with risk-aversion. Chapter 4 briefly describes how the possibility for firms to choose among different abatement technologies can affect the results. Chapter 5 summarises the findings. Chapter 6 discusses the assumptions made.

2 Integrating risk-aversion in firms' abatement decisions under tradable emission allowance schemes

2.1 Risk-aversion: the general theoretical concept

The most common concept of risk-aversion (as one form of risk-attitude) used in economic theory has been formalised in the *expected utility theory* by von Neumann and Morgenstern (1947) and builds on the distinction by Knight (1921), who differentiated between risk, under which decision makers assign (numerical) probabilities to the possible outcomes, and uncertainty, in which they do not.[1]

More precisely, the *expected utility theory* conceptualises the utility of an individual facing a choice set of alternative univariate probability distributions (Machina and Rothschild 1998, p. 201). This can be imagined, for example, as a choice between different investment possibilities which each have probability distributions of revenue streams. The individual's preferences must be complete over the relevant range, transitive and continuous (in an appropriate sense) (ibid., p. 202; Eisenführ and Weber 2002, p. 211). Furthermore, the risk(s) in question must be separate from other risks that the individual may be facing (regarding other economic goods) (Kolstad 2000, p. 243, fn 6; Eisenführ and Weber 2002, p. 211). The resulting expected utility function for each investment possibility has the following form:

$$E[u(x)] = \sum_N p_i \, u(x_i) \text{ with } i = 1, 2, ..., N \qquad (1)$$

where x_i = stream of revenue in outcome i; $u(x_i)$ = utility assigned to this revenue outcome i by the specific individual; p_i = probability of the outcome i; $E[u(x)]$ = expected utility (= expectation value of all possible utility outcomes) that the individual assigns to the investment possibility.

A *risk-averse* economic agent has a decreasing marginal utility of the revenue streams from an investment. In other words, the higher the revenue streams, the higher his utility ($U' > 0$), but the *additional* utility of each additional stream of

---

[1] Kolstad (2000, p. 243, fn.6) notes that there are also other approaches using different assumptions about how people evaluate risks. Mas-Colell et al. (1995) handle this issue more deeply.

revenue is lower than the additional utility obtained from each respective previous increase in revenue. The second derivative of the utility is thus negative ($U'' < 0$). Figure 1 illustrates this: For example, the additional utility from obtaining x_3 instead of x_2 is not as high as the additional utility when receiving x_2 instead of x_1. When conceptualising this discrete example as a continuous function, we obtain the von Neumann-Morgenstern utility function of a risk-averse individual.

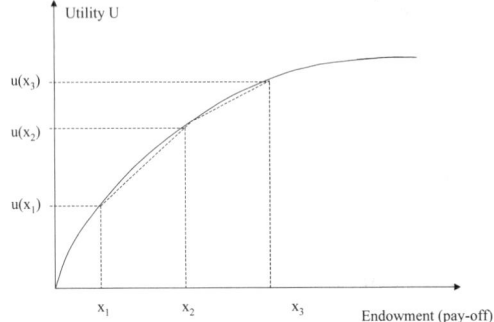

Fig. 1. Risk-aversion in a von Neumann-Morgenstern utility function
Source: Eisenführ and Weber 2002, p. 222

From this function, the certainty equivalent and the risk premium of a given investment (with expected pay-offs and their probabilities) can be derived. The certainty equivalent is that pay-off x^* which, when obtainable with certainty, gives the risk-averse agent the same utility as the expected utility of the uncertain alternative: $u(x^*) = E[u(x)]$. For a risk-averse agent, the certainty equivalent x^* is smaller than the expectation value $E(x)$, since the agent would be willing to "give up" a part of the pay-off if he could obtain certainty "in exchange". The risk premium denotes exactly how much the agent would be willing to give up, i.e., the difference between the expectation value of the pay-offs $E(x)$ and the certainty equivalent x^* (see Figure 2).

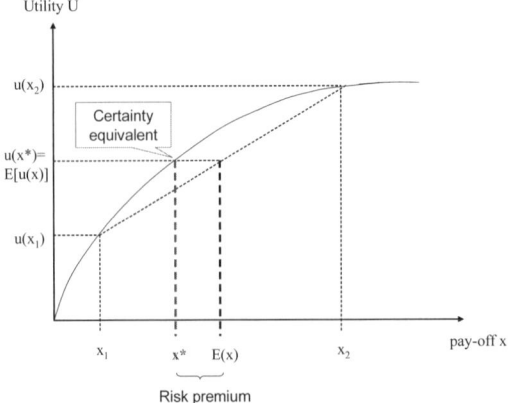

Fig. 2. Certainty equivalent and risk-premium of a given asset/stream of revenues in a von Neumann-Morgenstern utility function
Source: Machina and Rothschild 1998

In the following, it is assumed that that the risk-attitude of the investors in question is constant over time.[2]

2.2 Application of the theoretical concept to emission abatement decisions by economic agents facing volatile allowance prices

When transferring the above-mentioned concept to decisions on emission abatement investments under tradable emission allowances, one first has to note that such risk-aversion here plays a role only when investments are, to a certain extent, irreversible. This means that no significant "scrap value" could be obtained when closing, and selling, the installation.

How can we find the appropriate risk premium under tradable allowances? Three approaches are presented and discussed below: an increased discount rate in NPV calculations, an explicit derivation of the expected utility from different possible NPV scenarios, and an adjustment of the expected allowance price.

Increased discount rate

One common way to measure risk premiums in investment theory is to apply a higher discount rate (i.e., "required interest" for the capital invested) for the Net Present Value (NPV) calculation[3]. This results in some potential investments being dropped that would be profitable under certainty. If one imagines an investment in equipment that reduces emissions (for example, a more fuel-efficient boiler that leads to lower CO_2 emissions, or a Flue Gas Desulfurisation Unit for SO_2 abatement[4]), the potential investment would be characterised as follows:

$$\prod_0 = R - G - I - C_0 - P(e_0^* - e^a)$$ [5] (2)

---

[2] For deeper insights into the discussion about whether firm executives do indeed act risk-averse, see for example MacCrimmon and Wehrung (1984). The presumption of risk-aversion as the predominant risk-attitude is supported by the fact that "equities have historically had higher average returns than bonds, suggesting that investors have had to be induced with higher rewards in order to get them to make riskier pruchases." (Sharpe et al. 1995, p. 167f).

[3] This would be based on alternative assets with comparable (diversifiable and non-diversifiable) risk, and can be modelled by, e.g., CAPM models.

[4] Instead, one could also imagine the cost differential between two differently expensive and differently emission-efficient plants; or between two differently advanced types of Flue Gas Desulfurisation Units. In all the latter cases, and the case of boilder replacement, the calculation must be for cost differentials between different investment alternatives or between the old and the new technique. Only under the example of "one" possible Flue Gas Desulfurisation Unit, absolute streams (costs of the Unit) are sufficient.

[5] For a complete representation, one would have to add the opportunity cost pe* to the above-mentioned actions. However, this makes the formal representation more complicated, not least since in practice (e.g., most EU Member States in the EU-ETS for CO_2), regulators often grant free allocations to new installations and capacity extensions, and they withdraw allowances from closed plants. Both rules eliminate a part of the opportu-

$$\prod_{1...n} = R - G - C_{1...n}[I] - P(e_{1...n}*[I] - e^a) \text{ and} \tag{3}$$

$$NPV = \prod_0 + \sum_{i=1}^{n} \prod_{1...n}(1+r)^{-i} \tag{4}$$

whereas
$\prod_t$ = profits in period t,
R = revenues (assumed to be independent of the abatement decision here),
G = general (non-emission) costs (for fuels, raw materials, labour, etc.),
I = investment in abatement technology (assumed to be effective one period after the investment),
C_t = Abatement costs in period t,
P = allowance price,
e* = optimal emission level at price P,
e^a = emissions level corresponding to the individual allowance allocation
r = discount rate
n = number of periods considered.

The use of the discount rate to formalise abatement investments under tradable allowances has two principal limitations:

First, it implicitly assumes that uncertainty increases from year to year. This is plausible under the assumption that prices fluctuate according to certain types of Markov chains, such as "geometric Brownian motion processes" in which the variance increases with time[6]. However, markets for tradable allowances usually experience a lot of learning on the side of the participants about what the "real" abatement costs – and therefore allowance prices – are; this applies especially to the first year(s) after implementation of such a scheme.

Second, and more importantly, increasing the discount rate seems to be not universally appropriate for abatement investment appraisal under tradable allowances: setting a higher discount rate compared to the case of certainty implicitly assumes that the higher the investment, the higher is the "value at risk", the "exposure", of the profits. However, when a firm is a buyer of allowances at the prevailing market prices (i.e., its optimal emissions e* are higher than its allocation e^a), it can reduce its exposure (e* – e^a) – rather than increase it – by abating more than it would do under certainty.

Therefore, under freely allocated tradable allowances[7], risk-averse *sellers* of allowances will indeed reduce investment in abatement when the allowance price risk is higher. However, risk-averse *buyers* of allowances behave the contrary and *increase* their abatement investments under higher allowance price risk. To properly reflect this in a hypothetical "risk-adjusted" NPV calculation, risk-averse buyers would have to apply a higher, not lower, discount rate.

---

nity cost – for example, production reduction that leads to a plant being completely shut down results in no spare allowances to sell.

[6] See Dixit and Pindyck (1994).

[7] And when allocation rules do not base a firm's future allocation on its present (e.g., investment) actions. This assumption is discussed in Chapter 6.

In general it looks questionable whether an increase in the discount rate is an appropriate representation of risk-aversion in all cases other than stochastic fluctuations of product sales revenues. For fluctuations of input prices, it seems generally inappropriate: after all, by reducing exposure to stochastic input factor prices, a firm can stabilise its profits and thereby increase its expected utility.

Explicit calculation of the expected utility of each investment alternative, based on each alternative's NPV probability distribution

It seems more realistic that a potential risk-averse investor would not adjust the discount rate, but instead calculate, for each investment alternative, the "unadjusted" NPV for several allowance price *scenarios*, assign a utility to each scenario outcome, and thus derive the expected utility he assigns to each investment alternative. Simplified, he could compare mean and variance of the resulting profits of the different alternatives. But for both he would need a utility function or a comparable measure of the degree of his risk-aversion. In general, this approach requires comprehensive calculations of assumed price outcomes and their probabilities, as well as explicit formalisation of the investors' degree of risk-aversion.[8] Formally, this would be done by inserting the NPV_i of the respective investment alternative (instead of the "revenue stream" x_i) under outcome i in equation (1), and thereby calculating the expected utility:

$$E[u(x)] = \sum_N p_i\, u(NPV_i) \text{ with } i = 1, 2, ..., N \qquad (5)$$

Adjustment of the assumed allowance price

As an alternative to constructing NPV probability distributions for each investment alternative and comparing their expected utility, a simpler method is to set a "threshold value" for the allowance price which the investor considers likely. For example, he can estimate that two thirds of all expected price outcomes of an assumed price distribution function (with given specific expectation value and variance) will be above a certain price level X €, and use this value instead of the expectation value. When adding a certain price level Y € which will, with probability y, not be exceeded, he can also derive a "corridor" of the expected allowance price. This approach seems the most likely "hands-on approach" in light of the above-mentioned difficulties. Interview answers by two allowance trading practioners from electricity generating firms indicate that it is indeed used. A similar approach is also suggested by Spangardt (2005).

---

[8] The expected utility concept by Neumann and Morgenstern is, in principle, a static concept, while investments are inherently intertemporal. Therefore, when using on an aggregate level the approach presented here, one must implicitly assume firm homogeneity in terms of the (not risk-adjusted) discount rate, or at least the absence of any correlation between this and the actors' risk-attitude (and the seller or buyer position, which are discussed in chapter 6). When there are such correlations, it is possible that the aggregation leads to analytical problems. To analyse these problems would require deep formal treatment of this issue, which is beyond the scope of this paper.

Fraunhofer-ISI (2005, p. 258) describe such a corridor of an "expected allowance price" by means of a precise example for an abatement cost function. This is slightly modified here for illustration purposes (taking away the names of the abatement measures and re-arranging the precise shape of the curve). In addition, the idea of an explicit derivation of the corridor from a confidence interval of 95% of an assumed normal price distribution has been inserted by this author.

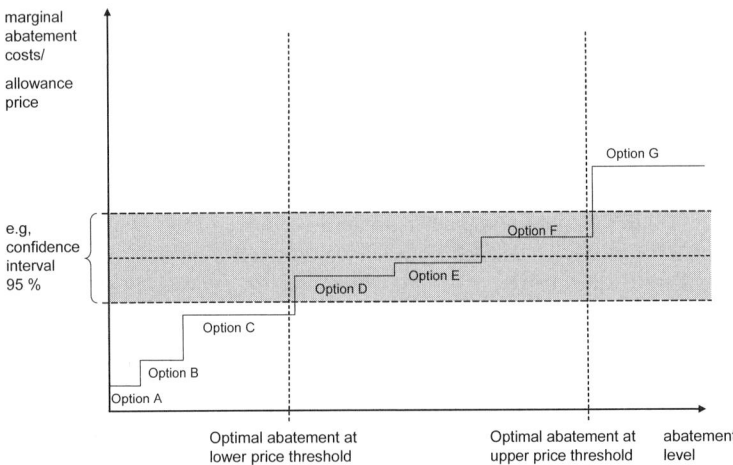

Fig. 3. Derivation of optimal abatement level under a corridor of possible price outcomes
Source: Fraunhofer-ISI (2005:258) with modifications by this author

Fraunhofer-ISI (ibid.) suggests that a risk-neutral investor would invest up to the allowance price expectation value (the central dotted line) and therefore implement all abatement measures A through E. A risk-averse investor, they suggest, would implement only measures A through C.

This description covers, in the view of this author, only the perspective of a potential *seller* who aims to limit his/her exposure from uncertain sales revenues. For a risk-averse buyer, in contrast, the relevant threshold would not be the lower price bound, but the upper bound, since the "worst" that can realistically happen to him/her is a high price of the level indicated by the upper bound.

The question is now whether Zhao's suggestion that risk-attitude does not affect the real option valuation can be maintained when a representation of risk-aversion is used that is in line with the latter two approaches presented here. This is discussed in the following chapter 3.

2.3 The impact of allowance price risk on aggregate investment levels and aggregate innovation rates

Impact on aggregate investment

Whatever the precise conceptualisation of risk-aversion under tradable emission allowances, one result can be clearly proposed for risk-averse agents under freely allocated allowances (as long as a firm's abatement decisions do not influence their individual allocation): As Ben-David et al. (2000) suggest, when allowance prices fluctuate stochastically, risk-averse sellers invest less in abatement than they would under certainty, while risk-averse buyers invest more in abatement than under certainty. Both groups of actors do this to reduce their "exposure" to fluctuations of the allowance price and therefore reduce the variance of their profits. The overall effect on investmnts is ambiguous. As a result, under risk the trading volume is smaller than under certainty, with prices generally the same as under allowance price certainty (ibid., p. 593).

Interestingly, they find in an experiment they performed with students that the trading volume did not significantly decrease under price risk, and furthermore prices do not differ significantly between situations of price risk and of price certainty. As an explanation, they suggest that buyers of allowances do not increase their abatement investment as expected, because they wait for the uncertainty to be resolved. This explanation seems to address exactly the "real option value" of postponed investment which will be described in chapter 3. However, the use of this argument does not seem sufficient, namely since Ben-David et al. also report that the prices are not significantly different from the situation of certainty. Then, the question is first why sellers should not also wait with their investments, i.e., an explanation is missing as to why sellers should have a lower tendency to wait with their investments than buyers have. If both buyers and sellers would wait, then prices would have to go up. But this they did not do significantly in the experiment.

A similar, but more complex issue is the one of emission-reducing plant replacement, which Ben-David et al. (2000) do not address. The incentive for plant replacement results from the reduced allowance expenditure and/or spare allowances to sell. In addition, in terms of exposure, plant replacement normally brings the investor from a buyer position into a seller position at least for several periods, until overall cap reductions may render the new facility a buyer as well.

The replacement incentive then depends on the relative duration of the changing exposures, and their relative size – in other words, the integral of those periods where the firm in question is a buyer and those where it is a seller. If the (risk-free discounted) overall integral of the considered replacement investment suggests a "net" seller position, then increased price risk decreases the value of replacement. If, in contrast, the "net" effect is a buyer position, then increased price risk increases the value of replacement. Whether a net seller or buyer position over time applies depends, in part, on how quickly emission caps are reduced by the regulator – if they are reduced quickly, the agents could be buyers for a longer time (with the old plant) than they are sellers after replacement (with the new plant).

Ben-David et al. (2000) do not address *auctioned* allowances. For auctions, the suggestion that risk-averse *buyers* of allowances tend to increase abatement investment results in the following proposition: Under auctioning, all agents are buyers. Therefore, if the majority of agents in a tradable allowace market is risk-averse, auctioned allowances lead to more investment than freely allocated allowances. This difference between auctioned and freely allocated allowances is stronger when allowance price uncertainty is higher.

Impact on aggregate innovation

The impact of allowance price risk on aggregate innovation (in the sense of such abatement investments or new plant investments that use technologically advanced production methods, like highly efficient coal or gas based power production, or renewable energy forms) seems more unambiguous than the impact on aggregate plant construction investment: environmental innovation is in general defined as reducing marginal abatement costs. Thus, an innovative technology investment leads per se to higher optimal abatement than "off-the-shelf" investments (as long as the marginal abatement costs of the innovative alternative are lower over the whole relevant range of potential abatement levels). Therefore, on the whole, innovators should, in general, be more "on the sellers' side" than agents who employ "off-the-shelf" investments. Since sellers reduce their abatement level under risky prices compared to certain prices, one can expect at least a tendency that higher uncertainty reduces innovation. This suggestion relies, albeit, on the absence of psychological factors, such as high (negative) valuation of an allowance deficit.

To the extent that innovation is in the form of investment, the point made above for auctioned allowances and investment applies to the innovation impact as well: Under stochastic allowance prices, auctioned allowances lead to higher innovation than freely allocated allowances.

3 The interplay of real options and risk-averse economic agents

3.1 Real option valuation in general

Real options theory is based on first considerations by McDonald and Siegel (1985) and Brennan and Schwartz (1985), which have been extended especially by Dixit and Pindck (1994) as well as Trigeorgis (e.g., 1995). First applications to investment decisions under tradable allowances have been made by Herbelot (1994); Edleson and Reinhardt (1995); Lambie (2002); Spangardt et al. (2003); Laurikka (2003) and Weber and Swider (2004). (For a concise overview, see Laurikka 2005). Real options valuation is based on the principle that an investment decision is hardly ever a "now-or-never" decision – as would be implied by pure NPV calculations. Instead, investments can be *postponed* to gain more information on future developments (on markets, technologies, etc.). The benefits of additional information gained by waiting can be so important that they dominate the profits foregone when the investment is postponed, leading to the decision "to wait", even though investing instantaneously would have a positive NPV. (This can be, for example, if "waiting-in-line" yields more information on *how* to invest in terms of final product choice, production technology, fuels and raw materials used, etc.). The opportunity to invest can, in this respect, be regarded as an "option" to invest *at any time*. This option is "excercised", and loses its "option value", as soon as the investment is made (and the "loss" of option value is the higher, the less "scrap value" can be obtained from the invesmnt when it is abandoned again.). Since abatement investments, just like other production capital investments, are "real" investments, one speaks about "real options" valuation.

To illustrate the potential gains from such options, imagine the following expected pay-offs from a potential investment (with p denoting the allowance price and R the profits at time t). The numbers attached to the arrows denote the probabilities.

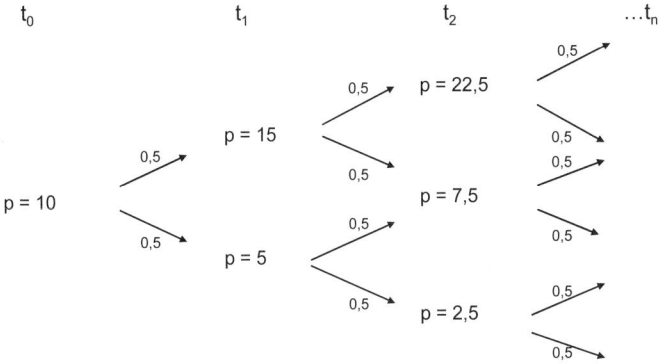

Fig. 4. Pay-offs from a hypothetical investment under different hypothetical outcomes (absolute returns, without consideration of risk attitude)
Source: Dixit and Pindyck (1994, p. 42), with modifications by the author[9]

As can be seen, the option value depends positively on the price volatility and on the share of allowance costs/abatement costs in total costs.[10]

It is important to note that the gain from waiting, for instance, for one year, is not only a prevented negative outcome in this first year – in fact, in the case shown above the avoided "unfavourable" outcome would still yield a positive return and its avoidance for itself would therefore not justify postponement. But the

---

[9] The price outcomes and probabilities in the Figures 4 through 6 have been taken from Dixit/Pindyck (1994, p. 42) and represent a 50% increase or 50% fall in the price compared to the previous value in each period, each with 0.5 probability. The return numbers for a buyer and for a seller of allowances in figures 5 and 6 are illustrative values chosen by this author. With this parameter choice the expectation value for the price (10 €) and the return (100 Mill €) is the same in each period.

[10] Note that a "real option" can also be "created", for instance by initiating an R&D program that *can* be continued if subsequent periods indicate favourable outcomes (due to exgoneous factors as well as factors endogenous to the R&D process), and can instead be cancelled otherwise. Thus creating a real option can make an R&D investment valuable even when its net present value, weighted with outcome probabilities, is negative (see Dixit and Pindyck 1994) – after all, the investment does not have to be continued "all the way through", so the calculation does not have to fully include unfavourable outcomes. Again, the higher the volatility, the higher is this real option value. "Creating" and "keeping open" a real option value are, in this way, "two sides of the same coin".

gain from waiting for one year results also from the fact that in many cases the outcomes in further years are more likely to be positive when they have been positive in the first year (and the same applies for negative outcomes). In this way, one year's development contributes to the development in future years. This is the case when the underlying processs is a "Markov" process (see Dixit and Pindyck 1994).

3.2 Real option valuation by risk-averse agents

We look at first at a risk-averse seller. She values the negative outcome (for her, a low allowance price) even higher than a risk-neutral agent (the drop in utility in the negative case is larger for her than for a risk-neutral agent).

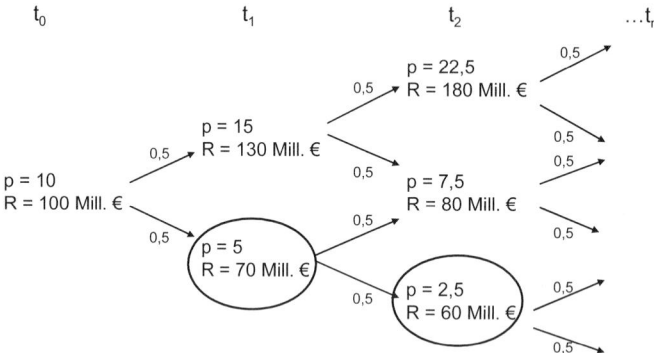

Fig. 5. Focus of a risk-averse seller
Source: Dixit and Pindyck (1994, p. 42), with modifications by this author

As one can see, also risk-neutral actors have a positive real option valuation, which means that not only risk-averse sellers postpone investments. However, risk-aversion increases the real option value of a potential seller. As a result, for sellers, real option value and risk-aversion point in the same direction.

What is the situation for a risk-averse buyer? She values the avoided negative outcome (in her case, an increase in prices) higher than a risk-neutral agent would. Therefore, risk-aversion means a tendency for postponement also for buyers.

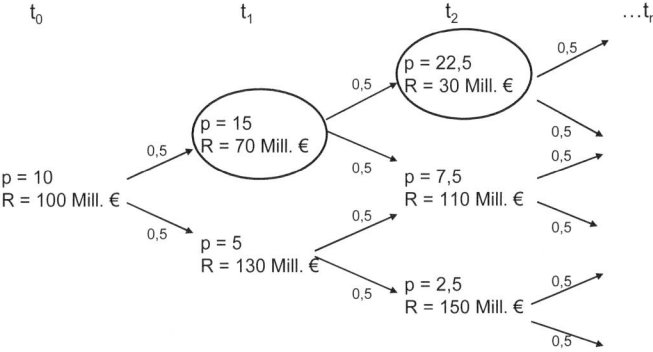

Fig. 6. Focus of a risk-averse buyer
Source: Dixit and Pindyck (1994, p. 42), with modifications by this author

But as mentioned in chapter 2, a risk-averse buyer also has a tendency to increase abatement in order to reduce his exposure. As a result, for a potential buyer, the "exposure effect" and the "real option effect" point in opposite directions.
In this case, the overall effect will depend on the relative importance of the option value of waiting versus the strength of the desire to limit the exposure. What does the size of the option value depend on?

First, the size of the option value depends on the loss avoided when waiting, but also on the benefits of an early investment (e.g., cost reductions already in early years, which would be given up in the event of waiting). Secondly, the size of the option value also depends on how much information on the relative likelihood of certain future developments the investor gains by waiting, i.e., how much of the potential unfavourable outcomes can be thus avoided. This depends on the degree to which prices at time t can stochastically predict prices at time $t + n$, i.e., on the size of the remaining "random factor".

If the profits foregone while waiting are small for a majority of the buyers, and if the stochastic function has a relatively small random component, then the option value of waiting should dominate on an aggregate level the "desire" to invest more resulting from the "exposure aspects". Conversely, if the profits foregone while waiting are relatively high as compared to the avoided potential unfavourable outcomes in future years (e.g., because the random factor is relatively large anyway), then on an aggregate level the desire to invest resulting more from "exposure aspects" is likely to dominate the option values.

As a result, we can state that:

- also risk-neutral actors have a positive real option valuation, so that risk-aversion is not required for creating a tendency for postponement;
- however, while for risk-averse sellers the risk-aversion reinforces the tendency to postpone, for risk-averse buyers it has an ambiguous effect.
- under auctioned allowances, all firms are potential buyers who want to reduce their exposure; on the other hand the risk–aversion enhances the option valuation for all of them. Therefore, as with buyers under freely allo-

cated allowances, risk-aversion has an ambiguous effect on the overall investment ratio.

4 Changes in the technologies of abatement

The analysis so far presumed that all potential abatement options were identical in their capital intensiveness and in terms of the impact ("leverage") exerted by an unfavourable future price outcome. However, apart from deciding on investments and their timing, uncertainty is likely to influence to some degree firms' decision on which abatement options to use (see Laurikka 2005). The influence may be only on the *abatement choices* in the narrow sense, but it is more likely to cover also production *technology decisions* – not least for the case of CO_2 emissions, where "end-of-pipe" abatement options are still far from being economically competitive.

Apart from the degree of abatement, there are two other possibilities for firms to react to increased allowance price uncertainty:

a) not to invest at any time and rely entirely on fuel shifts that are possible with the existing equipment.
b) change the type of investment.

Ad a) As the experiences under the Acid Rain Program have shown, this strategy can have considerable use in certain cases. Thus, nearly half of all abatement in the first five years has been achieved by fule shifts, or fuel mixings (see, e.g., Burtraw 2000). One of the largest US utilities, Southern Company, has largely relied on this strategy for a period of nearly ten years (personal information by Gary Hart, Senior Executive for Environmental Affairs at Southern Company, 25 July 2003).

However, this strategy is limited to such pollutants were fuel shifts are easily applicable without considerable (capital intensive) plant modification; and the strategy can achieve only limited emission reduction volumes.

Ad b) In terms of technology choice, several aspects matter (Laurikka 2005):

- employ technologies with a low exposure to the allowance price. This might favour gas to coal, unless a higher exposure on the fuel markets overrides this effect);
- employ technologies that have short pay-back periods (Yang and Blyth 2005), (e.g., because of a relatively low ratio of fixed costs to variable costs when calculated over 10–20 years),
- employ technologies that allow for "staged" investments (step-by-step), which favours technologies with small unit sizes (including certain renewable energy forms,
- employ "fuel-fuelled" plants that allow easy change of fuels (Laurikka 2005; Yang and Blyth 2005).

Effects on diffusion and innovation:

The firms' decisions on which technology they employ are likely to affect also the *direction* of technical progress: Firstly, the technology decisions of investors influence the relative rates of *diffusion* (market penetration) among different competing advanced technologies – such as advanced cycle gas turbines (CCGT), highly efficient coal combustion, carbon capture and storage (CCS) and renewable energy sources. In addition, the technology choices are likely to affect also firms' relative R&D spending between different advanced technologies, as well as cost reductions and performance improvements through "learning-by-doing" – thereby influencing the direction of innovation "as such", i.e., of the emergence of new or advanced technologies or technology variants.

5 Conclusions

When all actors are risk-averse, or actors are risk-averse on aggregate (and with homogeneous distribution of risk-averse and risk-loving actors among the potential sellers and buyers), and they all assign some degree of real option values, then

1. Under free allocation a clear *aggregate* tendency for postponement can be suggested, which increases with volatility. This is because there are three trends (exposure effect for sellers, option values for selles and buyers) that have a postponement effect, while there is only one trend (exposure effect for buyers) that has an accelerating effect;
2. Risk-aversion can change the composition of which firms abate first and which "wait" (since risk-averse potential sellers and buyers act differently);
3. Under free allocation it is plausible that risk-aversion leads to fewer investments in advanced ("innovative") abatement options at any given point in time than would be the case under certainty or with risk-neutral agents. This is because innovative technology is in general associated with lower marginal abatement costs and therefore with higher optimal abatement levels, resulting in relatively more innovative investors being sellers than buyers. The exact degree of this uncertainty effect on the innovativeness of investments is, however, difficult to determine;
4. Under auctioned allowances, the overall effect of risk-aversion on postponement is ambiguous because the "exposure effect" and the real option value point in opposite directions.

6 Discussion of the applied assumptions

6.1 Independency of a firms' allocation from its considered actions

The above analysis, like Ben-David et al. (2000), assumes implicitly that a pure "historical" principle in grandfathering is used – in other words, that allocation is based purely on historical data that any given firm can no longer influence. This principle is violated, for example, when free allocation is granted to new installations (and even more so when fuel-specific or other technology-specific allocation formulae are used for new installations); it is also violated under rules that foresee withdrawal of allowances from closed plants, and under allocation rules in which a regulator relates a firms' future allocation to its present emissions *after the emissions trading has started* (frequently called "updating").

When firm's individual abatement actions can influence their individual allocation, then this modifies the results in reality. However, it does not appear to qualitatively change the effects described here, at least as far as abatement measures in a plant (as opposed to plant replacements) are concerned.

6.2 No systematical differences between sellers and buyers regarding risk-attitude

The above analysis was based on the assumption of all agents being risk-averse, or at least of risk-aversion on an aggregate level, with homogeneous distribution of risk-averse and risk-loving agents among the potential sellers and buyers. Recall that this implies that firms differ in their abatement cost function not (systematically) because of different risk-attitudes, but because of cost factors such as different locations (and therefore fuel availabilities), different stages in their investment cycles (and therefore different sunk costs in the event of plant replacement), or because plants belong to different industry sectors that have different technological opportunities for emission reductions.

If, instead, all sellers are (or become) sellers because they are risk-loving and therefore have, e.g., a higher propensity to take the risks associated with new technologies (e.g., regarding functioning), then one might end up with a different result. Then both sellers and buyers would still have a certain incentive to postpone investment, in order to avoid the adverse pay-offs of unfavourable price outcomes (which are not valued as high by risk-loving actors, but still exist). On the other hand, both sellers and buyers now also have an unambiguous incentive from their "exposure side" to invest *more*: the risk-loving sellers want to increase their exposure (which they can do by investing more), while the risk-averse buyers want to reduce their exposure (which they can also do by investing more).[11]

In this case, the overall effect will depend on the relative importance of the option value of waiting, versus the degree of risk-love of potential sellers. We have argued in section 3.2 that the option value critically depends on the foregone prof-

---

[11] For the "missing link" in their explanation that buyers would apply a wait-and-see-strategy under uncertainty, leaving open the answer why sellers should not apply such a reaction as well, or even accelerate investment under uncertainty, which would be necessary to explain the observed price stability. However, in the experiment by Ben-David et al., the sellers and buyers were assigned their roles arbitrarily, so systematic "personal factors" can not be the reason.

its while waiting, and on the degree of prediction which the waiting time generates – in other words, on the relative importance of the random price factors.

Thus, if the profits lost while waiting are relatively high compared to the avoided potential unfavourable outcomes in future years (because the random factor is relatively large anyway), then on an aggregate level the desire to invest more resulting from "exposure aspects" is likely to dominate the option values.

Second, the result depends on the number of risk-loving agents among the sellers, and their "strength" of risk-love (as measured by the Arrow-Pratt-Index, see Eisenführ and Weber 2002). When the heterogeneity of firms' results, instead, mostly from technical factors (i.e, "given" marginal abatement cost functions) as listed above, and less on the risk-love of sellers, then the option value may still be the stronger factor.

For none of the firms covered by an Emission Trading Scheme, the emissions market is the core business activity, although for energy utilities it is an essential part of their cost and risk calculation. From this, one can presume that risk-aversion – or, at best, risk-neutraility - rather than risk-love is the dominant risk-attitude of the emitting agents. Therefore it looks plausible to presume that the option value effect is, in many cases, still the dominant factor, and postponement is a more likely reaction than an acceleration of adoption. However, more empirical research on the risk-attitude of agents and on firms' actual adoption actions seems vital.

6.3 No "flipping" of market positions

Irrespective of the risk-attitude, a "flipping" of some agents' market position (buyers becoming sellers) because of indivisibilities in the marginal abatement cost functions can complicate the issue. It is regarded plausible by the author that this still leads to more abatement by buyers than under certainty, but some investments may not be undertaken that would be done if abatement measures were more easily divisible. Thus, the overall investment may be even a bit lower when accounting for "flipping".

The case of plant replacements stated in section 2.3 is an example of "flipping".

6.4 Higher volatility does not raise the expectation value

If the assumption of a constant expectation value under higher volatility is given up, and instead it is assumed that higher volatility leads to a higher expectation value, then the effect of volatility on investment is also more complex. Such a higher expectation value can result if the possible range of expected prices (e.g., the confidence interval) is larger for price increases than it is for price drops.

After all, allowance prices can not drop below zero, so a skewed probability distribution is plausible. On the other hand, the upper price range is not infinite either, since it is limited by the penalty to be paid for non-compliance. Usually, emissions trading schemes set a defined "nominal" penalty value for non-compliance, and in addition require the missing allowances to be made up in the subsequent period, thereby requiring a higher "effective sanction".

When the effective sanction is considerably higher than double or triple the expected allowance price, then it is plausible that agents expect a skewed probability distribution function, in which a higher volatility goes together with a higher expectation value.

Under the EU Emissions Trading Scheme for CO_2, the nominal sanctions are 40 € in 2005–2007 and 100 € in 2008–2012; in addition, the missing allowances must be delivered in the following period. As the current allowance price is 25–30 € (December 2005), the effective sanction is rather 65–70 € in the current period. Since the actual price of 25–30 € is somewhat –not much – below half the effective sanction, a slight skewedness of the price probability function can be expected, albeit no large one. After 2008, the skewedness might be larger unless prices increase significantly, because the nominal sanction is then lifted to 100 €. Also under the US Acid Rain Program, where the nominal sanction is 2000 $ and current prices have risen to 700 $ (from previous levels below 200 $ before 2004), the actual price is below half the nominal sanction value, so a certain degree of skewedness is plausible.

6.5 The role of long-term emission target expectations

We have concluded above that a firm considering an irreversible abatement investment is likely to stand back from the most advanced alternatives when it is risk-averse, going to be a seller, and allowance prices are uncertain. However, when the investment is planned to be in operation for so many years that its owner, through the regulator's expected cap reduction path, is going to be a buyer during much of the latter years of the investment's lifetime, then the firm may even opt for a very advanced technology. The crucial factors in this context are the number of years that the investment will be in operation, the abatement possibilities through later modernisations within the respective plants, and the cap reduction path. These factors together determine for how many years the owner is a seller and for how many years he/she is a buyer.

6.6 Overall profitability effects

Incentives for abatement are, after all, "derived" incentives, since the ultimate objective of a polluting firm is not to reduce its overall compliance costs $(C + P\ (e^*-e^a))$, but to maximise its profit. Therefore not all effects that increase incentives to reduce compliance costs also increase incentives to abate or assign funds to R &D (see Grubb and Ulph 2002). This finding has, in the context of this paper, relevance especially to the comparison of auctioning versus free allocation: under auctioning, some actors might not find it profitable any more to produce at all, and therefore will also not abate. The same actors would not have seized production under free allocation when this involves also a rule that foresees withdrawal of allowances upon closure of a plant. Such a rule is applied by most EU Member States under the EU CO_2 Emissions Trading Scheme (ETS) (see DEHSt 2005), but is not used in the Acid Rain Program in the USA (see Ellerman et al. 2000).

References

Ben-David S, Brookshire D et al. (2000) Attitudes Toward Risk and Compliance in Emission Permit Markets. Land Economics 76 (4): 590–600
Burtraw D (2000) Innovation Under the Tradable Sulfur Dioxide Emission Permits Program in the U.S. Electricity Sector. RFF Discussion Paper 00–38
DEHSt (Deutsche Emissionshandelsstelle (2005) Implementation of Emissions Trading in the EU: National Allocation Plans of all EU States. Download: http://www.dehst.de/cln_007/nn_593634/SharedDocs/Downloads/EN/ETS/EU__NAP__Vergleich.html
Dixit AK, Pindyck RS (1994) Investment under Uncertainty. Princeton
Edleson ME, Reinhardt FL (1995) Investment in Pollution Compliance Options: the Case of Georgia Power. In: Trigeorgis L (ed) Real options in capital investment – models, strategies and applications. Praeger, Westport, Connecticut
Eisenführ F, Weber M (2002) Rationales Entscheiden. 4th edition. Springer, Berlin et al.
Ellerman AD, Joskow PL, Schmalensee R, Montero JP, Bailey EM (2000) Markets for Clean Air: The U.S. Acid Rain Program. Cambridge University Press, New York et al.
Fraunhofer-ISI (2005) Flexible Instrumente im Klimaschutz. Eine Anleitung für Unternehmen. In German language under http://www.isi.fhg.de
Grubb M, Ulph D (2002) Energy, the Environment, and Innovation. In: Oxford Review of Economic Policy 18 (1): 92–106
Herbelot O (1994) Option Valuation of Flexible Investments. The Case of a Scrubber for Coal-Fired Power Plants. Working Paper MIT-CEEPR 94–001WP, Massachusetts Institute of Technology, Cambridge/Mass
Knight FH (1921) Risk, uncertainty and profit. Hart, Schaffner & Marx, Houghton Mifflin Company. Boston, Massachusetts
Kolstad CD (2000) Environmental Economics. Oxford University Press, New York, Oxford
Lambie NR (2002) Analysing the effect of a distribution of carbon permits on firm investment. Paper presented at the 46th Annual Conference of the Australian Agricultural and Resource Economics Society, 13–15 February, Canberra
Laurikka H (2005) The impact of climate policy on heat and power capacity investment decisions. In Antes R, Hansjürgens B, Lethmathe P (eds) Emissions Trading and Business 1, Physica, Heidelberg/New York
MacCrimmon KR, Wehrung DA (1984) The Risk In-Basket. Journal of Business 57: 367–387
Machina MJ, Rothschild M (1998) Risk. In: The New Palgrave – A Dictionary of Economics. First paperback edition, pp 201–206. Macmillan Press, London; Stockton Press, New York
Mas-Colell A, Whinston MD, Green JR (1995) Microeconomic Theory. New York et al., Oxford University Press
von Neumann J, Morgenstern O (1947) The Theory of Games and Economic Behavior. 2nd Edition, Princeton University Press, NJ

Sharpe WF, Alexander GJ, Bailey JV (1995) Investments. International Edition, 6th ed Prentice-Hall International, London et al.

Spangardt G, Meyer J (2005) Risikomanagement im Emissionshandel. In: Lucht M, Spangardt G (ed) Emissionshandel. Springer

Trigeorgis L (ed) (1995) Real options in capital investment – models, strategies and applications. Praeger, Westport, CT

Weber C, Swider D (2004) Power plant investments under fuel and carbon price uncertainty. Paper presented at the 6th IAEE European Conference on Modelling in Energy Economics and Policy

Yang M, Blyth W (2005) Modelling impacts of climate change policy uncertainty on power investment. Paper presented at the Workshop "Climate Change, Sustainable Development and Risk – an Economic and Business View", organised by Martin-Luther-University Halle-Wittenberg and Gesellschaft für Operations Research e.V., 16th–18th November 2005, Leucorea, Lutherstadt Wittenberg/Gemany

Zhao J (2003) Irreversible Abatement Investment under Cost Uncertainties: Tradable Emission Permits and Emissions Charges. Journal of Public Economics 87: 2765–2789

Climate change, sustainable development and risk: realizing a financial fund within the TEM model as an economic and business opportunity

Stefan Pickl

Universität der Bundeswehr München
Faculty of Computer Science
Institute of Applied System Science and Information Systems
Werner Heisenberg Weg 39, 85577 Neubiberg-München, Germany
stefan.pickl@unibw.de

Abstract

The global climate is changing, and will continue to change, in ways that affect the planning and day-to-day operations of businesses, government agencies and other organisations. The possibility of establishing a fund supporting the aims of the Kyoto Protocol is currently one of the central topics for discussion. The key questions are: how can such a fund be realized and how can it be embedded in an optimal energy management?

We present a mathematical approach which is based on the Technology Emissions Means (TEM) model which was developed by the author. In addition it is shown how a fund structure can be designed. The economic and business opportunities are discussed within the framework of a climate change risk analysis.

Keywords: Climate change risk analysis, fund program, TEM model, certificates procurement management, International Monetary Fund, business intelligence

1 Introduction

In a first step we summarize the TEM model as a time-discrete model which models the economic behaviour within the Kyoto process. The simulated trajectories represent several strategies of the actors. We introduce a cost-game which is a basis for an optimal decision-making process. Within such a game the so-called core may be characterized. As the parameters may vary we may seek both stable and unstable coalitions. A risk analysis approach is presented. Finally we compare the approach with actual discussions of emission trading, balancing economic and business opportunities against sustainable development and risk categories.

2 Historical background

The conferences of Rio de Janeiro 1992 and Kyoto 1997 demanded new economic instruments which have a focus on economic and business opportunities. The Kyoto Protocol was agreed in 1997, enacted in 2005 and is valid until 2012. It contains a plan to decrease the emission of greenhouse gases. One of the objectives of the Kyoto Protocol is specified in ***article 2*** of *The United Nations Framework Convention on Climate Change* (retrieved on November 15, 2005):

"The ultimate objective of this Convention and any related legal instruments that the Conference of the Parties may adopt is to achieve, in accordance with the relevant provisions of the Convention, stabilization of greenhouse gas concentrations in the atmosphere at a level that would prevent dangerous anthropogenic interference with the climate system. Such a level should be achieved within a timeframe sufficient to allow ecosystems to adapt naturally to climate change, to ensure that food production is not threatened and to enable economic development to proceed in a sustainable manner."

This citation demonstrates the necessity of economic models which try to

- develop suitable benchmark processes
- examine disparities
- consider a climate risk analysis.

In summary, the Kyoto protocol tries to deal with climate change in order to realize sustainable development. The risks within that process should be understood and minimized. The approach corresponds to the fact that several organisations have begun to analyse the potential financial risk of climate policies for the electricity sector using more complex financial models. One example may be found in the Coalition for Environmentally Responsible Economies (CERES 2006).

This report stated that "American Electric Power did the most comprehensive job of analyzing future scenarios, quantifying their implications and providing and discussing them". In the following we start with an introduction of the TEM model as an analytical tool which examines specific processes within the Kyoto protocol.

An important economic tool forming part of the treaty of Kyoto in that area is Joint-Implementation. This is a program which intends to strengthen international cooperations between enterprises in order to reduce CO_2-emissions. A sustainable development can only be guaranteed if the instrument is embedded into an optimal energy management. For that reason, the Technology-Emissions-Means (TEM) model was developed, giving the possibility to simulate such an extraordinary market situation. The realization of Joint-Implementation (JI) is restricted by technical and financial constraints. In a JI program, the reduced emissions resulting from technical cooperations are recorded at the Clearing House. The TEM model integrates the simulation of both the technical and the financial parameters. In Pickl (1999) the TEM model is treated as a time-discrete control problem. Furthermore, the analysis of the feasible set is examined in Pickl (2001). In the following, a short introduction into the TEM model is given. A key to effective emission analysis is the accurate and standardized reporting of emissions. For that reason the TEM model refers directly to the emissions reduced. In addition to analysing an electric power company's corporate governance and management systems, actors may begin to quantify the financial risk provided by climate change in that they ask for emission pricing, i.e. a likely price for reducing carbon dioxide. This aspect will be considered after the TEM model is introduced.

3 The basic model

In order to provide a view of the behaviour of the key elements of the Kyoto process, the presented TEM model describes the economic interaction between several players (sometimes we say equivalently actors) which intend to maximize their reduction of emissions E_i caused by technologies T_i, by expenditures of money or by financial means M_i. The index stands for the i-th player. The players are linked by technical cooperations and by the market. The effectivity measure parameter em_{ij} describes the effect on the emissions of the i-th player if the j-th actor invests money for his technologies. We can say that it expresses how effective technology cooperations are (like an innovation factor), which is the central element of a JI Program.

The variable φ can be regarded as a memory parameter of the financial investigations, whereas the value λ acts as a growth parameter. For a deeper insight see Pickl (1999).

The TEM model is represented by the following two equations:

$$E_i(t+1) = E_i(t) + \sum_{j=1}^{n} em_{ij}(t) M_j(t), \tag{1}$$

$$M_i(t+1) = M_i(t) - \lambda_i M_i(t)[M_i^* - M_i(t)]\{E_i(t) + \varphi_i \Delta E_i(t)\} \tag{2}$$

Furthermore, we force that

$$0 \leq M_i(t) \leq M_i^*, \; i = 1,...,n \text{ and } t = 0,...,N. \tag{3}$$

Additionally we assume

$$-\lambda_i M_i(t)[M_i^* - M_i(t)] \leq 0 \text{ for } i = 1,...,n \text{ and } t = 0,...,N. \tag{4}$$

Then we have guaranteed that $M_i(t+1)$ increases if $E_i(t) + \varphi_i \Delta E_i(t) \leq 0$ and decreases if the term is positive. In the following section it is explained why this is necessary from a practical point of view. A detailed description is contained in Pickl (1999).

4 Empirical foundation

At the centre of the TEM model is the so-called em-matrix. It is a great advantage of the TEM model that we are able to determine the em$_{ij}$ parameters empirically. These parameters offer a quantitative measure of climate risk under a range of potential outcomes. This will be explained in the following in detail:

In the first equation of the TEM model, the level of the reduced emissions at the t + 1-th time-step depends on the previous value plus a market effect. This effect is represented by the additive terms which might be negative or positive. In general, $E_i > 0$ implies that the actors have yet reached the demanded value $E_i = 0$ (normalized *Kyoto-level*). A value $E_i < 0$ expresses that the emissions are less than the requirements of the treaty. In the second equation we see that for such a situation the financial means will increase, whereas $E_i > 0$ leads to a reduction of $M_i(t + 1)$.

The second equation contains the logistic functional dependence and the memory parameter φ which describes the effect of the preceding investment of financial means. The dynamics does not guarantee that the parameter $M_i(t)$ lies in the interval, which can be regarded as a budget for the i-th actor.

For that reason we have to impose the following additional restrictions on the dynamical representation:

$$0 \leq M_i(t) \leq M_i^*, \, i = 1,...,n \text{ and } t = 0,...,N.$$

These restrictions ensure that the financial investigations can neither be negative nor exceed the budget of each actor. Now, it is easy to show that

$$-\lambda_i M_i(t)[M_i^* - M_i(t)] \leq 0 \text{ for } i = 1,...,n \text{ and } t = 0,...,N.$$

We have guaranteed that $M_i(t + 1)$ increases if $E_i(t) + \varphi_i \Delta E_i(t) \leq 0$ and that it decreases if the term is positive. Applying the memory parameter φ, we have developed a reasonable model for the money expenditure-emission-interaction, whereby the influence of the technologies is integrated in the em-matrix of the system. We can use the TEM model as a time-discrete model where we start with a given set of parameters and observe the resulting trajectories. Usually, the actors start with a negative value, i.e. they lie under the baseline of the Kyoto Protocol.

They try to reach a positive value of E_i. By adding control parameters, we enforce this development by an additive financial term. For that reason the control parameters are added only to the second equation of our model:

$$M_i(t+1) = M_i(t) - \lambda_i M_i(t)[M_i^* - M_i(t)]\{E_i(t) + \varphi_i \Delta E_i(t)\} + u_i(t).$$

The introduction of the control parameter $u_i(t)$ implies that each actor makes an additional investigation at each time-step. In the sense of environmental protection, the aim is to reach the state specified in the treaty of Kyoto, by choosing the control parameters such that the emissions of each player become minimized. For details and the treatment as an approximation problem see Krabs (2004). The focus is the realization of the necessary optimal control parameters via a played cost game, which is determined by the way the actors cooperate. Let us first discuss where and how this aspect can be integrated into the TEM model. For analysts it might be necessary to integrate a qualitative measure under a range of potential outcomes. Because the em_{ij}-parameters can be varied, this approach considers the fact that analysts can use this model to simulate the potential financial behaviour and the risk of different policies on the electricity sector. Numerical examples are contained in Pickl (1999).

Risk management in that context combines the

- optimisation of energy efficiency
- establishment of an emission trading procedure
- divestiture of carbon-intensive assets.

5 Realizing a fund via a played cost-game

In order to reach steady states, which are determined in Pickl (1999), an independent institution may coordinate the trade relations between the actors (like a clearing-house mechanism). The trade relations are expressed by the em-matrix. In practice, the imposing of *taxes* or the giving of *incentives* means that in the TEM model the em_{ij}-parameters will change. It was assumed that the principle of JI implies that technical cooperation will be rewarded. If there is a cooperation between player 1 and player 2, an additional parameter ε, $\varepsilon > 0$, (which implies that the measure of effectivity increases) was introduced. The cooperation of the grand coalition is expressed by the parameter ω. We summarize this approach (which is the basis for the stochastic extension) in the following:

The coalition between player 1 and 2

$$\begin{pmatrix} em_{11} & em_{12} + \varepsilon & em_{13} \\ em_{21} + \varepsilon & em_{22} & em_{23} \\ em_{31} & em_{32} & em_{33} \end{pmatrix} \tag{5}$$

As well as the great coalition

$$\begin{pmatrix} em_{11} & em_{12}+\omega & em_{13}+\omega \\ em_{21}+\omega & em_{22} & em_{23}+\omega \\ em_{31}+\omega & em_{32}+\omega & em_{33} \end{pmatrix} \tag{6}$$

It is easy to construct a cost-game by

$$v_t(K) := \underbrace{\sum_{j \in K} M_j(t)}_{\text{without cooperation}} - \underbrace{M(K)}_{\text{cooperation}} \tag{7}$$

In Pickl (1999) this time-discrete game is described and analysed in detail. Furthermore it is shown that the difference between the cooperative and the non-cooperative case is always positive for the TEM model. Thus we have constructed a reasonable cost-game. The method was that at each time step (see parameter t in the equation above), this amount can be put into a central fund.

Up to now we have described in detail the possibilities for

- an emission analysis with the TEM model
- a market system analysis reflecting the em_{ij}-parameters
- several case-studies and benchmark scenarios.

6 Stochastic effects and emissions trading processes

The best and most reasonable method to reach the objectives of Kyoto is to reduce emissions where this is possible in the most cost-efficient way. Therefore every state defines a total quantity of emissions during a period of time. It assigns certificates of permission to emit a clearly defined volume of greenhouse gases to emitters. These certificates are tradable between the various emitters, so that as a result emitters with high costs of reducing emissions prefer to buy more certificates from other emitters who have lower-cost possibilities of reducing emissions. The following table shows the advantages and disadvantages of this way of trading certificates:

Table 1. Advantages and disadvantages of trading emission certificates

Advantages	Disadvantages
Caused by dynamic trends of market, the costs of reducing emissions are minimized. (dynamic efficiency)	The states are not forced to participate in the rules of Kyoto, so states with a high rate of emissions don't take part.
It is possible to accomplish a lasting long-term effect on reducing emissions of greenhouse gases.	The compliance with the Kyoto Protocol is not controllable enough, at best via random checks.
Emissions trading is applicable very flexibly across Europe and even globally.	Climate protection may be understood as a burden.
Emissions trading enables cost-saving possibilities for enterprises to reach the objectives of the Kyoto Protocol.	Some sectors, such as the chemical industry or nuclear energy, are not yet included.
	The very bureaucratic procedure for requesting permissions is a barrier to economic growth.

As shown by this table, a deterministic treatment of the Kyoto process might not be attractive. There are a lot of uncertainties in that procedure. For this reason there is a need for a stochastic analysis within a climate-risk approach.

The key idea is now that we interpret the variation of the em_{ij}-parameter (within the played cost-game) as variables which are described via a likelihood distribution. Then we are able to extend the approach.

Furthermore, the constitution of the fund is then directly connected with a risk analysis. A key to effective emission analysis and investment strategies should be at the centre of interest. The structure of such a possible fund (having a very close relationship to Keynes' ideas of a financial fund) is summarized in the following:

- globally valid
- local realization in certain projects (like JI)
- embedded in an international framework (financial system).

Nevertheless it is still an open question whether the Kyoto protocol and its inherent programs may solve the climate problem. As the price for one tonne CO_2 is below 1 € in 2007, the functional behaviour of the market is in discussion. At such a point, an orientation to the International Monetary Fund may help. Additionally, there is a need for certain analytical tools and instruments.

Additionally, such an analysis is entering more and more into the centre of interest of energy companies. An example can be found in CERES (2006) where the potential financial risk of climate policies on the electricity sector is analysed in detail. The extended (in a stochastic way) TEM model might be a useful mathematical tool to support such an analysis within a 'cap and trade' system such as the Kyoto Protocol.

7 Outlook and perspectives

The advantage of the approach presented here lies in the fact that we can combine a stochastic risk analysis with a game-theoretic bargaining approach. We can simulate a Joint Implementation Program as well as an emission trading process. Furthermore, we can describe certain scenarios and obtain benchmark processes which are connected with a certain likelihood for their appearance. Of course, the quality of the results is directly dependent upon the quality of the data.

According to Antes (2006) "Kyoto is a 'cap and trade' system that imposes national caps on the emissions of Annex I countries. On average, this cap requires countries to reduce their emissions 5.2% below their 1990 baseline over the 2008 to 2012 period. Although these caps are national-level commitments, in practice most countries will devolve their emissions targets to individual industrial entities, such as a power plant or paper factory.

An example of a 'cap and trade' system is the unrelated-to-Kyoto 'EU ETS'. Other countries may follow suit in time. This means that the ultimate buyers of Credits are often individual companies that expect their emissions to exceed their quota (their Assigned Allocation Units, Allowances for short). Typically, they will purchase Credits directly from another party with excess allowances, from a broker, from a JI/CDM developer, or on an exchange. National governments, some of whom may not have devolved responsibility for meeting Kyoto obligations to industry, and that have a net deficit of Allowances, will buy credits for their own account, mainly from JI/CDM developers."

As such deals can be made via a national fund, an agency, or collective funds, such as the World Bank's Prototype Carbon Fund (PCF), the approach presented here may be a useful tool for financial investors within that emerging market with banks, brokers, funds, arbitrageurs and private traders eventually participating. This stresses an orientation such as described in the previous section.

Apart from the above, several alternatives exist at the moment, such as "bubble"-structures, clusters of countries or "markets-within-markets". Some of those are related with the TEM model in earlier contributions like Pickl (2001) and Pickl (2004).

We conclude that the extension presented here allows – in an initial approach – the possibility to

- enumerate risks related to climate change impacts
- prioritise risks that require further attention
- establish a process for ensuring that these higher-priority risks are managed effectively
- advance market effectivity.

Details and first stochastic numerical results of such an approach will be contained in a forthcoming paper.

8 Conclusion

The possibility of establishing a fund supporting the aims of the Kyoto Protocol is currently one of the central topics for discussion. The key questions are: how can such a fund be realized and how can it be embedded in an optimal energy management? We present a mathematical approach which might accompany such an institutionalisation. With the TEM model a certain analytic approach is possible.

There is a need for further dynamic discrete models which deal with climate change, sustainable development and risk. The gap to economic and business opportunities should be better reflected and minimized – possibly via an international fund structure.

References

Antes R (2007) Nachhaltigkeit und Betriebswirtschaftslehre – Eine wissenschaftstheoretische und institutionelle Perspektive. (Sustainability and business administration science: a methodological and institutional perspective.) Metropolis Verlag, Marburg

Antes R (2006) Coroporate Greenhouse Gas Management in the Context of Emission Trading Regimes. In: Antes R, Hansjürgens B, Letmathe P (eds) Emission Trading and Business. Physica Verlag, Heidelberg, New York, pp 199–217

Betz R et al. (2002) Flexible Instrumente im Klimaschutz (Flexible Mechanisms of Climate Protection). Ministerium für Umwelt und Verkehr Baden-Württemberg, revised edition, Stuttgart

Boltjanski WG (1976) Optimale Steuerung diskreter Systeme. Leipzig Akademische Verlagsgesellschaft Geest & Portig K.-G., Leipzig

Center for Clean Air Policy (CCAP) (2001) Study on the monitoring and measurement of greenhouse gas emissions at the plant level in the context of the Kyoto mechanisms. Final Report, Washington, DC

Coalition for Environmentally Responsible Economies (CERES) (2006) Best Practice in Climate Change. Risk Analysis for the Energy Power Sector – The Results of the Ceres Electric Power/Investor Dialogue, http://www.ceres.org/pub/docs

Driessen TSH (1986) Cooperative Games, Solutions and Applications. Cambridge University Press, Cambridge, New York

European Environment Agency (EEA) (2001) Environmental signals 2001: European Environment Agency regular indicator report. Copenhagen

European Parliament (EP) (2003) Greenhouse gas emission allowance trading ***II. In: Texts adopted at the sitting of Wednesday 2 July 2003, provisional edition

Fichtner W, Enzensberger N, Rentz O (2003) CO_2-Emissionsrechtehandels-Regime und die Bedeutung des Zertifikatpreises. (CO_2 emissions trading regimes and the price of the certificates.) In: UmweltWirtschaftsForum 11 (3): 48–51

Grassl H (1993) Umwelt- und Klimaforschung. (Environmental and climate research.) In: Held M, Geißler KA (eds) Ökologie der Zeit. S. Hirzelverlag, Stuttgart, pp 75–84

Innovest (2002) Climate change and the financial services industry. Module 1 & 2, Finance Initiatives (UNEP), http://www.unepfi.net/

Kim Joy A (2003) Regime interplay: a case study of the climate change and trade regimes, submitted thesis to the School of Environmental Sciences of the University of East Anglia. Norwich

Krabs W (1997) Mathematische Modellierung. BG Teubner, Stuttgart

Krabs W, Pickl S (2004) Controllability of a time-discrete dynamical system with the aid of the solution of an approximation problem. Journal of Control Theory and Cybernetics

Kyoto (1997) Kyoto Contract. http://www.unfccc.org/resource/convkp.html

Knoedel P (2000) Der Emissionshandel bei BP Amoco. (Emissions Trading at BP.) In: UmweltWirtschaftsForum 1: 41–45

Kreikebaum H (1992) Zentralbereiche. (Central departments.) In: Frese E (ed) Handwörterbuch der Organisation. 3. edn, Stuttgart , pp 2603–2610

Meyer JW, Rowan B (1991, 1977) Institutionalized organizations: formal structure as myth and ceremony. In: Powell WW, DiMaggio PJ (eds) The new institutionalism in organizational analysis. Chicago/London, pp 41–62, reprinted: American Journal of Sociology 83 (2): 340–63

Organisation for Economic Co-Operation and Development (OECD) (2001) OECD Environmental Outlook. Paris Cedex

Ortmann G, Zimmer M (1998) Strategisches Management, Recht und Politik (Strategic management, law and policy). In: Die Betriebswirtschaft 6: 747–69

PCF Prototype Carbon Fund: Annual Report (2001) URL: http://prototypecarbonfund.org/docs/AR_download.htm

PCF Prototype Carbon Fund: Annual Report (2002) URL: http://prototypecarbonfund.org/docs/2002AnnualReport.htm

Pickl S (1999) Der τ-value als Kontrollparameter. Modellierung und Analyse eines Joint-Implementation Programmes mit Hilfe der kooperativen dynamischen Spieltheorie und der diskreten Optimierung. Shaker Verlag, Aachen

Pickl S (2001) Optimization of the TEM Model – co-funding and joint international emissions trading. Operations Research Proceedings 2000 (Selected Papers) 113–118. Berlin, Heidelberg, Springer Verlag

Pickl S (2001) Convex games and feasible sets in control theory. Mathematical Methods of Operations Research 53: 51–66

Pickl S (2004) An Algorithmic Solution for an Optimal Decision Making Process within Emission Trading Markets Annales du Lamsade No 3, Laboratoire d'Analyse et Modélisation de Systèmes pour l'Aide à la Décision, Proceedings of the DIMACS-LAMSADE. Workshop on Computer Science and Decision Theory: 267–278

Raiffa H (1968) Decision Analysis. Introductory Lectures on Choices under Uncertainty. Addison-Wesley Series in Behavioral Sciences: Quantitative Methods. Reading, Mass. etc.: Addison-Wesley Publishing Company

Richter R, Furubotn E (1996) Neue Institutionenökonomik (Institutions and economic theory). Tübingen

Rosenzweig R et al. (2002) The emerging international greenhouse gas market. Ed by PEW Center on Global Climate Change, Arlington, URL: http://www.pewclimate.org/ projects/trading.pdf

Royal Dutch/Shell Group (2003a) The Shell tradeable emission permit system. o.J., URL: http://www.shell.com/static/royalen/downloads/steps.pdf

Royal Dutch/Shell Group (2003b) Shell tradable emissions permit system (STEPS) – Learning and experience. URL: http://www.shell.com/static/royalen/downloads/steps_learning.pdf

RSU, Rat von Sachverständigen für Umweltfragen (2002) Umweltgutachten (Environmental report). Stuttgart

Sandhövel A (2003) Emissionshandel aus Bankensicht (Banks and emissions trading). In: UmweltWirtschaftsForum 11 (3): 39–43

Santarius T, Ott H E (2002) Attitudes of German companies regarding the implementation of an emissions trading scheme. Wuppertal Papers no 122e, Wuppertal

Scharte M (2002) Klimapolitik und Treibhausgas-Management (Climate policy and greenhouse gas management). St. Gallen, electronical dissertation at the University St. Gallen, URL: http://verdi.unisg.ch/www/edis.nsf/wwwDisplayIdentifier/2599/$FILE/dis2599.pdf

Schneidewind U (1998) Die Unternehmung als strukturpolitischer Akteur (The company as a political actor of structuration). Marburg

Schreiner M (1996) Umweltmanagement in 22 Lektionen (Corporate environmental management in 22 lessons). 4. edn, Wiesbaden

Sebastian HJ, Sieber N (1980) Diskrete dynamische Optimierung. Akademische Verlagsgesellschaft, Leipzig, 1980

Seo MG, Creed WED (2002) Institutional contradictions, praxis, and institutional change: a dialectical perspective. In: Academy of Management Review 27 (2): 222–247

Sorrell S (2002) The climate confusion: implications of the EU emissions trading directive for the UK climate change levy and climate change agreements. SPRU (Science and Technology Policy Research), University of Sussex, Brighton, URL: http://www.susex.ac.uk/spru/environment/research/ccfr.pdf

Tijs SH, Driessen TSH 1986: Game theory and cost allocation problems. Management Science 32: 1015–1028

UNEP Finance Initiatives (2002) Climate change and the financial services industry – Executive briefing paper. CEO briefing

Villiger A, Wüstenhagen R, Meyer A (2000) Jenseits der Öko-Nische (Beyond the econiche). Basel et al.

Vorholz F (2003) Was kostet die Luft? Der Markt hält Einzug im Umweltschutz – und sorgt in der Industrie für den großen Verteilungskampf. (What are the costs of clean air? The market moves into environmental protection and causes a great distribution fight in the industry.) In: Die ZEIT, No 29, 10.07.2003, p 17

WBCSD World Business Council for Sustainable Development, WIR World Resources Institute (2001) The greenhouse gas protocol: a corporate accounting and reporting standard. URL: http://www.wbcsd.org/DocRoot/NlH8ZRdvbUZhkGFQaOuj/ghg-protocol.pdf

Whittaker M, Kiernan M, Dickinson P (2003) Carbon finance and the global equity markets, (ed) by Carbon Disclosure Project. Februar 2003, URL: http://www.cdproject.net/

Wüstenhagen R (2000) Ökostrom – von der Nische zum Massenmarkt (Green electricity – from the niche to the mass market). Zürich

Wurzel R et al. (2003) Das britische Emissionshandelssystem (The UK emissions trading scheme). In: UmweltWirtschaftsForum 11 (3): pp 9–14

Young OR (2002) The institutional dimensions of environmental change: fit, interplay, and scale. Cambridge Ma., London

Zabel HU (2001) Ökologische Unternehmenspolitik im Verhaltenskontext (Ecological sound business policy – a behavioural approach). Berlin

Current evaluation practice of the Clean Development Mechanism

Felicia Müller-Pelzer

Von-Claer-Str. 4, 53639 Königswinter
felicia.muellerpelzer@googlemail.com

Abstract

The Clean Development Mechanism (CDM) is the most quickly evolving of the three flexibility mechanisms[1] of the Kyoto Protocol. The first CDM project activities were registered in 2004, i.e. before the Kyoto Protocol went into force, and the first emission reductions were issued in 2005.

This paper analyses the current evaluation practise of CDM project activities and comes to the conclusion that the two main CDM goals (the achievement of additional emission reductions and the contribution to sustainable development in the host country) are not awarded equal attention. As the mitigation of GHG emissions is closely linked to the fulfilment of the Kyoto Protocol, an externally accredited entity, the DOE (Designated Operational Entity) has to carry out validation, verification and certification activities. However, no transparent evaluation system has been established for sustainable development. The DNA (Designated National Authority) is in charge of assessing the contribution to sustainability of proposed project activities. As the host countries are sovereign in their decisions, the DNAs are free to assume this task as they find appropriate. Therefore, the quality of evaluation may vary considerably.

Although an external ex-ante evaluation is a prerequisite aimed at assuring the quality of the CDM project activities, it actually cannot guarantee a successful implementation. An approach is needed that enables the project developers to design and implement a sustainability strategy for their CDM project activities and to demonstrate their contribution to sustainable development. In addition, it should provide to the DNAs the information needed to evaluate the CDM project activity.

Keywords: Clean Development Mechanism (CDM), contribution to sustainable development, evaluation practices

---

[1] The three flexibility mechanisms: the emission trading (ET), the Joint Implementation (JI) and the Clean Development Mechanism (CDM).

1 Background

The United Nations Framework Convention of Climate Change (UNFCCC 1992) was established to stabilise greenhouse gas (GHG) concentration with the aim of preventing dangerous anthropogenic interference with the climate system. In pursuit of the ultimate objective of the Convention and guided by Article 3 of the Convention, the Parties established the Kyoto Protocol in 1997. Industrialised countries committed themselves to jointly reducing their GHG emissions from 2008 to 2012 by 5.2% below 1990 levels. Quantified emission limitation and reduction commitments are differentiated by country.

The Clean Development Mechanism (CDM) was established by the Kyoto Protocol together with two other Kyoto Mechanisms, the Joint Implementation (JI) and the emission trading (ET). During the UNFCCC Conference of the Parties in Marrakesh in 2001, modalities and procedures for the CDM were developed. Under the CDM ratifying Annex I Parties[2] can implement projects that either reduce emissions in non-Annex I Parties[3], or absorb carbon through afforestation or reforestation activities. In return, ratifying Annex I Parties receive Certified Emission Reductions (CERs), while the project activities should assist host Parties in achieving sustainable development and contributing to the ultimate objective of the Convention. The supervisory body of the CDM is the Executive Board.

As such, the CDM pursues three main goals: first, to assist non-Annex I Parties in achieving sustainable development, second, to support them in contributing to the ultimate objective of the Convention and third, to help Annex I Parties in achieving compliance with their Kyoto targets. However, these goals are not operational and possible trade offs can be identified: For instance a project activity strongly reducing emissions may have more negative impact than positive with regard to sustainable development. The Marrakesh Accords (UNFCCC 2001) do not provide detailed methodologies for CDM project activities as the Parties opted for a bottom-up approach. In addition, the subsidiary principle was adopted with regard to sustainable development: the host country was made sovereign to decide whether or not a CDM project activity contributed to the sustainable development of the country.

The CDM relies on the market mechanism, but is above all embedded in strategic political processes. Apart from diverging national interests, pressure groups try to influence the development of the CDM. The interests are manifold ranging from pursuing a cost-reduction for mitigation activities, to obtaining foreign currency and investments, to enhancing sustainable development. Market actors are often driven by the revenues from the emission reductions, but may also be interested in improving both their market presence and image, in enhancing their bargaining power, in building up business alliances as well as gathering experience with Kyoto mechanisms.

---

[2] Annex I countries are defined as Parties to the Convention (UNFCCC) which are also listed in Annex B of the Kyoto Protocol. These are the industrial countries and some of the countries with economies in transition (EITs).

[3] These are some of the EITs and the developing countries.

Many studies have been conducted (e.g. IISD 2005; Foot 2004) which evaluate the first achievements of the CDM; However, assessing the success of the CDM depends on the underlying value judgments.

The CDM cannot satisfy all these particular interests equally. Therefore, one has to abide by the goals stated in the Kyoto Protocol and the Marrakesh Accords, which reflect the consensus of the Parties. The contribution of non-Annex I Parties to the ultimate objective of the Convention is becoming prominent for post Kyoto scenarios. The two goals which are most discussed with regard to single CDM project activities are the achievement of additional emission reductions and the contribution to sustainable development. These goals are juxtaposed in the following to explain the concepts and to show the differences in evaluation practices under the CDM.

2 State of research

Two main strands of research can be observed: one is generally convinced of the benefits of the CDM as a Kyoto Mechanism. The other calls the CDM into question pointing out ecological and institutional shortcomings (to some extent CDM Watch 2004; Greenpeace 2005; Ott and Sachs 2004).

Among experts generally in favour of the CDM, attitudes range from defending the need for strict project-specific methodologies to pleading for a simplified assessment for enhanced participation in the CDM. For instance, the *CDM Gold Standard*[4] claims a demanding investment and barrier test as well as detailed sustainability requirements. As sustainable development is the main benefit of the CDM for developing countries (DCs), aid organisations and ecological NGOs such as the WWF[5] call for high quality certificates.

Some scientists, such as Michaelowa (1999), defend the additionality assessment of a CDM project activity relying on a stringent test of investment additionality to exclude free riders. On the contrary, other scientists and industry experts, such as Vanderborght[6], believe that financial analysis neglects risks and barriers to investment. They further criticise the assessment criteria as potentially becoming prohibitive, leading to very low participation. Instead, industry proposes to enhance participation to make the CDM workable.

Research has broadly dealt with the calculation of *emission reductions*, i.e. the development of baseline and monitoring methodologies, focusing on the *baseline selection* and the *additionality assessment* to guarantee the integrity of the Kyoto Protocol (Figure 1). In contrast, the contribution to *sustainable development* has had no evaluation procedure laid down. Sustainability is a very comprehensive concept composed of diverse perspectives (*constructivist paradigm*) which may be

---

[4] The original concept was pioneered by Helio International, promoted by SouthSouth-North and used for developing the CDM Gold Standard in a cooperation of 36 NGOs. The CDM Gold Standard is an accreditation standard for project activities.
[5] See http://www.panda.org/climate/goldstandard/ (2005–07–12).
[6] Dr. Bruno Vanderborght, Holcim Group Support (Brussels) SA.

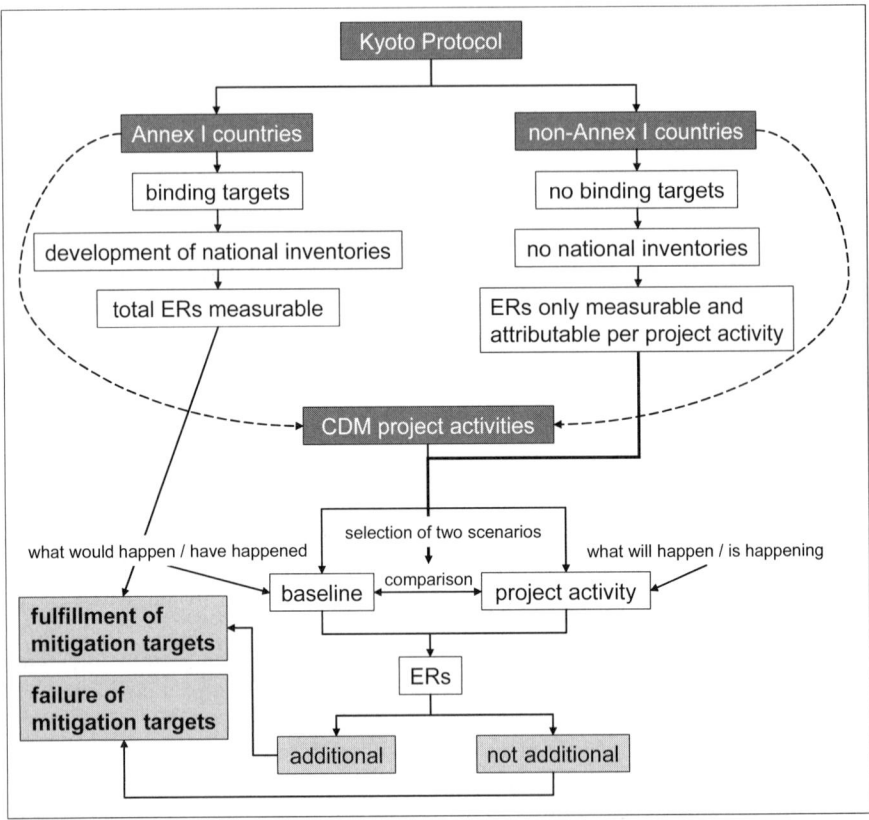

Fig. 1. Additionality of emission reductions to guarantee the integrity of the Kyoto Protocol

conflicting (Stockmann 2004). As the CDM is still in an early stage, the outcomes and long-term impacts of CDM project activities can so far only be estimated ex-ante and not evaluated ex-post.

2.1 Emission reductions

The Kyoto Protocol obliges ratifying Annex I countries to jointly reduce emissions by 5.2% from 1990 levels, but the Intergovernmental Panel on Climate Change (IPCC 2001 and 2007) considers this commitment insufficient.

2.1.1 Baseline concept and additionality concept

To measure the improvements generated by a project activity, i.e. the additional GHG emission reductions (ERs) and the contribution to sustainable development (SD), the setting of a baseline scenario (reference case) is necessary. The baseline does not necessarily represent the historical situation such as in a Before-After-

Comparison. A baseline should model the situation likely to take place if the project activity was not accepted under the CDM[7]. As this scenario is hypothetical it implies a certain percentage of uncertainty.

The additionality concept is closely related to that of the baseline. As the Marrakesh Accords left its interpretation open, two fundamental interpretations are now struggling against each other. The first is often referred to as *environmental additionality*, which requires demonstration of the fact that the emission reduction would not be achieved without the proposed project activity. The second interpretation goes further requiring additional emission reductions to be those which would not be achieved without the impact of the CDM. Thus, a project activity does not lead to additional emission reductions if it would have happened anyway. The first interpretation implicitly assumes that the baseline scenario is different from the one of the project activity; however, the second interpretation requires this assumption to be sustained appropriately. A project activity which would also take place without being accepted under the CDM has to be considered as business-as-usual, thus not additional.

At its 16th meeting, the EB published the *tool for the demonstration and assessment of additionality* (2005) which is the consolidation of the additionality tests provided by the approved methodologies. During the COP10 in Buenos Aires in December 2004, the EB clarified that use of the tool is not mandatory. However, the recent approval praxis of the EB indicates that the tool as in effect setting a minimal standard for additionality assessment.[8]

The baseline setting has far-reaching consequences for measurement of the additional emission reductions. If the baseline is not set conservatively, excess CERs are generated for business-as-usual project activities and the Kyoto abatement targets remain unmet. Subsequently, the number of carbon certificates is not equivalent to the achieved emission reductions which stay behind. A high quality baseline is therefore the prerequisite for the integrity of the Kyoto Protocol; however, the more refined the baseline setting gets, the more costly its development and application become. These transaction costs could themselves turn into a barrier for implementation of project activities, making them financially unattractive. To help small-scale project activities from being impeded by high costs of developing a methodology, the EB has established simplified methodologies for small-scale (SSC) CDM project activities.

2.1.2 Limits of additionality assessment

As the baseline is a theoretical concept, it is counterfactual and can never be proved. Additionality itself cannot be proved either. Additionality cannot be definitely demonstrated as nobody can look inside the minds of the project proponents. Even the project developers themselves might sometimes not be sure whether they would have commissioned the project activity without the CDM.

---

[7] This scenario can of course be equivalent to the historical situation if no change is expected.
[8] EB decisions since September 2004, http://cdm.unfccc.int/EB/Meetings (2005–07–12).

If the assessment is too lax, more free riders pass the EB approval procedure further, reducing the impact of the CDM. The risk of *cheating* might diminish when the procedures and control measures are refined. However, bad results at the initial stage could deeply damage the image of the CDM.

2.2 Sustainable development

The examination of the contribution to sustainable development by the project activity is the responsibility of the host countries, who act through their DNA. Neither proceedings nor quality standards are prescribed, therefore this may lead to perverse incentives. To attract foreign capital, the DNAs may be tempted to reduce their SD-requirements more and more- the so called "race to the bottom" (Thorne and Raubenheimer 2001).

Raised in discussion by the IISD paper "The development dividend" (IISD 2005), sustainable development seems to be turning into a secondary issue. IISD calls for a clear and strong set of criteria for sustainable development. Arguing if this is not politically feasible, at least common principles and guidelines for sustainable development under the CDM should be established.

DNAs and institutional CER-buyers have to decide which CDM project activities to select. As the Marrakesh Accords require a CDM project activity to contribute to sustainable development, many sets of criteria and evaluation approaches have been developed[9]; however, so far they have not, for the most part, been applied. In this paper, two prominent approaches will be discussed.

The *Gold Standard* is a quality label for CDM project activities. It establishes an enhanced control procedure by the DOEs. It supplements the Project Design Document (PDD) by 5 further Annexes. These contain a sustainable development assessment, a positive list of eligible project activities, an ODA screen[10], an obligatory Environmental Impact Assessment (EIA) as well as stricter criteria for the stakeholder consultation. The use of the Gold Standard PDD is voluntary.

The second approach, *MATA-CDM (Multi-Attributive Assessment CDM)*, was developed by Sutter (2003) to enable the DNAs to develop criteria and indicators which reflect the stakeholders' interests. MATA-CDM demonstrates how to develop sustainability criteria following a participative approach. The analysis was conducted in India, South Africa and Uruguay. Convinced by the tool, the DNA of Uruguay decided to officially adopt it to define its sustainability criteria and indicators.

Different theories may be used to analyse the processes taking place when a CDM project activity is implemented. The following illustrates how three strands of research (sustainability management, evaluation theory and principle agent theory) apply to the research topic:

---

[9] See e.g. Begg (2003), based on MCDA; Brown and Corbera (2003) based on MCA, Markandya and Halsnaes (2002).
[10] The ODA-Screen intends to prove that the CDM project activity is not implemented instead of Official Development Aid (ODA).

2.2.1 Sustainability management

The terms *sustainable development* and *sustainability* express the idea of a long term temporal scale. Sustainable development describes the process towards sustainability.

System theory further differentiates between strong and weak sustainability. The difference lays in the substitutability in a closed system. Strong sustainability does not permit any substitution, while weak sustainability assumes resources are inter-changeable, i.e. technological innovations can compensate for extinguished resources. In reality it is likely that a mixture of both interpretations will be encountered. For instance one may argue that the extinction of fossil resources can be compensated by innovations in alternative energy generation, while climate change and its consequences cannot be fully compensated by technical improvements (e.g. higher dams and stronger houses). (Hopwood et al. 2005)

The most well-known definition of sustainability is that of the Brundtland Commission from 1987 which expresses the concept of intergenerational equity: "meeting the needs of the present without compromising the ability of future generations to meet their needs" (WCED 1987). The challenging aspect of this definition is how to define the *needs*. Is this the lifestyle, the consumption pattern or just basic needs? It is already difficult to define the needs of the current generation, but what will the needs of the future generation be? How many future generations have to be taken into account? What would the discount rate be? Such considerations illustrate that sustainability represents a concept which requires interpretation.

Other attempts have also been made to clarify sustainability. In 1992, the 27 Principles of the Rio Declaration on Environment and Development (1992) were developed. They refer to questions of both inter- as well as intra-generational equity. Mostly, environmental issues are addressed. The precautionary principle and the polluter-pays-principle are adopted. A link between environmental protection and poverty eradication is established (Principle 6). Common but differentiated responsibilities of the States are recognised. Participation of the citizens is explicitly mentioned yet only in the context of environmental issues. The vital role of women (Principle 20), youth (Principle 21) as well as indigenous people and their communities and other local communities (Principle 22) in achieving sustainable development is recognised. Finally, cooperation and peace are strongly promoted.

During the 2000 United Nations Millennium Summit, the Millennium Declaration was adopted containing the eight Millennium Development Goals (2000) covering a range of developmental issues related to poverty eradication. Only Goal 7 addresses environmental issues. The aim of the Millennium Development Goals (MDGs) is to set concrete goals and quantifiable targets for governments.

However, the Rio principles as well as the MDGs stay rather general; above all the question as to how to achieve the targets, which concrete activities are required, cannot be answered in an international declaration. This leads to the conclusion that further interpretation of sustainability in each case is needed. Ethical and ideological concepts of the society are reflected in its interpretation of sustainability. The strategic as well as ethical implications of sustainability form the guiding principle, but there is no single operational definition.

Sustainable development has become a constituent part of economic and environmental policy in industrial and many developing countries. Increasingly, environment management systems and sustainability management systems are being implemented. Initially mainly operative, these approaches are becoming more and more integrated into strategic management. There is a variety of instruments available to integrate the ecological and social aspects into the conventional economic management. For instance the Sustainability Balanced Scorecard, early warning systems and checklists to name a few. However, most of them still focus on ecological aspects without paying due attention to social aspects, efficiency aspects and finally, to the integrative consideration of the economic, ecological and social goals. In addition, most of the instruments consist of quantitative or even monetary measurements. Thus, in case there is no common measure (e.g. money) for outcomes (incommensurability) and the outcomes of different options cannot be compared (incomparability), these instruments are not applicable (Spangenberg 2005). This is often the case when dealing with sustainable development.

2.2.2 Evaluation theory

At the beginning of evaluation theory (Stockmann 2004), the dominant strand was the *methodological rigorism* (e.g. Cook and Campbell 1979; Scriven 1980). Experimental designs were preferred to detect the *real* causal relationships in order not to base political decisions on wrong assumptions on the likely achievements of the programme. Assurance of internal validity was the main goal. However, the implementation of the experimental design often failed and a new strand aiming at enhancing the external validity came up. Evaluation mutated to a more political than scientific venture. The interests of the stakeholders entered into the focus of evaluation. Another strand known as the *constructivist paradigm* (e.g. Stake 1983; Patton 1987; Guba and Lincoln 1989) questioned the methodological rigorism rejecting the concept of a one and only *true* reality. On the contrary, the constructivist paradigm departs from the concept that reality is socially constructed of diverse perspectives which are in conflict with each other. From the constructivist paradigm then came the *transformative or participatory* (e.g. Mertens 1998, 2004) paradigm. Evaluation was then intended to actively involve the stakeholders, to give them a say and to overcome existing structures of influence and power. Today, broad consensus has been reached that evaluation is required to take into account the stakeholders' needs and that quantitative and qualitative methods can often be well combined (mixed-method-approach) to enhance validity (Stockmann 2004; Chen 2005). The focus no longer lays on the absolute merits of a method, but rather on whether and how to use the research techniques to come to the most conclusive findings. The type of question to be investigated is decisive for the appropriate choice of methods. Further, the results of an evaluation will be useful to the decision-making process in politics and management.

In comparison with other objects of evaluation, the *evaluation of sustainability* is faced with special challenges due to its comprehensiveness. Ideally, all aspects related to the ecological, economic and social pillars should be examined at all scopes, time scales and from different points of view, but the more stakeholders become involved, the more complex, time consuming and costly the process becomes. The impacts are often difficult to measure as a vast set of interdependencies and external influences has to be taken into account. Therefore, evaluations of sustainability are exposed to high subjectivity in assessment.

Seven core elements shape the requirements for evaluations of sustainable development: First of all, evaluations should follow a *holistic approach* (Hardi and Zdan 1997; Chen 2005), addressing all dimensions of sustainability. A generally accepted systematisation divides sustainability into three dimensions: the ecological, the economic and the social dimension. An evaluation should address them simultaneously and balance them against each other in decision making. This does not mean each evaluation has to deal with all three aspects equally. On the contrary, it is the key for a high quality evaluation to find the right balance between the three dimensions which is very context-specific. Under the expression *triple bottom-line*, the concept of the three pillars has become established in the area of sustainability management. Regarding the Clean Development Mechanism (CDM), no holistic evaluation approach has so far been established. The ongoing assessment focuses on the calculation of the emission reductions achieved by the project activity while neglecting other aspects of sustainable development. It is the responsibility of the host country to assess the sustainability of a project activity. As neither guidelines nor standards have to be formulated, the process is not transparent.

Sustainable development may be evaluated at different *time scales* (Hardi and Zdan 1997; Knoflacher et al. 2003). If the short-term time horizon is in the centre of interest, an evaluation concentrates on direct output and/or outcomes, i.e. it deals with the direct effects induced by the intervention. However, it is inherent to the concept of sustainable development that the final important question is whether or not effects may be noticed in the long-term. To measure this, the impacts have to be analysed. This can be done ex-ante to anticipate the likely contribution to sustainable development or ex-post to facilitate accountability. The long-term time horizon also incorporates the concept of inter-generational equity. The evaluation of long-term impacts very much depends on the quality and quantity of the data available. Regarding climate change, it is very difficult to predict the exact impact of emission reductions in the long run. It can only be estimated using scientific models. Although climate change can already be noticed today, future generations, especially in poorer countries will suffer most from the consequences.

Sustainable development may be measured at different *spatial scales* (Hardi and Zdan 1997; Knoflacher et al. 2003), i.e. at a local, regional, national and even global level. The concept of sustainable development in its perfection is not limited to national frontiers, which leads to improvements on a global scale. The area of climate change is a suitable example for impact at a global scale. As climate change is a global phenomenon, the impact on the world climate is the same wherever the reduction is achieved. This is why the CDM has been created. In addition, the CDM should lead to sustainable development in the host countries, i.e. at the local level.

Sustainable development is confronted with *system dynamics, risk and uncertainties* (Stirling 1999), first of all because the consequences of unsustainable development can be disastrous, and second because of the complex relationships between the elements leading to sustainable development. It is often difficult to evaluate the effects of an intervention, as external influences have to be taken into account. The evaluation has to be designed as such to measure whether the results can actually be attributed to the intervention or if they are caused by other influences. Climate change is a phenomenon extremely exposed to dynamics, risk and uncertainties. It is part of the *precautionary principle* to initiate mitigation measures now to prevent disastrous consequences. Regarding sustainable development in the host country, even if it has been forecasted that a CDM project activity will contribute to the sustainable development of the host country, this does not imply that the promised benefits will materialise, and if they do, it has still to be demonstrated whether this was due to the project activity.

When an evaluation is undertaken, the assessment is based on a *value* system (Stockmann 2004). A challenge for an evaluator who performs an evaluation for a third party is weighing the criteria, against which the intervention has to be assessed in order to reflect the value system of the client. The evaluation should not reflect the subjective appreciation of the evaluator; it should always be designed for the client. That the client is in most cases not one person, but a group of people with diverse priorities, complicates the identification of the values as well as their relative weight. The more heterogeneous the group representing the client, the more difficult it is to complete the normative component of an evaluation. In the context of CDM project activities, the *clients* of the evaluation are the DNAs and the Executive Board, but always in their function to preserve the stakeholders' interests. The participatory method MATA-CDM (Sutter 2003) allows the integration of different value systems into one system which is then applied to evaluate project activities.

This leads to the aspect of *participation* (Aarhus Convention 1998; Chambers and Mayoux 2004; Chen 2005). When the value system has to be set up, stakeholders' participation is crucial as they are the ones affected by the project activities and therefore, need to have a say. Stakeholders' participation is intended to influence the normative character of an evaluation. This does not mean the stakeholders' interpretation has to be adopted into the value system across the board. It is the role of the evaluator to critically assess the outcomes of the stakeholder consultation and to design the criteria and indicators in such a way the stakeholders' values are included appropriately. An evaluator therefore has to consult the stakeholders before setting up the evaluation system. Stakeholders' participation is a crucial element in the CDM process (stakeholder consultations are mandatory). Also MATA-CDM encourages participation: the sustainability criteria are set by a pool of experts and the country-specific weightings are determined in a stake-holder survey. The direct weighting and AHP are methods used to derive the weightings.

An evaluation is not only intended to measure effects, but also to provide guidance for *process* improvement (Stockmann 2004). Two kinds of evaluation may be distinguished here. A formative evaluation is carried out during the implementation of the intervention, and is intended to make timely trouble-shooting possible. Summative evaluations are intended to provide final conclusions on accomplished interventions which shall provide information on how to better design future interventions. These recommendations may include institutional changes. As the CDM host countries are autonomous in assessing the sustainability of the project activities carried out in their territory, no independent body exists which could function as a Quality Assurance Board. Therefore, formative as well as summative evaluations of sustainable development will only be carried out on a voluntary basis.

2.2.3 Principal Agent Theory

Information asymmetries between different actors can be illustrated using the Principal Agent Theory (Varian 2004). The principal hires the agent to perform tasks on his behalf, but he cannot control the agent if he performs them in the principal's interest, as the agent's incentives may differ from those of the principal (Fritsch et al. 2003). Problems may arise between the principal and the agent. These are known as *adverse selection*, if occurring before the conclusion of the contract, and *moral hazard* if occurring during fulfilment. To overcome these problems, either the principal or the agent has to become active. The control of the agent's performance by the principal is called *screening*. When the agent himself becomes active to reduce the information asymmetry, this is called *signalling*.

As climate change is a global phenomenon, the final principals are the stakeholders worldwide. However, as big groups are difficult to organise, representatives are needed to defend the principal's interests. Regarding the CDM, UN and governmental bodies as well as accredited experts (e.g. EB, DNA, and DOE) exist to assume these responsibilities. The following illustrates two main principal agent problems in the area of the CDM:

The main information asymmetry is between project developers and the EB when *additionality of emission reductions* has to be assessed. The project developers are in the position of the agent as they are required to report to the DOE and eventually to the EB which is the principal. It is obvious the project developers have a lead in information on the project activity and therefore plenty of possibilities for *cheating*. The project developers pursue their special interest to get their project activity accepted by the EB. If the project developers are free riders, the interests differ substantially: The project developers will try to present the project activity in a way making it acceptable for the CDM. They could submit financial data that suggests higher costs and try to construct a carbon-intensive baseline to make the project activity appear as environmentally friendly as possible. There is a similar information asymmetry between the host DNA being responsible for assessing whether a project activity is likely to *contribute to sustainable development* of the host country and the project developers. Apart from that, another principal agent constellation exists: the DNA itself may become agent when it has to defend the stakeholders' interests. DNAs are governmental bodies and thus, political feasibility may prevent the DNAs from fulfilling the principal's interests. This may manifest in opaque assessment procedures, absence of sustainability criteria and project activities with negative impact on sustainable development.

To fight against adverse selection and moral hazard, screening and signalling are carried out under the CDM. Project activities are screened by the DOEs regarding the emission reductions and by the DNA regarding the contribution to sustainable development. However, the DOE has to assess the project activity under sustainability aspects only if sustainability criteria have been included into the monitoring protocol (either to fulfil the CDM Gold Standard or voluntarily). Compliance with the Gold Standard criteria signals the CDM project activity's contribution to sustainable development. Alternatively, project developers can signal the quality of their project activities by clear documentation as well as by conservative estimations. (Müller-Pelzer 2004)

3 The present evaluation system

The present evaluation system of the CDM (Figure 2) is divided into two parts: the estimation of the emission reductions achieved by the project activity and the contribution to sustainable development by the project activity. These two aspects of the CDM are treated in very different ways.

3.1 The first goal: additional emission reductions

The evaluation process of the first goal of the CDM, the additional emission reductions, has achieved a high level of transparency. All obligatory documentation has to be made available to the public on the CDM website (http://cdm.unfccc.int). The public has the opportunity thereby to actively participate in the evolution of the CDM. Comments can be provided during the public input phases for new proposed methodologies and project activities. Unfortunately, this opportunity is generally not taken advantage of demonstrated, by the ever declining public input on methodologies.[11]

By creating baseline and monitoring methodologies, the first step towards making the emission reduction goal operational has been taken. In the baseline methodology, the project developers have to demonstrate the additionality of the expected emission reductions. In the monitoring methodology, they have to specify how they are going to monitor the emissions reductions during the implementation and operation of the project activity. The information is provided in the PDD (Project Design Document).

---

[11] As of June 2005, less than one comment per new proposed methodology is submitted on average, more than 20% of the new proposed methodologies do not get any comments from the public, 30% of all comments have been provided by the Öko-Institut and 17% by the Federation of Electric Power Companies. This picture shows that the involvement of the stakeholders is rather limited and unbalanced, http://cdm.unfccc.int/methodologies/PAmethodologies (2005–07–12).

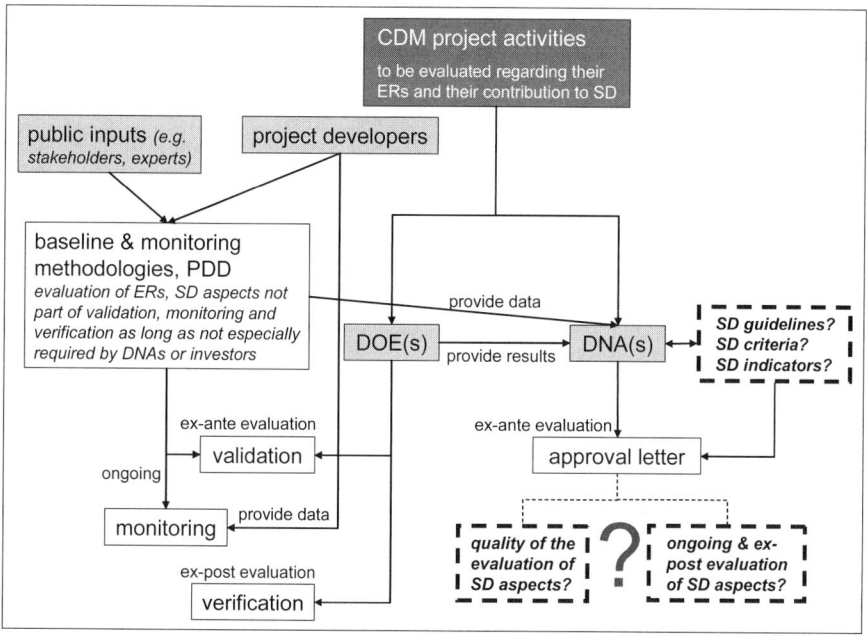

Fig. 2. Evaluation design of the CDM

The project activity has to be validated by a Designated Operational Entity (DOE)[12] before its registration can be requested. During the operation of the CDM project activity, the emission reductions have to be verified by a second DOE (for small-scale CDM project activities, this task can be adopted by the same DOE which carried out the validation).

3.2 The second goal: sustainable development

For the second goal, the contribution to sustainable development, transparency has not yet been realised. A list of the DNAs is available on the CDM website providing contact addresses and link to websites if existent, but there is no website or document providing an overview on the sustainable development requirements of the respective countries. In the PDD, project developers have to provide a verbal explanation of the project activity's contribution to sustainable development, but sustainability aspects are only addressed by the DOEs if included in the monitoring protocol. An external evaluation of the second goal is therefore hindered by the low data availability and transparency.

Due to the sovereignty of the non-Annex I countries, an evaluation by internationally accredited entities is not pursued. The responsibility for assessing sustain-

---

[12] A Designated Operational Entity (DOE) is an institution accredited by the EB, which undertakes validation as well as verification and certification activities for CDM project activities.

ability is put into the hands of the host country, i.e. the DNA. On the other hand it has to be taken into account that the requirements for sustainable development differ from country to country. Therefore, each country is autonomous in setting sustainable development criteria. As sustainable development should respond to the stakeholders' needs, the value generated by a project activity may only be estimated by the host country itself. A universal scheme for all countries and all project types to determine the contribution to sustainable development has therefore been rejected.

Today, some DNAs have published general guidelines and criteria, but few have published an operational set of criteria and indicators.[13]

3.2.1 CDM Gold Standard

The Gold Standard tries to compensate this lack. However, it is faced with several problems: The main point of criticism is that DOEs are not necessarily familiar with issues related to sustainable development. They are accredited by the EB and designated by the COP for *technical and sectoral scopes*[14] for which they can prove the necessary expertise. However, it is assumed they are competent in assessing sustainability *without their expertise being assessed*. Furthermore, the assessment of sustainability is not binding in contrast to the calculation of emission reductions. In case of not meeting the minimal requirements of the Gold Standard, the project participants (PPs) do not have to fear further consequences apart from not being awarded the quality label. The Gold Standard is therefore not an instrument to improve the quality of a CDM project activity. Although the criteria of the Gold Standard have been set comprehensively, they may not perfectly match with the single project activity, because sustainability is complex and difficult to standardise. The function of this quality label for CERs is to signal the higher quality of a CDM Gold Standard project activity in comparison to ordinary CDM project activities. Therefore, PPs hope to obtain a premium price on the market; however, this price effect is uncertain. Thus, PPs using the CDM for image reasons and having little doubts of not meeting the Gold Standard requirements bear the additional effort, but most PPs are likely to go for the easier solution.

3.2.2 MATA-CDM

In contrast, MATA-CDM takes the sovereignty of the DNAs into account. It is a participatory approach and supports the decision making of the DNA, which consists of approval or rejection of project activities. Two points have to be raised on

---

[13] 37 of the 80 DNAs interviewed between 11 January 2007 and 5 April 2007 apply an approach to demonstrate the contribution to sustainable development in the host country. Nevertheless, this number does not tell anything about the quality of the approach. The impression gained from the description provided by the DNAs was that most used a set of criteria, some required only a verbal description of the contribution to sustainable development, but there were also a few (3) who used real evaluation tools.

[14] 15 "sectoral scopes" were identified by the CDM Accreditation Panel, http://cdm.unfccc.int/DOE/scopes.html#11 (2005–07–11).

the applicability of MATA-CDM in practice: First, as in a number of developing countries the democratic culture is not well established, it is likely the benefits of this participatory approach will not be largely recognised (MATA CDM is only applied in one country to date). Second, DNAs may not be willing to take a clear stand on their criteria, as strict requirements may reduce the host country's attractiveness to investors. As the opportunity for obtaining foreign currency is a very short-term and concrete objective, it may have a higher influence on decision making than the long-term and vaguely perceived benefits of sustainable development.

Both the CDM Gold Standard and MATA-CDM are methodologically limited. The Gold Standard relies on the subjective judgement of an auditor rating a project activity on each criterion between + and −2. This illustrates the attempt to standardise the judgement, but at the price of reduced expressiveness. MATA-CDM uses a utility function and thus assumes indicators are at least weakly commensurable and options are at least weakly comparable, but this is not necessarily the case when sustainable development is the object of evaluation.

4 Conclusion

An independent entity is not available to assess the impact of CDM project activities on sustainable development. The DNAs are not necessarily active in defending the interests of the stakeholders due to competition for CDM project activities. As a result a global political solution for this problem seems to be rather unlikely. Thus, stakeholders in developing countries can only safeguard their interests in sustainable development if the local project developers commit to running their companies in a sustainable way. What sustainable development means in practice is highly context-specific and has to be determined for each case individually. It is therefore crucial to encourage companies to embed sustainable development into their strategic management and to contribute to it according to their capacities. If the awareness of sustainability issues is high, sustainability management can even become a market advantage.

Therefore, the need for a project-specific management approach arises, which allows project developers to develop and implement their sustainability strategy as well as to demonstrate their contribution to sustainable development. This implies an ongoing evaluation of CDM project activity by the project developer with the aim of unveiling conflicts among goals, recognising, early on, failures in the achievement of said goals, and introducing prompt corrective measures if required.

References

Aarhus Convention (1998) Convention on access to information, public participation in decision-making and access to justice in environmental matters. Aarhus, Denmark, UNECE

Begg K et al. (2003) Encouraging CDM energy projects to aid poverty alleviation. Attachment 3. Assessment of Sustainability Benefits from small-scale community projects. June 2003, http://www.iesd.dmu.ac.uk/contract_research/publications/kb4.pdf (2005–12–20)

Brown K, Corbera E (2003) A Multi-Criteria Assessment Framework for Carbon-Mitigation projects: Putting "development" in the centre of decision-making. Tyndall Centre for Climate Change Research Working Paper 29, February 2003, http://www.tyndall.ac.uk/publications/working_papers/wp29.pdf (2006–01–24)

CDM Watch (2004) Market failure – Why the Clean Development Mechanism won't promote clean development. By Pearson B, November 2004, http://www.cdmwatch.org/files/Market%20failure.pdf (2005–07–12)

CDM Website http://cdm.unfccc.int (2005–10–07)

Chambers R, Mayoux L (2004) Reversing the Paradigm: Quantification and Participatory Methods. http://www.enterprise-im-pact.org.uk/informationresources/toolbox/quantify cationandparticipatorymethods.shtml (2007–03–28)

Chen HT (2005) Practical Program Evaluation: Assessing and improving Planning, Implementation, and Effectiveness. Thousand Oaks, Sage Publications, Inc

Cook TD, Campbell DT (1979) Quasi-Experimentation. Design and Analysis Issues for Field Settings. Chicago, Rand McNally

Foot S (2004) Evaluation of the present Clean Development Mechanism. In: Environmental Law and Management, vol 16, issue 3, 2004: 125–134, http://www.icfconsulting.com/Markets/Environment/doc_files/cdm-evaluation.pdf (2004–11–04)

Foot S (2004) An evaluation of the present Clean Development Mechanism (Part 2). In: Environmental Law and Management, vol 16, issue 3, 2004: 193–199, http://www.Icfconsulting.com/Markets/Environment/doc_files/cdm-evaluation-2.pdf (2005–02–15)

Fritsch M et al. (2003) Marktversagen und Wirtschaftspolitik – Mikroökonomische Grundlagen staatlichen Handelns. 5. Aufl., Verlag Franz Vahlen, München 2003

Greenpeace (2005) Klimarappen: Prima für Ölprofiteure aber sicher nicht fürs Klima! Bern 2005–01–05, http://info.greenpeace.ch/de/klima/pressreleases/pr050105klimarappen (2005–07–12)

Guba EG, Lincoln YS (1989) Fourth Generation Evaluation. Newbury Park, London, New Delhi, Sage

Hardi P, Zdan T (1997) Assessing Sustainable Development: Principles in Practice. IISD

Hopwood, B., et al. (2005) Sustainable Development: Mapping Different Approaches. Sustainable Development. Wiley InterScience (13): 38–52

IISD (2005) Realizing the Development Dividend: Making the CDM Work for Developing Countries. Phase 1 Report – Pre-publication Version, by Cosbey A, Parry JE, Browne J, Babu YD, Bahandari P, Drexhage J, Murphy D, International Institute for Sustainable Development (IISD), May 2005, http://www.iisd.org/pdf/2005/climate_realizing_dividend_sum.pdf (2005–07–14)

IPCC (2001) Technical Summary – Climate Change 2001: Mitigation – A Report of Working Group III of the Intergovernmental Panel on Climate Change. p 24, http://www.ipcc.ch/pub/wg3TARtechsum.pdf (2005–07–14)

IPCC (2007) Contribution of Working Group III to the Fourth Assessment Report of the Intergovernmental Panel on Climate Change – Technical Summary. http://www.mnp.nl/ipcc/pages_media/FAR4docs/chapters/TS_WGIII_220607.pdf (2007–08–26)

Knoflacher M et al. (2003) Assessment of Sustainability – Can it be standardised? EASY-ECO-2 Evaluation of Sustainability – European Conferences, Vienna

Kyoto Protocol (1997) The Kyoto Protocol. Kyoto 1997, http://unfccc.int/resource/docs/convkp/kpeng.html (2005–07–10)

Markandya A, Halsnaes K (2002): Climate Change and sustainable development: prospects for developing countries. Earthscan, London 2002

Marrakesh Accords (2001) The Marrakesh Accords and the Marrakesh Declaration. Marrakesh 2001, http://unfccc.int/cop7/documents/accords_draft.pdf (2005–07–10)

Mertens DM (1998) Research methods in education and psychology: Integrating diversity with quantitative and qualitative approaches. Thousand Oaks, CA, Sage

Mertens DM (2004) Institutionalizing Evaluation in the United States of America. Evaluationsforschung: Grundlagen und ausgewählte Forschungsfelder. Stockmann, R. Opladen, Leske & Bulrich: pp 45–60

Michaelowa A (1999) Baseline methodologies for the CDM – which road to take? Presentation at the Institute for Global Environmental Strategies (IGES), IGES Working Paper, Shonan Village, June 1999

Millennium Development Goals (2000) http://www.un.org/millenniumgoals/ (2006–01–24)

Müller-Pelzer F (2004) The Clean Development Mechanism – a comparative analysis of chosen methodologies. HWWA-Report 244, Hamburg 2004,http://www.hwwa.de/Publikationen/Report/2004/Report244.pdf (2005–10–31)

Ott HE, Sachs W (2004) Ethical Aspects of Emission Trading. Wuppertal Papers No 110, September 2000, http://www.wupperinst.org/Publikationen/WP/WP110.pdf (2005–07–14)

Patton MQ (1987) Evaluation's Political Inheritency: Practical Implications for Design and Use. The Politics of Program Theory. Palumbo DJ, Thousand Oaks, CA, Sage: 100–145

Rio Declaration on Environment and Development (1992) Report of the United Nations Conference on Environment and Development. Rio de Janeiro, 3–14 June 2002, Annex I, http://www.un.org/documents/ga/confl51/aconfl5126-1annex1.htm (2006–01–24)

Scriven M (1980) The Logic of Evaluation. California, Edgepress

Spangenberg JH (2005) Economic sustainability of the economy: concepts and indicators. International Journal of Sustainable Development, 8 (1/2): 47–64

Stake RE (1983) The Case Study Method in Social Inquiry. Evaluation Models. In: Madaus GF, Scriven M, Stufflebeam DL, Boston, Kluwer-Nijhoff: pp 279–286

Stirling A (1999) On Science and Precaution: In the Management of Technological Risk – An ESTO Project Report Prepared for the European Commission – JRC. Seville, Institute Prospective Technological Studies

Stockmann R (2004) Evaluation in Deutschland. In: Stockmann R (ed) Evaluationsforschung: Grundlagen und ausgewählte Forschungsfelder. 2. Aufl., Leske + Bulrich, Opladen, 2004: pp 31–43

Sutter C (2003) Sustainability Check-Up for CDM Projects – How to assess the sustainability of international projects under the Kyoto Protocol. Wissenschaftlicher Verlag Berlin, 2003, http://www.climnet.org/EUenergy/CDM.htm (2004–06–27)

Thorne S, Raubenheimer S (2001) Sustainable Development (SD) appraisal of Clean Development Mechanism (CDM) projects – experiences from the SouthSouthNorth (SSN) project. Forum for Economics and Environment – First Conference Proceedings, http://www.econ4env.co.za/archives/ecodivide/Theme3a.pdf (2006–06–27)

Tool for the demonstration and assessment of additionality (version 2) 28 November 2005, http://cdm.unfccc.int/methodologies/PAmethodologies/AdditionalityTools/Additionality_tool.pdf (2006–02–08)

UNFCCC (1992) United Nations Framework Convention on Climate Change, Rio de Janeiro 1992, http://unfccc.int/ (2005–07–10)

UNFCCC (1997) Kyoto Protocol

Varian HR (2004) Grundzüge der Mikroökonomik. 6. Aufl., Oldenbourg Wissenschaftsverlag GmbH, 2004, aus dem Amerikanischen übersetzt von Prof. Dr. Reiner Buchegger, Originaltitel: Intermediate Microeconomics. Sixth Edition, W. W. Norton & Company, Inc., New York, USA

WCED (1987) Our Common Future. World Commision on Environment and Development, Oxford University Press

Sustainable development, the Clean Development Mechanism, and business accounting

Pala Molisa[I], Bettina Wittneben[II]

[I] School of Accounting and Commercial Law
Victoria University of Wellington
PO Box 600, Wellington, New Zealand
pala.molisa@vuw.ac.nz

[II] Rotterdam School of Management
Erasmus University Rotterdam
Burg, Oudlaan 50, 3062 PA Rotterdam, Netherlands
BWittneben@rsm.nl

Abstracts

With the institution of international responses to the imperatives of sustainable development and climate change mitigation, such as the UN-based Clean Development Mechanism (CDM) as part of the Kyoto Protocol, practical concerns such as how to value, measure, quantify and account for such issues have come to the fore. This paper argues that if such institutional mechanisms are to adequately respond to these imperatives, they will have to address their political as well as technical dimensions. There is an ongoing political struggle over the meaning of Sustainable development between business and other stakeholders and this has influenced how institutional mechanisms such as the CDM have come to be structured. This paper examines some of the structural limitations of the CDM and the extent to which accounting techniques and processes are able to address and overcome them.

Keywords: Clean Development Mechanism, climate change, sustainability accounting, sustainable development

1 Introduction

With the Kyoto Protocol coming into force in early 2005, it is an opportune time to consider the challenges of realising its prime objectives: cost-effective mitigation of greenhouse gas (GHG) emissions, and sustainable development (SD). The Protocol's Clean Development Mechanism (CDM) and the emergence of the carbon market have effectively brought business centre-stage as the sector through which SD and climate change mitigation are pursued (Wittneben 2007). The expanded role of business necessitates a closer look at how it can fulfil these objectives.

Against this background this paper explores some of the challenges facing business in meeting the CDM's climate change mitigation and SD objectives. We argue that the major challenges to meeting these objectives are not primarily technical, but political. Addressing SD politically will mean that the CDM and business will need to develop more critical approaches to accounting than have been traditionally adopted in the past. In order to illustrate this claim the paper examines the technical accounting needs produced by the CDM and contrasts these needs with the political issues that will need to be addressed if it is to realize its goal of promoting SD. By outlining some of the available sustainability accounting methods, we show that accounting techniques already exist by which business(es) could undertake comprehensive carbon accounts and SD accounts for CDM projects. However, their uptake by business has been poor – a trend which betrays a political interest on the part of business to prevent societal transitions toward SD. We outline some of the developments within the accounting and business literature which shows a political struggle unfolding between business and other stakeholders over the meaning of SD. These struggles have a bearing on the CDM's ability to promote SD. Is the CDM's current market structure adequate as a mechanism for promoting SD, or does it have to be transformed to deal more effectively with the substantive issues which SD entails?

The paper proceeds as follows. The second section provides an outline of the CDM's institutional structure and the role business plays in its processes. This section also identifies some of the accounting needs that the CDM generates for business. The third section explores the meaning of SD in order to establish a context for examining the suitability of the CDM's institutional structure for promoting this objective. The fourth section then outlines some of the accounting techniques business has available in order to provide substantial carbon accounts and sustainability accounts for CDM projects. Here we demonstrate how accounting's obstacles to addressing SD are not primarily technical ones but political ones. This exploration points to a need for business to re-conceptualize SD not as a technical problem to be solved but rather as a political one. The paper therefore concludes by exploring accounting research which offers some ways forward for overcoming this technical-political contradiction.

2 The Clean Development Mechanism

The CDM is one of the cooperative implementation mechanisms introduced by the Kyoto Protocol of the United Nations Framework Convention on Climate Change (UNFCCC) to enable cost-effective emissions of greenhouse gases (GHG). Other mechanisms are Joint Implementation (JI, Article 6) and emissions trading (ET, Article 17). As part of the framework established by the Kyoto Protocol, the CDM has the dual objective of mitigating GHG emissions and sustainable development. The mechanism's objective is defined in Article 2 of the Kyoto Protocol as follows:

"The purpose of the Clean Development Mechanism shall be to assist countries not included in Annex I achieving sustainable development and in contributing to the ultimate objectives of the Convention, and to assist countries included in Annex I in achieving compliance with their quantified emission limitation and reduction commitments under Article 13" (Kyoto Protocol to the United Nations Framework Convention on Climate Change 1998, p. 12, English version).

The CDM is supposed to create incentives for industrialised countries to support SD initiatives in so-called 'developing' countries and to simultaneously reduce global GHG emissions. This recognises that while per capita emissions of 'developing' countries remain far below that of industrialised countries, the aggregate emissions of Non-Annex I countries are growing at a much faster rate and are projected to surpass Annex B countries by 2020 (Banuri and Gupta 2000, p. 74). The CDM therefore attempts to strike a balance between the global need to reduce GHG emissions and the developmental aspirations of Non-Annex I countries. From the beginning, the climate discussion has been framed in terms of the differential responsibilities, vulnerabilities, obligations and capacities between industrialised (Annex I) and 'developing' (Non-Annex I) countries (Banuri and Gupta 2000, p. 74). The UNFCCC implies a concern for equity between industrialised nations that have caused most of the GHG emissions and developing countries that are most vulnerable to climate change (Wittneben 2007, pp. 11–12). Along this line, the CDM provides incentives to Annex I countries to undertake emission mitigation initiatives beyond their own borders, namely in Non-Annex I countries. Under current rules, Annex I business organisations are able to submit CDM project ideas to the designated operational entity (DOE) and the CDM executive board. By placing more emphasis and agency on business as the sector responsible for driving initiatives, the CDM represents a shift in the institutional power relations of entities within the climate change mitigation and SD arenas; it has perhaps decreased the influence of Non-Annex I countries in setting up mitigation projects while bringing to the fore the increased role of business in advancing SD and climate change mitigation in 'developing' countries (Wittneben 2007, pp. 21–22). Together with the UNFCCC secretariat, the CDM Executive Board is responsible for legislating, validating and registering climate change mitigation projects. If the CDM project idea is approved, it can be implemented by a corporation or NGO of the North or South, but is based in Non-Annex I nations. Upon completion of the project, the Designated Organizational Entity, an independent evaluator, and CDM Executive Board validate its contribution to reducing GHGs and offer the country of the North an emission credit certificate (Wittneben 2007, p. 16). In validating a CDM project, an important consideration is whether the project can demonstrate "additionality". Additionality is met if it can be shown that anthropogenic emissions of greenhouse gases by sources are reduced below those that would have occurred in the absence of the registered CDM project activity. With respect to SD, the role of defining it and assessing the extent to which CDM projects advance it remains with the host country. The CDM is therefore a balancing act, on the one hand, a policy instrument designed to offer economic incentives to (business) organisations for pursuing climate change mitigation strategies, and on the other, an instrument designed to further the developmental aspirations of Non-Annex I nations.

With this sort of institutional structure, one of the key challenges facing the CDM is how to measure the efficacy and effectiveness of CDM projects in meeting its dual objectives. On the one hand, organizations will have to develop new accounting reporting practices capable of reliably measuring the carbon credits generated by a CDM project; an area that is increasingly coming to be called *carbon accounting*. On the other hand, organizations will also have to develop techniques capable of *accounting for SD*. As a way of categorizing these forms of accounting, it is perhaps more useful to see the former (carbon accounting) as a subset of the latter (accounting for SD). This is because climate change mitigation is only one of SD – a concept that, although used increasingly by business – is still little understood (Gray and Bebbington 2000; Gray and Milne 2002; Milne and Ball 2005). Thus, before we outline some of the accounting methods that could be used to evaluate the SD impact of CDM projects, let us take a necessary detour into what this politically contestable and often vague notion may actually entail.

3 Sustainable development and sustainability

The *Brundtland Report* provides the starting point for any discussion on SD. It defines SD as development which meets the needs of the present generation without compromising the ability of future generations to meet their own needs. It contains two key concepts (UNWCED 1987, p. 43 cited in Milne and Ball 2005, p. 22):

- the concept of 'needs', in particular the essential needs of the world's poor, to which overriding priority should be given; and
- the idea of limitations imposed by the state of technology and social organization on the environment's ability to meet present and future needs.

This definition of SD relates to both present *and* future generations and it also requires that needs of all people are met, especially (in terms of priority) those people most severely impoverished.[1] These needs are both social and environmental.

---

[1] The Brundtland Report was commissioned against an increasing awareness throughout the 60s, 70s and 80s that current modes of socio-economic organisation had failed to de-

It is therefore an *intra-* and *inter-generational* concept which incorporates notions of *fairness* and *justice* relating to access and distribution of social and environmental wealth. Sometimes, these social and environmental needs are referred to as *eco-justice*[2] and *eco-efficiency*[3], respectively. Eco-efficiency, whilst a useful concept also needs to be distinguished from *eco-effectiveness* – the notion which captures the idea of reducing our *overall* ecological footprints (Wackernagel and Rees 1996). To some, SD is merely a precursor for the more important concept of *sustainability* (Milne and Ball 2005). Wackernagel and Rees (1996, pp. 32–40) argue that sustainability simply means, "living in material comfort and peacefully within the means of nature". Many commentators emphasise that sustainability requires a *scale of economic activity relative to its ecological life support system* (Norton 1991; Daly and Cobb 1989; Daly 1992; Dobson 1998). Sustainability is therefore seen as a primarily global concept, operating at the level of socio-economic systems and ecological systems. Moreover, given its link to SD, sustainability also contains eco-justice and eco-efficiency considerations. As such, Gray and Milne (2002) argue "sustainability emphases not just an *efficient* allocation of resources over time, but also a *fair* distribution of resources and opportunities [within] the current generation and between present and future generations, and a *scale* of economic activity relative to its ecological life support systems" (p. 69, emphases in original). Daly (1992) from an environmental economics perspective outlines three criteria for defining sustainability: (1) rates of use for renewable resources that do not exceed their rates of regeneration; (2) rates of use for non-renewable resources that do not exceed the rate at which sustainable renewable substitutes are developed; and (3) rates of pollution emission that do not exceed the assimilative capacity of the environment.

One of the most important points to stress from all these definitions of SD and sustainability is that they require an interrogation of *biophysical* and *social limits* and the constraints they impose on meeting human needs (both social and environmental). In fact, with its reference to 'needs', 'equity' and 'fairness' it also requires a critical examination of 'development' itself (Bebbington 2001). Bebbington (2001, p. 132) suggests that the key structural question inherent within SD appears to be the question of how we organize our economic systems so that development (under some revised definition) takes place without undermining the environment, on which all present and any future development rests. In this sense, SD is a profoundly relational concept that only becomes meaningful when the interdependencies between people (our socio-economic system) and between people and the environment (our socio-ecological complex) are recognised. The implication of this is that if SD is to be substantively addressed, it will require dialogic or

---

liver acceptable standards of living for the poorest parts of the developing world and was incurring unacceptable environmental costs. It also, therefore, includes the objective of poverty alleviation, especially for those parts of the world most severely afflicted.

[2] Eco-justice tries to capture the idea of equity between present peoples and between generations, and in particular, the access for all peoples to environmental resources (Gray and Bebbington 2000).

[3] Eco-efficiency captures the notion of reducing, for instance, material and energy units per unit of output (Gray and Bebbington 2000).

deliberative processes that can enable people to undertake a critical examination of the social relationships which determine their relations to each other and to nature.[4] Recognition of these insights brings to light some of the major obstacles facing the CDM in promoting SD.

First, the CDM is a *market* mechanism; it is *not* an institutional mechanism designed for facilitating critical dialogue over the meaning of SD and its relations to existing socio-economic and socio-ecological arrangements. It therefore takes for granted the very logic and rationality of the socio-economic system (that of a market mechanism) that a more authentic discourse on SD requires examining and challenging in the first place (Banerjee, 2003; Lehman 1996, 1999). Its market structure represents an instrumental or technical 'solution' to the perceived climate crisis and the wider crisis of unsustainability. It does not provide the opportunity to construct the dialogic or deliberative processes for addressing the fundamentally political (and social) issue of how people do and should relate to each other and to their wider environment. It is important to realise that this criticism goes to the heart of business existence – it claims that addressing SD is not simply that the CDM's market structure fails in providing the incentives for incorporating non-monetary aspects of SD into the decision-making processes of business. It is not simply that the various co-benefits of CDM projects that contribute to SD are not commodified, do not produce revenues, and hence are not incorporated in the market when considering potential investment projects – that "no part of the CDM architecture specifically monetises those [SD] benefits and as such they play a limited role, if at all, in directing investment" (Pearson 2007, p. 251). What it claims is that for SD to be realized, the CDM (or any institutional mechanism for that matter charged with promoting SD) will require deliberative processes that can place the very shape, rationality and function of business under scrutiny. This sort of scrutiny is what the notion of SD implies by referring to the limits imposed by social organization on meeting human needs. To simply assume that the CDM's market rationality is adequate to meeting the aims of SD is to fall foul of the problem of *juridification* – using the very same process which caused the problem to solve it (Power 1991, pp. 30–37).

There is some evidence that the consequences of juridification are already starting to show. We are seeing, for example, how the CDM's market structure is privileging inexpensive climate change mitigation projects (sometimes called the 'low-hanging fruit') over the pursuit of SD by considering how it is promoting renewables in Non-Annex I countries. A cursory review of current CDM projects highlights the scarcity of renewables (see Table 1). Ellis et al. (2004, p. 32) note a concerning trend towards projects that deliver large volumes of cheap credits that contribute very little to SD:

> "a large and rapidly growing portion of the CDM project portfolio has few direct environmental, economic and social effects other than GHG mitigation, and pro-

---

[4] This is why many commentators on SD and sustainability emphasise more participatory forms of democracy and the need for a more empowered public sphere in which to discuss, debate and examine, for example, our fundamental relationships to each other and to nature (i.e. Gray 1992; Lehman 1999).

duces few outputs other than emission credits. These projects types generally involve an incremental investment to an already-existing system in order to reduce emissions of a waste stream of GHG [...] without increasing other outputs of the system".

Table 1. Certified emission reductions until 2012 (%) in each category

CERs Until 2012 (%) in each category

- Demand-side EE 1%
- Afforestation & Reforestation 0.3%
- Fuel switch 7%
- Transport 0.1%
- HFCs, PFCs & N2O reduction 31%
- Supply-side EE 11%
- CH4 reduction & Cement & Coal mine/bed 20%
- Renewables 30%

Source: Fenhann 2008

Wara (2006, p. 23) points out that CO_2 reduction projects only make up 29% of the CER pipelines of projects to be realized by 2012. Renewable energy projects make up a mere 18% of the credits approved. He concludes that non-CO_2 gases dominate the CDM pipeline with over 70% of projects at the time of writing. This trend is not likely to be reversed until all such project development is exhausted. Vallette et al. (2004, p. 8) suggest that the world price of carbon will have to reach at least $ 50/tonne before significant funds are diverted into renewables. It has also been argued that existing (inequitable) distributions of wealth will only be exacerbated by the CDM, because the introduction of new investment markets – whether they be CDM projects or the certified emission reduction units (CERs) resulting – does nothing in equalising the power inequalities that determine the parties in the best position to appropriate economic rents (surpluses) that accrue (see for example, Banuri and Gupta 2000, p. 81).

The CDM is a market, not a development fund nor a renewable promotion mechanism. And it is certainly not a deliberative mechanism that can enable public discourse on the direction in which our societies are headed. The problems, then, are not simply institutional but also systemic. The CDM as an institutional structure adopts the wider shape, function and rationality of the market economy that it (and business) is embedded within. It is the inability of the CDM to place these more systemic considerations under scrutiny which is perhaps its greatest obstacle in promoting SD.

Second, and following on from the first point, the CDM simplifies what is essentially a global and holistic issue (SD) to a more 'reduced' level by not providing a definition of SD, as well as by leaving responsibility to business which lacks the mandate governments do for acting in the international arena on behalf of communities. Gray and Milne (2002, p. 5) point out that it is only at the level of communities and eco-systems (as opposed to the level of organizations) that an understanding of SD – cumulative environmental and economic changes – are best understood. At first sight, the CDM appears to be a good mechanism for promoting SD since it operates at the appropriate (international) level with which to properly account for SD. However, although the CDM has as its objective the pursuit of SD, neither it nor the Kyoto Protocol actually offers a substantive definition of it. Instead, it leaves it to developing country governments hosting CDM projects to define SD, even if these governments are not publicly elected. Now in a way, this was a political decision which emerges during climate change negotiations, and is meant to safeguard the sovereignty of developing countries to define their own developmental aspirations. On the other hand, it has also effectively removed from the international level (i.e. the United Nations) the need to discuss collectively the nature of SD. Since CDM projects may move to the countries that have the lowest SD requirements, this also means a race to the bottom for securing meaningful SD criteria and rules.

Instead of coordinated collaboration between governments to work towards the principles and objectives of SD, the CDM leaves responsibility to business agencies for driving societal transitions toward SD in the developing world. This is a major obstacle for realizing SD because the notion as a systems concept needs to be grasped at the global level. It is arguable whether the concept has any application at the corporate or even regional level or perhaps more accurately, that it makes little sense for it to be applied at the corporate or even regional levels unless the wider global view is attained (Gray and Milne 2002; Milne & Ball 2005).

While there are other shortcomings of the CDM that will have to be addressed, we suggest that it is these two – its lack of critical mechanisms by which to deliberate internationally the meaning of SD as well as its inability to address the nature of SD as a global and systemic (socio-economic and socio-ecological) concept – which pose the most important constraints on its ability to realize its objective of promoting SD. It is important to recognize that these are not primarily *technical* constraints but *political* ones and they arise, to an important extent, because of the vested interests of business in preventing the substantive issues of SD from being addressed.

It is important to remember that ever since the 1992 Rio Earth Summit which first thrust SD onto the world stage and cemented it as an important international policy objective, business – which operated in this instance through the powerful lobbies of the International Chamber of Commerce (ICC) and the Business Council for Sustainable Development (now WBCSD – World Business Council for Sustainable Development) – has manoeuvred to remove itself from the SD agenda. For those such as Hildyard (1995), Rio could not address substantive issues of sustainability because it failed to examine key structural features of unsustainability:

"The net outcome was to minimise the status quo [...] Unwilling to question the desirability of economic growth, the market economy or the development process itself, UNCED never had a chance of addressing the real problems of 'environment and development'. Its Secretariat provided delegates with materials for a convention on biodiversity, but not on free trade; on forests but not on agribusiness; on climate but not on automobiles. Agenda 21 [...] featured clauses on 'enabling the poor to achieve sustainable livelihoods' but none on enabling the rich to do so; a section on women but none on men. By such deliberate evasion of the central issues which economic expansion poses for human societies, UNCED condemned itself to irrelevance even before the first preparatory meeting got underway" (pp. 22–23).

The inability to address these substantive issues which the notion of SD throws up is reflected in how the CDM too is set up. Despite these constraints, the potential for accounting in assisting businesses in evaluating and reporting the SD impact of CDM projects still needs to be addressed. We cannot just assume that simply because the CDM exhibits these limitations that it, with the help of accounting, will not be able to contribute significantly to SD (whether this be via climate change mitigation or, for example, through poverty reduction and equitable distributions of wealth) in some way. The next section considers available sustainability accounting techniques which businesses could use for evaluating the SD impact of CDM and it explores whether or not they provide avenues for substantively addressing SD and for overcoming some of the CDM's identified limitations.

4 Accounting for sustainability and SD

Gray (1993) identifies three methods of sustainability accounting that business or organizations could use to evaluate the SD impact of CDM projects:

1. Sustainable cost
2. Natural capital inventory accounting
3. Input-output analysis

Another form of sustainability accounting that needs mentioning even if only because of its popularity is:

1. Triple Bottom Line accounting (TBL) (Elkington 1999)

Ecological accounting methods include (continuing the previous list of sustainability accountings) (see Factor 10 1999; Wackernagel and Rees 1996; Milne and Ball 2005):

1. The ecological footprint
2. Material inputs per Unit of Service (MIPS)
3. Surface Area per Unit of Service (FIPS)
4. Eco-toxic exposure equivalent per Unit of Service (TOPS)

4.1 Sustainable cost and full-cost accounting

Sustainable cost refers to the hypothetical cost of restoring the Earth to the state it was in prior to an organization's impact. It is "… the amount of money an organization would have to spend at an end of an accounting period in order to place the biosphere back into the position it was at the start of the accounting period" (Gray 1994, p. 33). This method works with the notion of capital maintenance and applies it to the biosphere. A sustainable organization or CDM project, then, would be one that maintains natural capital intact for future generations. Sustainable cost is deducted from the accounting profit (calculated using generally accepted accounting principles or GAAP) to arrive at a notional level of sustainable profit or loss (Bebbington and Gray 2001, p. 563; Lamberton 2005, p. 8). Preliminary projects have begun which explore the practical issues of applying sustainable cost frameworks in organizational contexts (i.e. Bebbington and Gray 2001; Bebbington and Tan 1996, 1997; Howes 1999).

Another method that operates within the sustainable costing framework is the Sustainability Assessment Model (SAM). SAM is an accounting technique that quantifies the social, financial, environmental impacts as well as the resource usages of specific projects and their components. Alongside other forms of full cost accounting, such as the Sustainable Cost Calculation, SAM highlights the social and environmental impacts of economic activity and incorporates them into the full cost of the project (Baxter et al. 2004). SAM assesses the social, environmental, economic and resource use impacts over the project's lifecycle. SAM differs from other sustainability evaluative tools by taking the project as its focus. SAM may provide a useful starting point for discussing sustainability issues. The main categories that SAM uses to calculate impacts are *financial flows, resource usages, environmental impacts* and *social impacts*. Financial flows are the economic benefits the project generates for the organisation and its stakeholders. Resource usages are the value of resources that are not accounted for in financial flows. Environmental impacts arise primarily from the environmental damage produced by project activities. Social impacts capture both positive and negative aspects of the project not captured by the other categories (i.e. indirect employment or redundancies associated with project).

The great advantage of sustainable cost methods is that they work within a monetisation framework. They are therefore more easily adaptable to the decision-making processes of business organizations than methods which adopt non-monetary measures.[5] Another strength is that it attempts to incorporate intergenerational considerations with the idea of biospherical limitations into the decision-making calculus. It does not, however, deal adequately with issues of wealth distribution (see Bebbington et al. 2007a, pp. 227–228). It also embodies a more fundamental limitation already alluded to above. The dangers of accounting for

---

[5] This, however, should not be taken to imply that such an approach is not deeply challenging to "business as usual". Gray (1992), for example, notes that such calculations may show that business has actually not made a profit for a very long time – that in fact it has been treating as an expense what is actually capital: the biosphere's natural capital.

natural capital using an established accounting principle, in this case *capital maintenance*, is that it still works within a price-driven framework. Like the CDM's institutional framework, it takes for granted precisely what needs to be questioned in any sustainability accounting – the market structure and the forms of rationality and relations of power that constitute it (Banerjee 2003; Cooper 1992; Hines 1991; Lehman 1996; Maunders and Burritt 1991).

4.2 Natural capital inventory accounting

Natural capital inventory accounting involves the recording of stocks of natural capital over time. Changes in stock levels are used as an indicator of the quality of the natural environment. Distinctions are made between various types of natural capital stocks in order to enable the recording, monitoring and reporting of depletions of enhancements within distinct categories (Lamberton 2005, p. 9). Gray (1994, pp. 17–45) suggests four categories of natural capital:

1. Critical capital (i.e. ozone layer, tropical hardwood, biodiversity)
2. Non-renewable/non-substitutable capital (i.e. oil, petroleum, mineral products)
3. Non-renewable/substitutable capital (i.e. waste disposal, energy usage)
4. Renewable capital (i.e. plantation timber, fisheries)

Natural capital inventory accounting could be predominantly non-financial. It could track resource flows in quantitative but non-monetary units (Gray 1992). On the other hand, it could value natural assets using financial units (e.g., Jones 1996). Accounting for natural inventories is still in its exploratory stages. Like sustainable costing methods, the influence of conventional accounting over natural capital inventory accounting is evident in the application of the capital maintenance concept. It is also evident in the way it uses the management accounting tool of inventory control. It also shares similar limitations to sustainable costing as a *SD* tool. It does not adequately deal with issues of wealth distribution (and hence with the notions of eco-justice that SD entails) and insofar as it adopts monetisation as a way of valuation it does not fundamentally challenge the socio-economic system that financial figures represent. It can, however, be one of the most promising methods for addressing eco-efficiency concerns. The major challenge, though, involves the identification of the accounting entity to which to apply this method. This could be at either the community or regional level (Gray 1992; Lehman 1999). It is doubtful, given that we are dealing with ecological *system* flows when it comes to SD, that this method would have much application at only the level of organizational activity. The implication this has for CDM projects is that they will have to be situated within the ecological systems with which they interact in order to ascertain its cumulative impact on resource flows. It is also doubtful, notwithstanding the preceding discussion, that natural capital inventory accounts could meaningfully reflect nature's interconnectedness and enormous diversity (Lamberton 2005, p. 10).

4.3 Input-output analysis

Input-out analysis accounts for the physical flow of raw materials and energy inputs and product and waste outputs in physical units (Lamberton 2005, p. 10). It attempts to measure all material inputs into the process and outputs of finished goods, emissions, recycled materials and waste for disposal. Resource flows are accounted for using units of volume, although Gray (1994) considers accounting in financial units feasible. Input-output analysis uses a balancing concept familiar to accountants, that of "what goes in must come out". It thus provides a disciplined approach to providing environmental information.

The advantages of input-output analysis are that it is able to identify potential resource and energy savings; it is often the first step in an environmental audit process, and it can facilitate product innovation and pollution prevention strategies, particularly when it forms part of a product and/or process life cycle analysis (Lamberton 2005; Jasch 1993). However, input-output analysis does not measure sustainability or unsustainability. Instead, it provides a transparent account of resource flows into and out of a process. This enables further analysis of environmental impact and can aid in informing sustainability strategies (Gray 1994; Jasch 1993).

4.4 Triple Bottom Line accounting (TBL)

TBL is a form of sustainability accounting that reports on an organization's economic, social and environmental impacts. Underpinning TBL accounts is the evolving three-dimensional definition of sustainable development (e.g., Van den Bergh 1996; UNWCED 1987).

Some versions of TBL use monetary units to measure economic, social and environmental performance whilst other versions such as those of the Global Reporting Initiative (GRI) use a wide array of indicators to measure performance toward the goal of SD. A major advantage of such a version of TBL is that it is part of an overall (albeit voluntary) attempt to systematize accounts. The GRI Sustainability Accounting Guidelines draw on the three-dimensional definition of sustainability using a series of performance indicators to measure the economic, social and environmental dimensions as well as a set of integrated indicators which capture multiple dimensions (see Table 2).

Table 2. GRI Framework for performance indicators

	Category	Aspect
Economic	Economic	Economic Performance
		Market Presence
		Indirect Economic Impacts
Environmental	Environmental	Materials
		Energy
		Water
		Biodiversity
		Emissions, effluents, and waste
		Suppliers
		Products and services
		Compliance
		Transport
		Overall
Social	Labour Practices and Decent Work	Employment
		Labour/Management Relations
		Occupational Health and Safety
		Training and Education
		Diversity and opportunity
	Human Rights	Investment and Procurement Practices
		Non-discrimination
		Freedom of Association and Collective Bargaining
		Abolition of Child Labour
		Prevention of Forced and Compulsory Labour
		Complaints and Grievance Practices
		Security Practices
		Indigenous Rights
	Society	Community
		Corruption
		Public Policy
		Anti-Competitive Behaviour
		Compliance
	Product Responsibility	Customer Health and Safety
		Products and Services Labelling
		Marketing Communications
		Customer Privacy
		Compliance

Source: GRI 2006

Major strengths of the TBL GRI framework is that it provides a coherent system to recording, monitoring, reporting and auditing an organization's impact in economic, social and environmental terms. The disadvantages, however, include difficulties in observing a clear link between the indicators and sustainability (Baker, 2002). It also does not deal with the problem of poverty directly and its voluntary status lends no real impetus for businesses in making use of its framework. Like other sustainability accounting frameworks, it lacks the ability to account for the long-term economic and ecological impacts that SD as a global concept requires. This would necessitate an evaluation of societies and communities rather than organizations (Milne and Ball 2005) and, as alluded to above, it would require a critical analysis and a public discourse about the socio-economic system within which business takes place (Bebbington 2001). By making it a voluntary scheme it also seems to assume that the role of, *inter alia*, poverty alleviation is primarily for the government to deal with rather than business. This, however, ignores the question of whether business activities create poverty and also the extent to which business lobbies governments in order to remove this issue from national and international agendas. Only a critical dialogic or deliberative process such as the one imagined by Lehman (1999, pp. 217–241) would be able to move such issues into public discourse.

Milne and Ball (2005) provide a compelling analysis of the way business has manipulated the use of TBL accounts in order to marginalize the radical nature of SD. They show that rather than exhibiting a systematic uptake of TBL reporting, currents trends show TBL use as sporadic, unsystematic and, in fact, slowing down. They also point out how business's use of TBL significantly underplays the eco-justice, equity fairness, and eco-effectiveness aspects of SD such that eco-efficiency is coming to act as a stand in for business for the more radical notion of SD that reports such as the UNCWED (1987) articulate. They argue that in order for business to move away from the over-emphasis on eco-efficiency – an emphasis which works within the same system that caused the problem in the first place (McDonough and Braungart 1998) – businesses will need to develop greater ecological literature and to make use of *ecological accounting* methods (i.e. Birkin 1996).

4.5 The ecological footprint

The ecological footprint estimates the "area of productive land and water ecosystems required to produce resources that the population consumes and assimilate the wastes that the population produces" (Rees 2000, p. 371). In ecosystemic terms, the power of footprints lies in its ability to show the extent to which existing populations are living beyond the assimilative capacities of the biosphere or as Bebbington et al. (2007b, p. 371) put it "the extent to which the human population is living off ecological capital rather than ecological surplus". Implicit to the footprint is a vision of a one-planet. It is therefore also a problematizing tool in the sense that it can show the extent to which some populations (primarily Western) are consuming far in excess of their "share" of biological resources. In doing so, it may serve as a point of departure for discussing issues of inter- and intra-generational equity. One of the strengths of the ecological footprint is its ability to take the global perspective that SD demands. By showing disparities between populations in terms of consumption shares, it is also able to provide indications of wealth distribution and the issues for equity and fairness that such distributions imply. It is important to note, however, that the ecological footprints only provide *indications* of (un)sustainability; it does not explain *why* it is that such distributions occur. In order to answer this, focus will have to be brought back to the nature of socio-economic arrangements and why they lead to such distributions.

4.6 MIPS, FIPS and TOPS

The ecological footprint is related to the concept of Material Inputs per Unit of Service (MIPS) and the "ecological rucksack" it generates; Surface Area per Unit of Service (FIPS); and Eco-toxic exposure equivalent per Unit of Service (TOPS) (Factor-10 1999 cited in Milne and Ball 2005). Ecological rucksacks are calculated for end-use products. They seek to assess the amount of raw materials (including energy) used in producing, transporting, consuming, and finally disposing of an end product less the weight of the rucksack (Milne and Ball 2005). MIPS calculations relate material inputs to "service" outputs. Improvements in MIPS ratios can come from either less material inputs used or from improved services from given inputs. FIPS and TOPS operate on similar principles; FIPS relates surface area used up to "service" outputs whilst TOPS relates the amount of eco-toxic substances used to "service" outputs.

The goal of these methods is to foster dematerialisation and also to provide ways for understanding that environmental harm is not simply caused by pollution but is also significantly caused by resource extraction and determined by resource productivity. These methods are able to convey changes in terms of the amount of material used in service activities. It is important to note, however, that the per-unit measures of dematerialization are ecologically irrelevant since they do not account for possible absolute increases in material flows from the ecosphere to the economy (Milne and Ball 2005, p. 20). Because of this Milne and Ball (2005, p. 19) argue that "improvements in dematerialization need to come at the level of economies and ultimately, the planet and not simply on the basis of individual products – a tenfold decrease in material inputs per computer is little use, if it coincides with a greater than tenfold increase in consumption of these products". This insight recognizes, again, the global nature of SD as a concept and the need to critically consider how current socio-economic arrangements may encourage eco-efficiency gains only to produce absolute increases in material consumption.

5 Discussion and conclusions

Fortunately, despite the limitations evinced in existing approaches to SD accounting covered above, there are developments in accounting research which attempt to overcome these limitations. The common characteristic to all these approaches is that they all treat accounting, not primarily as a technical practice, but rather as a political or social phenomenon (Tinker 1980, p. 158).

Lehman (1999), for instance, seeks to develop notions of accounting beyond conceptions which view it simply as an instrumental or procedural approach for establishing the accountability of organizations by calling for the creation of social practices which can place critical dialogue or discourse at the heart of accountability. Such an approach has implications for the CDM. If businesses are to have a greater chance of promoting SD through CDM projects, they will have to make use of deliberative mechanisms in which SD can be critically, creatively and collectively examined and explored by all the communities concerned. It should be clear from the preceding discussion that this approach is not simply advocating a weak form of stakeholder engagement. It is, more importantly, premised on the ability of communities to call into question the very shape, rationality and function of business itself and in doing so, to enable them to explore where it is they fit within this socio-economic context and whether or not this is to be changed. Other research in accounting has developed these dialogic themes by theorizing how accounting researchers could engage more effectively with organizations and other stakeholders in order to develop more adequate forms of sustainability accountings – theorizations that have been informed by actual engagements with organizations (i.e. Bebbington et al., 2007a, b; Thomson and Bebbington 2005). What is implied by such research, is the need to see sustainability accounts as just one part (and perhaps a small one at that) of an organization's or a society's transition to more sustainable forms of development.

A sustainability accounting method on its own is severely inadequate in terms of capturing all that is entailed by the notion of SD. Combined with other tools, however, may begin to present a more complete picture of CDM project's impact on society and on the environment. Nevertheless, this may not be enough. What they have to be a part of is a more critical and collective discussion on the direction our societies are heading. This would necessarily entail an examination of not only the shape, function and rationality of our socio-economic arrangements but also an explicit examination of the vested interests in maintaining or transforming it. Only then will the CDM, business and accounting be able to address the issue of what SD is, and whether or not our current trajectories are taking us closer to or farther away from it. If this is done, we will no longer be within the anaesthetised realm of technical accounting discourse. Instead, we will be firmly dealing with the politics of SD.

REFERENCES

Baker M (2002) The GRI – the will to succeed is not enough. Corporate Social Responsibility News and Resources. http://www.mallenbaker.net/csr/CSRfiles/GRI.html

Banerjee SB (2003) Who sustains whose development? Sustainable development and the reinvention of nature. Organization Studies 24 (1): 143–180

Banuri T, Gupta S (2000) The Clean Development Mechanism and sustainable development: an economic analysis. In Ghosh P (ed) Asian Development Bank: Implementation of the Kyoto Protocol: Opportunities and Pitfalls for Developing Countries 73–101

Baxter T, Bebbington J, Cutteridge D (2004) Sustainability assessment model: modelling economic, resource, environmental and social flows of a project. In: Henriques A, Richardson J (eds) 'The Triple Bottom Line–Does It All Add Up?' Earthscan, London

Bebbington J (2001) Sustainable development: a review of the international development, business and accounting literature. Accounting Forum 25 (2): 128–158

Bebbington J, Gray RH (2001) An account of sustainability: failure, success and a Reconceptualization. Critical Perspectives on Accounting 12: 557–587

Bebbington J, Brown J, Frame B (2007a) Accounting technological and sustainability assessment models. Ecological Economics 61: 224–236

Bebbington J, Brown J, Frame B, Thomson I (2007b) Theorising engagement: the potential of a critical dialogic approach. Accounting, Auditing and Accountability Journal 20 (3): 356–381

Bebbington J, Tan J (1996) Accounting for sustainability. Chartered Accountants Journal July: 75–76

Bebbington J, Tan J (1997) Accounting for sustainability. Chartered Accountants Journal February: 37–40

Birkin F (1996) The ecological accountant: from cogito to thinking like a mountain. Critical Perspectives on Accounting 7: 231–257

CDM Watch (2005) The World Bank and the carbon market: rhetoric and reality. http://cdmwatch.org/files/World%20Bank%20paper%20final.pdf

Cooper C (1992) The non and nom of accounting for (m)other nature. Accounting, Auditing and Accountability Journal 5 (3): 16–39

Daly HE (1992) Allocation, distribution and scale: towards and economics that is efficient, just and sustainable. Ecological Economics 6: 185–194

Daly H, Cobb JB (1989) For the common good. Boston: Beacon

Dobson A (1998) 'Justice and the Environment: Conceptions of Environmental Sustainability and Theories of Distributive Justice'. Oxford University Press: Oxford

Elkington J (1999) Triple Bottom Line Reporting: looking for balance. Australian CPA March: 19–21

Ellis J, Winkler H, Corfee Morlot J (2004) 'Taking Stock of Progress under the CDM', OECD and IEA Information Paper, COM/ENV/EPOC/IEA/SLT (2004): 4

Factor-10 Club (1999) Factor 10: Making sustainability accountable – putting resource productivity into praxis. Factor 10 Club: http://www.factor10-institute.org/Pdf-Files.htm

Fenhann J (2008) CDM Pipeline Overview, January 2008. UNEP Risoe Centre. http://www.CDMPipeline.org/Publications/CDMPipeline.xls (accessed January 2008)

Global Report Initiative (GRI) (2006) Sustainability reporting guidelines. http://www.global reporting.org

Gray RH (1992) Accounting and environmentalism: an exploration of the challenge of gently accounting for accountability, transparency and sustainability. Accounting, Organizations and Society 17 (5): 399–425

Gray RH (1993) 'Accounting for the Environment'. London: Paul Chapman

Gray RH (1994) Corporate reporting for sustainable development: Accounting for sustainability in 2000AD Environmental Values: 17–45

Gray RH, Bebbington J (2000) Environmental accounting, managerialism and sustainability: is the planet safe in the hands of business and accounting? Advances in Environmental Accounting & Management 1: 1–44

Gray RH, Milne MJ (2002) Sustainability Reporting: who's kidding whom? Chartered Accountants Journal of New Zealand 83 (7): 16–18

Hildyard N (1995) Foxes in charge of the chickens. In: Sachs, W. (1995) 'Global ecology: a new arena of political conflict'. London: Zed Books, pp 22–35

Hines R (1991) On valuing nature. Accounting, Auditing and Accountability Journal 4 (3): 27–29

Howes R (1999. Accounting for environmentally sustainable profits. Management Accounting 77 (11): 32–33

Jasch C (1993) Environmental information systems in Austria. Social and Environmental Accounting 13 (2): 7–9

Jones MJ (1996) Accounting for biodiversity: a pilot study. British Accounting Review 28: 281–303

Lamberton G (2005) Sustainability accounting – a brief history and conceptual framework. Accounting Forum 29: 7–26

Lehman G (1996) Environmental accounting: pollution permits or selling the environment. Critical Perspectives on Accounting 7: 667–676

Lehman G (1999) Disclosing New World: Social and Environmental Accounting. Accounting, Organizations and Society 24 (3): 217–241

Maunders K, Burritt R (1991) Accounting and ecological crisis. Accounting, Auditing and Accountability Journal 4 (3): 9–26

McDonough W, Braungart M (1998) The next industrial revolution. The Atlantic Monthly Digital Edition: http://www.theatlantic.com/issues/98oct/industry.htm

Milne MJ, Ball A (2005) 'Business and sustainability: agenda for change or soothing palliatives?' Inaugural Professorial Lecture. Victoria University of Wellington

Norton BG (1991) 'Toward unity among environmentalists'. New York, Oxford University Press

Pearson B (2007) Market failure: why the Clean Development Mechanism won't promote clean development. Journal of Cleaner Production 15 (2): 247–252

Power M (1991) Auditing and environmental expertise: between protest and professionalism. Accounting, Auditing and Accountability Journal: 30–42

Rees W (2000) Eco-footprint analysis: merits and brickbats. Ecological Economics 32 (3): 371–374

Thomson I, Bebbington J (2005) Social and environmental reporting in the UK: a pedagogic evaluation. Critical Perspectives on Accounting 16 (5): 507–533

Tinker AM (1980) Towards a Political Economy of Accounting: An Empirical Illustration of the Cambridge Controversies. Accounting Organizations and Society 5 (1): 147–160

United Nations World Commission on Environment and Development (1987) 'Our Common Future' Oxford, Oxford University Press

United Nations, Kyoto Protocol to the United Nations Framework Convention on Climate Change (1998) http://unfccc.int/resource/docs/convkp/kpeng.pdf

Vallette J, Wysham D, Martinez N (2004) 'A wrong turn from Rio: The World Bank's Road to Climate Catastrophe', Sustainable Energy and Economy Network, Institute for Policy Studies, Transnational Institute, Research and Policy Brief. 10th Session of the Conference of the Parties, Buenos Aires

Van den Bergh J (1996) 'Ecological economics and sustainable development'. Cheltenham, Edward Elgar

Wackernagel M, Rees W (1996) 'Our Ecological Footprint: Reducing Human Impact on the Earth'. New Society Publishers, Canada

Wara M (2006) 'Measuring the Clean Development Mechanism's Performance and Potential' Working Paper #56, Program on Energy and Sustainable Development. Stanford University

Wittneben B (2007) The Clean Development Mechanism: Institutionalising New Power Relations. Erasmus Research Institute of Management. Report Series Research in Management ERS-2007-004-ORG

Risk management in the Clean Development Mechanism (CDM) – the potential of sustainability labels[1]

Adrian Muller

Socio-economic Institute
University of Zurich
Blümlisalpstrasse 10, 8006 Zurich, Switzerland
adrian.mueller@soi.uzh.ch

Abstract

There is a danger that the CDM will fail to achieve its goals, namely reduction of greenhouse gas emissions and enhanced sustainable development. Sustainability labelling is a promising strategy to hedge against such failures. Labels could also serve as a business risk-hedging tool. The existing labels for the CDM are not comprehensive enough, however. A two-tiered stakeholder participatory approach with national flexibility under an international umbrella could be a promising option. Due to the necessary bureaucracy this might not be feasible. Labels in the spirit of the existing approaches – addressing only restricted aspects of sustainability or not applicable to all sectors – may be a second-best option. Other instruments for the further regulation of the CDM, such as a profit tax, should therefore be discussed as well.

Keywords: CDM, labels, sustainability indicators, risk, equity

---

[1] Many thanks to Reimund Schwarze for comments and to Simon Mason for detailed comments and for the correction of the English and to the Workshop participants for fruitful discussions and comments. Financial support from the Swiss National Science Foundation is gratefully acknowledged. The usual disclaimer applies.

1 Introduction

The Clean Development Mechanism (CDM) allows the so-called Annex-I countries facing greenhouse gas (GHG) emissions caps under the Kyoto Protocol (KP) to achieve emission reductions in countries not subjected to emissions caps.[2] Governmental and private entities can invest in such activities to produce "Certified Emission Reductions" (CER) that directly account for domestic reduction goals in Annex-I countries or that can be sold on the carbon market. Investment in the CDM is a business activity and subject to corresponding risks such as volatile prices (i.e. of the CERs). Besides business risks, the CDM leads to a range of sustainability risks, for example, of only achieving uncertain emission reductions or of resulting in a sell-out of cheap reduction possibilities without further gains for the host countries (Rose et al. 1999; Ott and Sachs 2002; see Molisa and Wittneben in this volume for a general discussion of the sustainability-business trade-off in the CDM in a societal context and Müller-Pelzer for a somewhat more project-related discussion).

In order to tackle the business risks, classical risk-hedging mechanisms such as portfolio approaches can be applied. In addition, it can be expected that the carbon market itself will offer a wide range of products already common in financial markets (Nolles 2004). As the metric in the market is monetary, however, sustainability is not likely to be of decisive importance in such considerations.

The sustainability risks can be framed as external costs of the CDM: internalization of these costs would be the socially optimal strategy. Internalization could, for example, be achieved by a tax on CDM projects. Given the potential for considerable rents in CDM projects, taxing by rent sharing mechanisms is a promising suggestion (Denne 2000; Muller 2007). Tax proposals, however, usually face considerable opposition and an implementation in the CDM might not be politically viable on a global level. China, as a first nation, however imposes a tax on certain CDM projects expected to have low sustainability benefits but high profitability (CDM China 2005; Point Carbon 2005).

Currently, the sustainability risks – if tackled at all – are usually addressed by another, not market-based policy instrument, namely by standards or labels. There are various institutions (mainly NGOs and governmental agencies, but also some of the Carbon Funds; cf. Sutter 2003; CDCF 2005) providing sets of sustainability indicators to assess, and labels to communicate certain standards to be achieved by CDM projects. This is mainly done on a voluntarily basis, but some funds or countries require all CDM projects they fund or host to comply with the standards they set. The UNEP has also issued a series of "Environmental Due Diligence" documents for risk assessment in several renewable energy techniques (UNEP 2003).

Such labels or standards can also act as instruments to hedge against business risks as they could be combined with assured minimal CER prices, for example. They can also build on a certain segment of customers that are likely to buy CERs standing for reliably verified emission reductions. Even if CERs from unsustain-

---

[2] Annex-I countries are basically developed countries and economies in transition, and the Non-Annex-I countries are primarily developing countries.

able projects dominate the market and even if there are no stringent reduction plans for the post-Kyoto phase, such labels and standards could be a way forward on a voluntary basis.

Sustainability labelling is thus an approach that could address both business and sustainability risks and that would break with the monetary focus of other business risks hedging strategies. In this paper, the potential of labelling for the CDM is assessed. Some shortcomings of current labels can be identified; one goal of this paper is to identify critical issues in CDM labelling and to suggest more optimal labelling strategies.

The CDM is one of the so-called flexibility mechanisms of the KP, broadening the supply side of CERs for the carbon market. As such it is a crucial part of the global carbon policy instruments designed in the KP. Labels perform well as a voluntary consumer information instrument, as experience from other sectors shows, where there are increasing market shares of sustainability-labelled products (e.g. Wüstenhagen 1998; Bird et al. 2002). The greatest challenge for voluntary labels is how to gain large market penetration. This is one of the important marketing issues for sustainability labels for goods such as food (organic) or electricity (green power) (Wüstenhagen 1998).

Going one step further, labels for the CDM can be considered from a global policy instruments perspective, i.e. regarding their potential to increase the sustainability of the CDM in general by requiring all projects to comply with some additional standards. There are, however, theoretical results indicating that such mandatory standards are less efficient than other instruments such as guaranteed higher CER prices for sustainable projects (Fischer and Newell 2004). Interesting is also the theoretical economic literature on labels (relevant key concepts are "green products" or "impure public goods", and "warm glow" as one explanation for buying such goods) and whether they have positive or negative environmental (or general welfare) effects. Results usually depend on modelling details. Effects are not always positive and a detailed analysis of the concrete situation at hand is necessary (see e.g. Kotchen 2005, 2006 or Althammer and Dröge 2006). An additional complication comes from the importance of foreign direct investment (FDI) for many developing countries. Mandatory high standards could hinder some FDI projects that might not be very sustainable or that would not achieve true emissions reductions but that definitely would be advantageous from a standard economic development perspective.

The organisation of this chapter is as follows. Section 2 presents the CDM with a particular focus on its risks. Section 3 discusses sustainability indicators and labelling on a more general and abstract level. Section 4 brings these lines together and addresses labelling for the CDM, pointing out potential pitfalls and options for improvement. Section 5 concludes.

2 The Clean Development Mechanism

The CDM is expected to lower compliance costs under the KP as the marginal GHG reduction costs are lower in developing countries. To qualify for the CDM, a project has to reduce emissions below a baseline representing "business as usual". Following the current understanding, this basically means that the project has to be profitable only with the additional revenues from CER sales (UNFCCC 2004). Setting this baseline is a crucial, complex and controversial task (Sutter 2003; Müller-Pelzer 2004), but once settled, the GHG reduction aspect of the CDM project is well-defined. Besides this, it is an explicit goal of the CDM to foster sustainable development in the host countries (UNFCCC 1997). There is clearly a large potential for sustainable CDM project activities (in rural energy supply or waste management, for example; see also Pal and Sethi in this publication), but the sustainability goal lacks further concretization in the legal documents and allows considerable leeway in interpretation.

The partners in a CDM project are a public or private CER buyer, an institution in the host country and often also a consultant facilitating the transactions between buyers and host country institutions. Currently, many projects seem not to involve direct investment from Annex-I partners but rather CER purchase agreements only, thus disproportionally shifting the risk to Non-Annex-I parties (Sutter 2005). Since February 2005, unilateral CDM projects are possible, i.e. an institution in a developing country can submit a project for registration under the CDM without involvement of other parties (UNFCCC 2005a). The CERs can then be sold on the market.

The CDM sector is growing rapidly (Fenhann 2006, 2007). Following is a short summary of the current (May 2007) situation, although changes are fast and certain aspects of the situation will probably look different within a matter of months. On the other hand, some trends or characteristics have now been stable for more than a year and thus may continue. As of May 2007, there are 645 CDM projects registered, 1221 are under examination and 19 withdrawn or rejected. In accumulated (till 2012) CER production, 38% are Hydrofluorocarbons (HFC) destruction (26%) or N_2O capture/destruction (12%) projects that are not likely to deliver any additional sustainability benefits (Schwank 2004; Cosbey et al. 2005). These percentages have remained relatively stable up to January 2006, although the number of projects and accumulated CER production has almost doubled between October 2005 and then (mid-January: HFC 40%, N_2O 11%). The percentage for HFC has however dropped since then and criticism of HFC projects has intensified (ENB 2006; Point Carbon 2007). Landfill gas (Methane reduction) projects account for 9% of accumulated CER production. They are likely to deliver sustainability benefits if combined with energy use, but there are some large projects that only flare the gas, thus foregoing these additional returns. The accumulated volume of CERs from the 1,866 projects is about 1,910 million t CO_2 equivalents by 2012. The number of countries hosting CDM is steadily increasing but is still characterised by a skewed distribution: only India (33%), China (24%), Brazil (12%) and Mexico (8%) have more than 3% of the total number of projects. The inequality in distribution is decreasing, however (in January 2006, India and Brazil together accounted for 60% of the projects). Regarding CER generation, the distribution is even more skewed: China accounts for 49%, India for 17%, Brazil for 8%; other than these, only South Korea (5.3%) reaches over 3%. Only 2.2% of the projects are located in Africa, mostly in South Africa (17), Egypt (7) and Morocco (5), a share that has declined since mid-January 2006 (2.6%). The biggest buyers are currently the UK, the Netherlands and Japan with 376, 137 and 135 projects, respectively, whereas for 979 projects this information seems not to be available (i.e. it is likely that a CER purchase agreement has not yet been reached or that the projects are unilateral).

The current experience, with almost 38% of CER production not being accompanied by increased sustainability together with the unspecific formulation of this goal in the KP, as well as the questions regarding additionality, emphasize the danger that the CDM may fail to achieve its goals. One specific aspect is that the CDM may lead to the sell-out of cheap reduction possibilities in developing countries, leaving them with only the more expensive ones in the future, when they may face caps themselves. This could theoretically be offset by technology transfer or learning effects that would keep marginal abatement costs down (Rose et al. 1999). At least part of the experience, however, indicates that such actions might not take place (as observed in South Africa, for example, according to Steve Thorne in UNFCCC (2005b)). These reservations regarding the sustainability performance of the CDM initiated the discussion on sustainability indicators or labels for the CDM, aiming firstly to establish indicators for measurement of the sustainability performance and development of CDM projects, and secondly to identify levels of good performance and to define standards and labels for this (e.g. Sutter 2003; Goldstandard 2005). The COP11/MOP1 in December 2005 reached decisions for improvement of the CDM, but no decisions have been taken to strengthen the sustainability goal or to address the lack of projects in the least-developed countries (Wittneben et al. 2006). This situation still prevailed after the COP12/MOP2 in November 2006, where the skewed regional distribution was still an issue of primary importance (ENB 2006; UNFCCC 2006). Besides this, prime topics identified for 2007 were the role of several specific and partly new project types for the CDM (Carbon Capture and Storage, Afforestation/Reforestation, switching from non-renewable to renewable biomass), the ongoing discussion on additionality and the discussion on how to deal with HFC projects under the CDM.

Besides being an instrument of climate policy, the CDM provides a framework for new business opportunities. It is potentially profitable to invest in GHG reduction projects in developing countries and to sell the CERs on the carbon market. The mechanism also opens up opportunities for additional profit in ordinary investments due to the CERs produced. This can, for example, be the case for renewable energy projects. However, the opportunities of the CDM are also associated with several risks. First, there is the uncertainty of the prices of CERs. This is mainly due to still unclear details of rules and allocation plans for the first commitment period of the Kyoto Protocol (2008–2012) to succeed the "trial period" of the European Emissions Trading System (EU-ETS) now under operation (2005–

2007). The general political and institutional uncertainty regarding the KP or any successor agreement after 2012 further adds to this uncertain business environment. However, the COP11/MOP1 in Montreal, December 2005, decided formally to take up discussions on regimes beyond 2012 and initiated several processes in this context (Wittneben et al. 2006). This discussion was continued at the COP12/MOP2 in Nairobi, November 2006. Concrete decisions have not yet been reached, but the importance of proceeding rapidly towards such decisions is widely acknowledged (ENB 2006). In any case, the EU-ETS is likely to proceed and it can be expected that in particular some NGO initiatives will also be willing to buy CERs outside a stringent post-Kyoto agreement (such as the NGO "My-Climate" investing in projects to offset carbon emissions from air travel). The total market volume of CERs, however, is heavily dependent on global participation. On this level, the role and volume of "hot air" credits in the market will also be decisive for CER prices.

A specific CDM risk is the possibility that a project is not accepted by the CDM Executive Board, the institution in charge of the whole CDM registration process. This bears the danger of sunk costs, especially for small projects, as transaction costs to file a CDM project are still considered to be high. This problem together with the insufficient funding of the CDM Executive Board were addressed at the COP11/MOP1 in December 2005 (Wittneben et al. 2006). Additional funding was allocated in order to overcome the bottleneck in processing of project proposals and some "sectoral" CDM will be possible, allowing for certain CDM activities bundling several projects at different sites. Problems with monitoring and verification of the CERs might occur and there is some potential that political opposition against specific CDM projects could arise. There is also the danger that a project may not deliver as many reductions as it claimed regarding the specific base-line chosen. Still changing or newly implemented national policies regarding revenue sharing also affect the profitability of projects. CER revenue shares are explicitly a matter of negotiation between the partners for each project (UNFCCC 2005c), but it is the prerogative of national governments to set additional rules, such as China did in 2005 by imposing a tax on CDM projects (CDM China 2005, Point Carbon 2005: e.g. 65% of CER revenues for HFC projects, 30% for N_2O).

Due account has also to be paid to the peculiar characteristics of CERs: they are no physical necessity for any firm, but the political decision to cap GHG emissions make them a valuable input for any process emitting GHGs. The demand for CERs is thus generated politically and the commodity itself (i.e. CERs) as well as the market are basically designed in a political process. In general, neither buyers nor sellers are interested in the quality of CERs (for example measured by the reliability of the reductions achieved or by assured sustainability effects). The quality-based self-enforcing control mechanism present in usual goods markets is thus missing (Repetto 2001).

3 Sustainability indicators and labelling

Sustainability indicators are designed to capture the state and development of a system regarding sustainability. A fundamental problem with the concept of "sustainability" is its overly general nature that hinders its operational use without further specification. Any set of sustainability indicators frames some aspects on a concrete and measurable or evaluable level, thus allowing for a more or less complete description of a system regarding its sustainability in current state and in evolution. Sustainability is most often framed by discerning economic, environmental and social aspects on an equal footing, a structure many indicator systems follow. This makes the implicit trade-offs in any sustainability assessment explicit. Another common approach is more hierarchical, defining the social aspects as final goals, the environment as an indispensable basis and the economic issues as efficient means to reach the goals given the boundary conditions according to the physical basis. The importance of long-term time horizons and inter-generational equity further increase the complexity of the concept.

Sustainability indicators have become common since the UN Conference on Environment and Development 1992 – the "Earth Summit". With the "Agenda 21", a programme of action for sustainable development was launched and countries were asked to develop national strategies for sustainable development. This boosted a wide range of initiatives to make sustainability a topic in various contexts, such as for local authorities (local Agenda 21), investment (sustainability funds), firms (ISO 14000) and energy production (environmental fiscal reform), to name just a few. In parallel, indicator systems for assessing the sustainability of the effects of such activities were developed.

The current situation regarding the application of sustainability is characterized by a lack of coordination and implementation of strategic action in national contexts (Swanson et al. 2004). While measuring indicators for the classical spheres of economics, environment and society is widespread and relatively well understood, this is not the case for more integrated aspects of sustainability. The aggregation of multiple indicators and the balance between generality/comparability and case-specific adequacy also remains a challenge. Implementation of strategic actions addressing integrated issues of sustainability and coordination between national and local strategies, and between sustainability policies and the general national budgeting process, is lacking. Furthermore, there is a still largely unused potential to implement classical economic policy instruments such as the environmental fiscal reform (Swanson et al. 2004).

While indicators measure state and development on a descriptive level, labels can be used prescriptively. They predominantly apply to single products, processes or producers. To specify what sustainability ideally means on such a concrete level is notoriously difficult. Basically two pragmatic strategies to deal with this problem can be discerned. Either one defines a limited set of ideals that should be achieved (e.g. recycling quotas), accepting the only partial view adopted, or one identifies some crucial negative aspects to be avoided (e.g. pesticide use), thus aiming at achieving a certain minimal standard. Furthermore, it is useful to discern between broad sustainability labels such as the EU eco-label and more restricted but correspondingly more detailed sector-specific labels such as green electricity (de Boer 2003). The main goal of labels is to provide simple, clear and reliable information on aspects of the goods purchased that consumers cannot otherwise verify. Labelling as a means of general and environmental information provision to the customer is not new (the "Blaue Engel" – blue angel – in Germany, for example, was established in 1977), but environmental labelling was boosted in the wake of Agenda 21, and social labelling became widespread at the same time.

The aggregation challenge is particularly important as labelling usually involves integrating a range of different indicators in a single measure. This is one point of criticism towards the EU eco-labelling scheme, and there is the danger that it undermines credibility (Karl and Orwat 1999). The problem is somewhat alleviated for the many labels that are specific for certain sectors and certain aspects of sustainability only, such as labels focusing on organic farming practices ("Organic", "Bio"), fair-trade aspects ("Max-Havelaar"), or key environmental and social aspects in forestry and fisheries ("Marine" and "Forest Stewardship Council" – MSC resp. FSC).

Current labelling schemes involve basically four groups of actors (de Boer 2003). Producers are interested in labelling as it can give them a competitive advantage – working somewhat like a brand – whereby the market can be segmented according to the new qualities the label stands for. It signals compliance with certain sustainability standards and thus directly addresses a certain consumer segment. Consumers gain access to information they did not have before or which it would be prohibitively costly to collect. Governmental institutions see labelling as having the potential to combat free-riding by reducing the asymmetry of information availability prevailing in any producer-consumer relationship. In addition, labels are one instrument that can be employed in order to achieve governmental sustainability goals. Other parties such as NGOs have mainly a type of lobbying interest as labels can support their case. This applies for example to the process of newly defining a label, starting a discussion in this context and working towards its regulation.

The recent development of labelling has been characterized by the emergence of a new type of labels that are no longer linked to any increase in individual material benefit (such as it is the case for organic food, for example, that is free of pesticide residuals). "Green" electricity (i.e. non-fossil, non-nuclear and no large hydro) is the paramount example. Labelling of "conflict-free" diamonds under the Kimberley-Process or the discussions on "fair-trade" oil are others. For all these products, there is no physical difference between the labelled and the conventional. The direct consumer gain is thus not tangible but wholly ideational. However, indirect differences with a public good/bad character probably arise, e.g. through the negative effects of destabilizations due to environmental conflicts in the countries of origin of many natural resources (Mason and Muller 2007).

Marketing is a crucial aspect for labelling schemes, especially for goods with only non-tangible ideational individual benefits. Bird et al. (2002) identify consumer information and aggressive marketing as a crucial factor in increasing market shares for green electricity. The presence of reliable labels and differentiated products, also offered in the low-price segment, are decisive as well. There does indeed seem to be a considerable willingness to pay for labelled products, e.g. in energy-efficiency, which even exceeds the expected cost savings over the lifetime of the product (Sammer and Wüstenhagen 2006). New aspects of sustainability marketing include the relevance of non-product-specific issues such as social and environmental problems and how they intersect with consumer wants and demands. The potential for and importance of affecting macro-conditions for the (new) markets on the policy level is also a new aspect because these conditions are considered to be externally given in conventional marketing (Belz 2005). Sustainability marketing has also a longer time-horizon than conventional marketing and puts less emphasis on the sales and transactions than on a lasting relationship with the customer.

4 Sustainability labels for the CDM

Section 4.1 presents a short overview of some of the existing indicator and labelling schemes for the CDM. In many aspects, labels for CDM take labelling to its limits – the related problems but also the potential are outlined in Section 4.2. Based on this, Section 4.3 investigates what an ideal label for the CDM could look like.

4.1 Existing sustainability labels and indicators for the CDM

Various indicator systems and labels for the CDM exist, provided by NGOs (e.g. the SSN sustainable development tool, the Gold Standard, the CCB standards), by governmental institutions (SUSAC, an EU institution to support CDM projects in African, Caribbean and Pacific Countries), and also by the carbon funds themselves (e.g. the Prototype Carbon Fund PCF) (Sutter 2003). Common to all these approaches is a set of criteria and often some (explicit or implicit) weighting scheme to make them comparable and to aggregate them to a one-dimensional measure allowing ranking and labelling of projects. Alternatively, or in addition, there are often minimal performance requirements for each criterion that have to be fulfilled by a project in order for it to qualify for the label.

Sutter (2003) identifies four key properties that should be fulfilled by any sustainability assessment tool for the CDM: adjustability regarding (the various stakeholders') preferences, possibility for relative measurements in the context of the sustainability of the larger surrounding system (e.g. the host country), reproducibility of results, and an assessment that is as comprehensive as possible. His multi attribute utility theory (MAUT) based approach provides a framework for assessing CDM projects fulfilling all these criteria. It is implemented by the government of Uruguay to assess all CDM projects that it may host (Heuberger and Sutter 2003). A recent assessment of all registered CDM projects as of autumn

2005 with a simplified version of this tool shows that none of the projects fulfil both the additionality and the sustainability requirements (Sutter and Parreno 2005).

Currently one of the most prominent CDM labels is the Gold Standard (Goldstandard 2005). For the first time, CERs have been issued for a Gold Standard certified project in March 2007 and more projects are in the pipeline (Goldstandard 2005, 2007). It mainly consists of a sustainability assessment based on three groups of 3 to 6 indicators. The projects are assigned a performance measure between -2 (major negative impacts) and +2 (major positive impacts) for each indicator. Within each sub-group, the sum of the indicator values has to be non-negative and the over-all sum over the sub-groups has to be positive. No indicator is permitted to score -2. The indicators in sub-group one are 1) water quality, 2) air quality, 3) other pollutants, 4) soil quality, 5) biodiversity effects and 6) employment (qualitative aspects); in sub-group two 7) livelihood of the poor, 8) access to energy services and 9) human and institutional capacity; in sub-group three 10) employment (numbers), 11) balance of payments and 12) technological self-reliance. As a second main point, the Gold Standard requires two main public consultations during the design phase in addition to the ordinary CDM stakeholder participation criteria: one at the beginning and another, more comprehensive one, before the project is validated. The Gold Standard states several rules for these consultations, such as information and publication requirements and at least one meeting to be held in the local languages. The third main point of the Gold Standard is conformity with the host countries' or the CDM Executive Board's requirements regarding an environmental impact analysis (EIA). If no such requirements are given for the project type, the project has to be checked against the requirements of the Gold Standard. The Standard requires an EIA if the first stakeholder consultation (see above) identifies significant environmental impacts.

Another recently developed set of standards are the CCB standards (CCB 2005). They have four groups of criteria (general, climate related, community related, biodiversity) with 6 indicators for the first and 3 each for the others. Fulfilment of these 15 criteria leads to accordance with the CCB standards. Each group provides two additional voluntary criteria, the fulfilment of which leads to a more stringent "silver"or "gold" label. Regarding sustainability, these criteria remain, however, rather general (mainly assessment of off-site negative impacts) and they refer in part more to ordinary good business practices than to specific sustainability requirements. Indeed, this applies to all general criteria other than the requirement of no significant land tenure disputes. These other general criteria refer to legality and management capacity together with information on the base-line, original situation and goals of the project.

4.2 Problems and potential of CDM labels

Labelling the CDM has several peculiar characteristics compared with common labelling schemes. It addresses not individual end consumers but rather governmental institutions, funds and productive industries emitting GHGs. It is a label for an end product with only immaterial qualities and with demand generated by politics. It is topically very broad, in principle providing comprehensive sustainability labels to the whole range of different activities eligible under the CDM. The trade-offs intrinsic in the sustainability concept are thus transformed into similar trade-offs in the labels. Most existing labels claim general validity without differentiation according to countries and their respective national sustainable development plans. They thus face the classical problem of any sustainability label: the breakdown to project level of the sustainable development ideal that has to be seen in a wider geographical, cultural and temporal context.

If Sutter's four criteria are taken as a starting point, none of the existing labels fulfil all of them. His MAUT-approach does fulfil them, but at present it defines an indicator set for assessment rather than a specific label. However, it could easily be further developed into a basis for CDM standards or labels, for example by setting requirements such as that the aggregate utility has to be larger, or no criterion must score less, than a certain value (cf. Heuberger and Sutter 2003). Due to the stakeholder participation, the MAUT approach is sensitive to case- and potentially also nation-specific issues. It does not, however, automatically assure incorporation of long-term aspects and coordination with national sustainability strategies.

While CDM sustainability labels thus face some problems in reporting the sustainability of CDM projects and working towards its increase in a comprehensive manner, they at least clearly capture a partial picture and can also be applied as risk hedging tools. This is so if a label succeeds in meeting consumers' information demands and can thereby establish credibility regarding the claims it makes. Sustainability being often too vague a goal to strive for, a label can for example claim to stand for high quality reductions (in particular verified, reliably achieved reductions) or for other specific aspects of sustainability. Comparison with existing labels from other areas suggests that such restrictions make it easier to establish credibility.

A particular potential for CDM labels exists in cases where sustainability risks and business risks coincide. This can be the case for projects involving smallholders, for example, where business risk hedging aspects such as guaranteed output prices can also play a crucial role in hedging against risks threatening local livelihoods. It is also the case for the above-mentioned high quality reductions, as they would lower the risk that the project actually generates fewer reductions than projected and thus leads to fewer revenues from CERs.

In this context, marketing issues related to CDM labels may become more important. This aspect is crucial to classical labelling initiatives and is the object of increasing awareness in this field (cf. Belz 2005). It has been largely neglected for CDM labels so far. This might be partly because of the different type of actors involved and the absence of individuals as consumers. The incorporation of marketing aspects may also take some time because the CDM labels were initiated and developed to reflect sustainability rather than business risk concerns.

4.3 Options for improving CDM labels

The discussion above suggests that an improved labelling approach for sustainable CDM should combine four key aspects: 1) a MAUT-based strategy, 2) clear labelling rules tied to the indicator values such as provided by the Gold Standard 3) inclusion of additional tools to assure coordination with national sustainable development goals (assuming that those national strategies make sense) and – as a particular aspect thereof – 4) sensitivity to long-term aspects.

Such an approach could be implemented employing a two-tiered instrument. Concrete specification for a MAUT-based procedure could be agreed on national levels, containing some further specifications for regions or sectors. An international institution could then accredit the national strategies and provide labels that thus would be credibly applicable on a global level. This international institution might provide some guidelines to facilitate comparability of national strategies. It could also provide some minimal requirements on certain indicators that are globally felt to be a necessity for any sustainability label, e.g. on how to include long-term aspects. Such a comprehensive approach, however, bears the danger of further and disproportionately increasing the administrative costs of CDM projects.

Another strategy would be to abstain from any more comprehensive approach and to concentrate on some issues considered as being particularly crucial for the situation at hand. The Gold Standard is such an example – but somewhat more flexibility regarding national situations and individual cases could be advantageous. Instead of restricting the indicators to be considered, a topically more specific approach could be taken, aiming to establish labels for more narrow national or regional contexts or for more narrow sectors only – such as for hydropower schemes or waste management.

Such restricted labels would work as information provision tools for consumers and they could also be risk-hedging tools for producers. They are however less adequate to assure sustainability of the CDM in a comprehensive manner. In particular, if any national or international policy is to be based on sustainability performance of CDM projects according to a set of indicators, a comprehensive approach with stakeholder participation is likely to be indispensable.

The peculiar properties of the CER as a commodity are important regarding marketing of labelled CERs, in particular because of the different characteristics of the buyers of CERs and the consumers of more conventional labelled goods. The influence on the wider economic context that is a topic in sustainability marketing in general is particularly important for the CDM, where rules for the mechanism itself are still under development. This changes marketing strategies, especially towards governments that could make labels a legal issue, thus settling the decision to buy once and for all. This is very different from individuals as consumers who are usually free to decide in favour of or against sustainability labelled goods with each consumption decision.

It has to be kept in mind that the comprehensive labelling approach as suggested here is likely to further add to the bureaucracy involved in the CDM that is already plagued by high transaction costs. This is definitely an argument for employing more restricted labels. It could also be an argument for embarking on a totally different strategy to further regulate the CDM, such as a profit tax (Denne 2000; Muller 2007). A tax, however, usually faces more opposition from lobbying groups than label-based and other command-and-control approaches (Dijkstra 1999). Nevertheless, a tax could also be combined with some simple labelling scheme defining several tax levels according to a basic notion of sustainability performance. This would be in line with the general philosophy of the CDM and the KP as frameworks for market-based approaches to reducing GHG emissions. In contrast, an overly bureaucratised CDM is clearly not in this spirit.

5 Conclusions

Labelling the CDM takes labelling in general to a new level – both regarding the commodity labelled and the actors involved. In particular the types of consumers addressed are different to those of other labels, such as for organic food or green electricity. The labelling strategies identified for the CDM can thus provide fruitful input to labels for other goods with similar types of target consumers, such as natural resources not sold to individual consumers. In particular, stakeholder participation in framing and applying the labels and a two-tiered approach setting general guidelines on an international level and more specific ones on a national level are new to labelling. Marketing aspects are likely to play a crucial role in this context – they are to be understood in a broad sense, including their effects upon the economic environment.

References

Althammer W, Dröge S (2006) Ecological Labelling in North-South Trade. DIW Discussion Paper 604, German Institute for Economic Research DIW, Berlin

Belz FM (2005) Sustainability Marketing: Blueprint of a Research Agenda. Marketing and Management in the Food Industry Discussion Paper No 1, TU München, http://www.food.wi.tum.de/upload/Diskussionsbeitrag/DP1SMBelz.pdf (24.2.2006)

Bird L, Wüstenhagen R, Aabakken J (2002) Green Power Marketing Abroad: Recent Experience and Trends. Technical Report TP-620-32155, National Renewable Energy Laboratory (NREL)

CCB (2005) The Climate, Community and Biodiversity Project Design Standards (CCB Standards). http://www.climate-standards.org/index.html (24.2.2006)

CDCF (2005) Community Development Carbon Fund – Community Benefits. http://carbonfinance.org/cdcf/router.cfm?Page=Projects (24.2.2006)

CDM China (2005) Measures for operation and management of clean development mechanism projects in China. Office of National Coordination Committee on Climate Change, http://cdm.ccchina.gov.cn/english/NewsInfo.asp?NewsId=100 (24.2.2006)

Cosbey A, Parry J, Browne J, Babu Y, Bhandari P, Drexhage J, Murphy D (2005) Realising the Development Dividend: Making the CDM Work for Developing Countries. Report of the International Institute for Sustainable Development IISD, Canada, http:// www.iisd.org/publications/pub.aspx?id=694 (24.2.2006) de Boer J (2003) Sustainability Labelling Schemes: The Logic of Their Claims and Their Functions for Stakeholders. Business Strategy and the Environment 12: 254–264

Denne T (2000) Sharing the Benefits: Mechanisms to Ensure the Capture of Clean Development Mechanism Project Surpluses. CDM Dialogue Paper, Center for Clean Air Policy, http://www.ccap.org/pdf/Benefits.pdf (24.2.2006)

Dijkstra B (1999) The Political Economy of environmental Policy. Edward Elgar, Cheltenham

ENB (2006) Summary of the Twelfth Conference of the Parties to the UNFCCC and Second Meeting of the Parties to the Kyoto Protocol: 6–17 November 2006, Earth Negotiations Bulletin, http://www.iisd.ca/vol12/enb12318e.html (25.5.2007)

Fenhann J (2006) Information on the CDM compiled at the UNEP Risoe Centre, 17.1.2006

Fenhann J (2007) UNEP Risoe CDM/JI Pipeline Analysis and Database. http://cdmpipeline.org/ (28.5.2007)

Fischer C, Newell R (2004) Environmental and Technology Policies for Climate Change and Renewable Energy. RFF Discussion Paper 04-05 (REV), Resources for the Future, Washington DC

Goldstandard (2005) The Gold Standard – Premium Quality Carbon Credits. http://www.cdmgoldstandard.org/ (24.2.2006)

Goldstandard (2007) Press Release March 14, 2007, http://www.cdmgoldstandard.org/uploads/file/Press%20release%20First%20CERs%20from%20GS%20Validated%20Project.pdf (24.5.2007)

Heuberger R, Sutter C (2003) Host Country Approval for CDM Projects in Uruguay: Application of a Sustainability Assessment Tool. http://www.cambioclimatico.gub.uy/~fpacheco/cambio_climatico/html/modules.php?op=modload&name=DownloadsPlus&file=index&req=viewdownload&cid=14 (24.2.2006)

Karl H, Orwat C (1999) Environmental Labelling in Europe: European And National Tasks. European Environment 9: 212–220

Kotchen MJ (2005) Impure public goods and the comparative statics of environmentally friendly consumption, Journal of Environmental Economics and Management 49 (2): 281–300

Kotchen MJ (2006) Green Markets and Private Provision of Public Goods. Journal of Political Economy 114 (4): 816–834

Mason S, Muller A (2007) Transforming Environmental and Natural Resource Use Conflicts. In: Cogoy M and Steininger K (eds) The Economics of Global Environmental Change. Edward Elgar, Northampton

Muller A (2007) How to Make the CDM Sustainable – The Potential of Rent Extraction. Energy Policy 35 (6): 3203–3212. A condensed version is published in Carbon Finance (3): 25

Müller-Pelzer F (2004) The Clean Development Mechanism. HWWA-Report 244, Hamburg Institute of International Economics, Hamburg, Germany

Nolles K (2004) Lessons on the Design and Implementation of Renewable Energy, Greenpower and Greenhouse Emissions Abatement Markets from the Financial Markets and Experimental Economics. IAEE Annual Conference 2004, Washington DC, http://www.ceem.unsw.edu.au/ceem_docs/IAEE%20Conference%20Paper%20-%20Nolles%202004.pdf (24.2.2006)

Ott H, Sachs W (2002) The Ethics of International Emissions Trading. In: Pinguelli-Rosa L, Monasinghe M (eds) Ethics, Equity and International Negotiations on Climate Change. Edgar Elward, Northampton. – Similar to Ott H, Sachs W (2000) Ethical Aspects of Emission Trading, Wuppertal Papers Nr 110, http://www.wupperinst.org/Publikationen/WP/WP110.pdf (24.2.2006)

Point Carbon (2005) News from the 31.10.2005, http://www.pointcarbon.com/Home/News/All+news/CDM+%26+JI/Host+countries/article11906-875.html (2.24.2006)
Point Carbon (2007) HFC-23 reduction from the CDM. Carbon Policy Update, 22 January 2007, see also the Report from the 28th Meeting of the Executive Board of the CDM, 15 December 2006, http://cdm.unfccc.int/EB/028/eb28rep.pdf (24.9.2007)
Repetto R (2001) The Clean Development Mechanism: Institutional Breakthrough or Institutional Nightmare? Policy Sciences 34: 303–327
Rose A, Bulte E, Folmer H (1999) Long-Run Implications for Developing Countries of Joint Implementation of Greenhouse Gas Mitigation. Environmental and Resource Economics 14: 19–31
Sammer K, Wüstenhagen R (2006) The Influence of Eco-Labelling on Consumer Behaviour – Results of a Discrete Choice Analysis. Business Strategy and the Environment 15: 185–199
Schwank O (2004) Concerns About CDM Projects Based on Decomposition of HFC-23 Emissions from 22 HCFC Production Sites. INFRAS, Switzerland, http://www.cdm.unfccc.int/pulic_inputs/inputam0001/Comment_AM0001_Schwank_081004.pdf (24.5.2007)
Sutter C (2003) Sustainability Check-Up for CDM Projects: Multi-Criteria Assessment of Energy Related Projects Under the Clean Development Mechanism of the Kyoto Protocol. Wissenschaftlicher Verlag Berlin, Berlin
Sutter C (2005) Personal Communication (31.10.2005)
Sutter C, Parreno J (2005) Does the current Clean Development Mechanism deliver its sustainable development claim? International conference: Climate or development? October 28–29, 2005, Hamburg Institute of International Economics (HWWA), Hamburg, Germany
Swanson D, Pinter L, Bregha F, Volkery A, Jacob K (2004) National Strategies for Sustainable Development. The International Institute for Sustainable Development (IISD)
UNEP (2003) Environmental Due Diligence (EDD) of Renewable Energy Projects – Guidelines. http://www.sefi.unep.org/index.php?id=40 (24.2.2006)
UNFCCC (1997) Kyoto Protocol to the UNFCCC. http://unfccc.int/essential_background/kyoto_protocol/background/items/1351.php (29.4. 2005)
UNFCCC (2004) Tool for the demonstration and assessment of additionality. http://www.cdm.unfccc.int/methodologies/PAmethodologies/AdditionalityTools/Aditionality_tool. pdf (24.2.2006)
UNFCCC (2005a) CDM-Executive Board Meeting Report of the 18th Meeting, Agenda sub-item 3(e), 57, http://cdm.unfccc.int/ EB/Meetings/018/eb18rep.pdf (24.2.2006)
UNFCCC (2005b) Steve Thorne at the UNFCCC SB 22 Meeting, May 2005, Bonn, http://www.iisd.ca/vol12/enb12266e.html (24.2.2006)
UNFCCC (2005c) Guidelines for Completing CDM-PDD, CDM-NMB and CDM-NMM, version 4, p 11, http://cdm.unfccc.int/Reference/Documents/Guidel_Pdd/English/Guidelines_CDMPDD_NMB_NMM.pdf (24.2.2006)
UNFCCC (2006) Report of the COP12/MOP2, Nairobi from 6 to 17 November 2006, Decision 1/CMP.2: Further guidance relating to the clean development mechanism, http://unfccc.int/resource/docs/2006/cmp2/eng/10a01.pdf#page=3 (25.5.2007)
Wittneben B, Sterk W, Ott H, Bround B (2006) In from the Cold: The Climate Conference in Montreal Breathes New Life into the Kyoto Protocol. Wuppertal Institute for Climate, environment and Energy, http://www.wupperinst.org/download/COP11MOP1-report.pdf (24.2.2006)
Wüstenhagen R (1998) Pricing Strategies on the Way to Ecological Mass Markets. In: Partnership and Leadership: Building Alliances for a Sustainable Future. Proceedings of the 7th International Conference of the Greening of Industry Network, Rome, November 15–18, 1998

Economic and social risks associated with implementing CDM projects among SME – a case study of foundry industry in India

Prosanto Pal[1], Girish Sethi[1]

[1] The Energy & Resources Institute (TERI)
Habitat Place, Lodhi Road, New Delhi 110003, India
prosanto@teri.res.in
girishs@teri.res.in

Abstract

The sustainable development of the potential host countries is an explicit objective of the Clean Development Mechanism (CDM), one of the financial mechanisms under the Kyoto Protocol. The three dimensions of sustainability which need to be met by CDM projects are environmental, social and economic sustainability. Energy efficiency projects in the small and medium enterprise (SME) sector directly contribute to sustainable development of the developing country's economy. There are a large number of products manufactured by SMEs using outmoded energy-inefficient processes. Some examples of such products are clay bricks, glass items, iron castings (foundries), steel re-rolling, textiles and a variety of food items. The total energy consumption of each of these sectors, taken as a whole, is very significant, simply because of the large number of SMEs. Although individual CDM projects are not possible for SMEs, it is technically possible to bundle multiple small projects under large regional "umbrella" projects. However there are several economic and social risks associated with the implementation of small-scale CDM projects. These are presented in this paper by means of a case study on the bundling of energy efficiency projects in a small-scale foundry cluster in Rajkot, Western India.

Keywords: Clean Development Mechanism (CDM), sustainable development, small and medium enterprises (SME), foundry, divided blast cupola (DBC), energy-efficiency improvement, economic and social risks

1 The SME sector in India

In India, small and medium enterprises (SMEs) are an important part of the economy in terms of their contribution to gross domestic production (GDP), industrial production, employment generation and exports. SMEs, or SSIs (small-scale industries), as they are commonly called in India, generate almost 40% of India's manufacturing sector output and 35% of total exports. As per the most recent government Census of SSIs there were about 11 million SSI units in the year 2002–03, 87% of which are unregistered (Ministry of Small-Scale Industries 2004). Almost 95% of the total industrial units are in the SME sector. The contribution of the SME sector to the industrial production was Rs 3,120 billion (USD 69 billion) and they collectively exported Rs 860 billion (USD 19 billion) of manufactured goods (Ministry of Finance 2005). The sector is also one of the major employers and provided employment to about 26 million people, which is second only to the agricultural sector.

Traditionally, India has had strong policies to promote the SME sector. For example, a large number of items are exclusively reserved for manufacturing in SMEs. At one time, there were more than 850 items in the reserved list, which has been gradually brought down to about 600 at present. This is because the government has recently realized that reservation of items for manufacturing in SMEs in some of the sectors impacted their growth specifically with regard to modernization and technology upgrading. The economic reforms undertaken since 1991 have been particularly harsh for SMEs. Economic sickness is widespread; official figures estimate that 10% of total SMEs are ailing, but unofficial figures put the proportion at over 40%. Closure of many SMEs has contributed to a decline in total employment growth within the country. The sector troubles are mainly blamed on technological stagnation, inadequate access to finance, poor marketing skills, infrastructure gaps and regulatory stringency. These issues pose serious threats to the long-term sustainability of the sector and to the livelihood of millions of workers.

Many of the SME activities are geographically clustered. Within a cluster, there is a great deal of similarity in the level of technology, the operating practices and even the trade practices among the individual units, which means that the potential to develop standard solutions is large.

2 Technology intervention in energy-intensive SME sectors

There are many energy-intensive sub-sectors in SMEs. Energy accounts for a major share of the operating cost in these sub-sectors, which means that to remain competitive, it is absolutely essential for them to improve their energy performance. Since 1994, TERI, with the support of SDC (Swiss Agency for Development and Cooperation), has initiated a programme aimed at achieving energy savings, with consequent reduction of carbon-dioxide emissions, in selected energy-intensive SSI sectors. The programme was developed as part of the initiative of the Swiss government to support developing countries in implementing UN conventions concerned with the global environment.

An analysis of 41 SME sectors was performed by TERI in 1994–95 in order to identify the most energy-intensive sub-sectors among them (TERI 1995). Statistical data were taken from government records (Ministry of Small-Scale Industries 1998). The glass industry cluster at Firozabad was excluded from the analysis since it had already been pre-selected for intervention on the basis of its importance among SME clusters in India and the high energy consumption during glass melting operations. Based on the analysis, it was found that the casting and forging sector accounted for the largest share of the total energy consumption (about 44%) among the chosen sub-sectors. Foundry and forging, together with the production of edible oils & fats and glass tiles, accounted for more than 70% of the energy usage in the SME sector.

Based on the analysis and subsequent deliberations with experts during a screening workshop held in 1995, TERI and SDC selected the following four sub-sectors for improving energy efficiency by demonstration and replication of energy-efficient technology:

- Foundries (energy-efficient cupola)
- Glass units (gas-fired pot and muffle furnaces)
- Brick kilns (Vertical Shaft Brick Kilns)
- Biomass gasifier for thermal applications.

For each of the above sub-sectors, a demonstration plant was built in order to promote the cleaner technology option among other units. Since the Clean Development Mechanism (CDM) is an instrument to promote cleaner technologies in developing countries under the Kyoto Protocol, the project team has recently been exploring the possibility of leveraging CDM for large-scale promotion of the demonstrated technology options among SME clusters in India.

A case study of a potential CDM project aimed at adoption of the demonstrated energy-efficient furnace technology, with consequent reduction of carbon-dioxide emissions, among 190 foundry units located in a geographical cluster in India is presented in subsequent sections of this paper. Some of the economic and social barriers to implementation of the "bundled" small-scale CDM project are also discussed, thus making explicit the current limitations of the CDM in promoting sustainable development of SMEs within developing economies.

3 Energy-efficient cupola demonstration

There are about 5,000 foundry units in India, most of which are small-scale. The majority of the foundry units are spread over 20 geographically disparate clusters in various parts of the country. The products (castings) manufactured by foundry units include manhole covers, motor bodies, pump parts, automotive parts, electric motor bodies, water pipes and a range of industrial and machinery items. Typical operations in a foundry unit are outlined in Fig. 1. In India, most of the operations such as moulding, molten metal transfer, pouring, knockout and finishing are still performed manually and hence are highly labour-intensive.

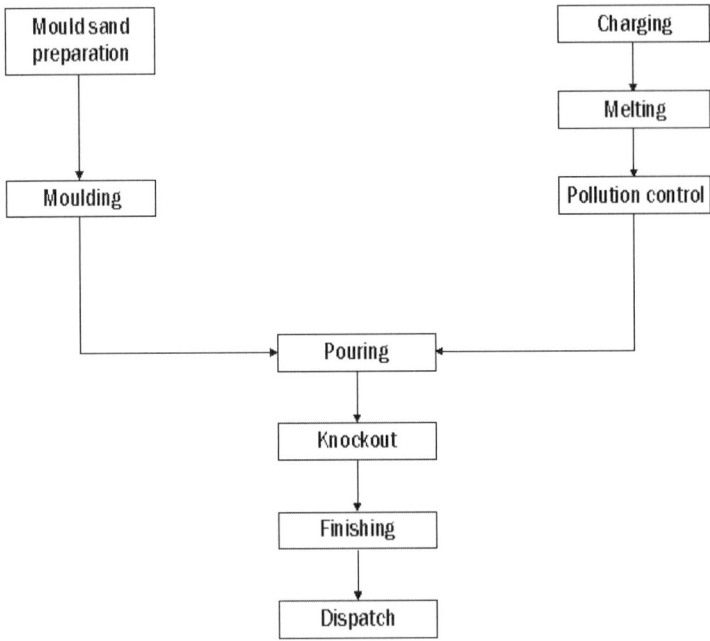

Fig. 1. Typical foundry operations

Most of the small-scale foundry units are family-owned and -managed. The general level of awareness among them about energy conservation and new technologies is low. Although some of the entrepreneurs are interested in energy efficiency and technological improvements, they are constrained by lack of technical know-how and finances. Melting in cupola furnaces is the most energy-intensive operation in a foundry unit.

The majority of the cupolas currently being operated by the small-scale foundry units are highly energy-inefficient and hence there is a large potential, in the range of 25 to 65 percent, to save energy through adoption of improved cupola designs. After discussions with foundry industry associations, it was decided that the best way of promoting a new technology or design is through the setting-up of a demonstration plant in one or two units within a cluster. Successful demonstration would help in motivating other units in the cluster to undertake similar projects within their units.

Consultations with foundry experts from the British Cast Iron Research Association (BCIRA) revealed that the DBC (divided blast cupola) is a well-proven technology for improving the energy performance of a cupola furnace at a modest investment. Hence within the framework of the project, TERI, in association with a British expert from BCIRA, designed and installed a full-scale demonstration DBC melting furnace at a foundry unit in Howrah cluster, in the state of West Bengal in Eastern India. An attractive energy saving of 35% in coke consumption was achieved in the demonstrated plant (Dugar 1999). The economics of operation of a DBC was found to be very attractive and the total investment in a new DBC was recovered within 1 to 2 years from the savings in coke alone. Further benefits of DBC, such as reduction in oxidation losses and more consistent metal quality, make the investment even more attractive for a foundry unit.

4 Barriers to replication of DBC among Indian foundries

Although the return on investment in a new DBC was demonstrated to be very attractive, replications of the technology did not happen immediately as was envisaged. Some of the reasons for the slow rate of adoption of the demonstrated technology by other foundry units could be attributed to one or more of the following factors:

- Technology barrier: the DBC concept, as adopted for the demonstration unit, is not perceived by the entrepreneurs as a new or novel concept/technology. Several foundry units had in the past already constructed a DBC, either on their own or with the aid of local consultants, although most of these furnaces are not optimally designed. Hence, convincing them of the benefits of the improved version of the DBC is a major barrier.
- Investment barrier: most foundry units have only limited capacity to evaluate a new technology and often choose those requiring the least up-front investment. The design specifications of the demonstrated DBC require a higher up-front investment due to stringent material specifications and standard auxiliaries such as blowers and the bucket-charging system. Since SMEs typically want to minimise advance costs, a higher initial cost of the optimally designed DBC is a barrier to adoption of the technology.

Since CDM offers a route to promote energy-efficient DBC among foundry units in India on a large scale, TERI helped Rajkot Engineering Association (REA) to develop a CDM project design document (PDD) for local foundry units as part of the National Strategy Study (NSS) project for CDM implementation in India (TERI 2005).

5 CDM project in small-scale foundries

A CDM project aimed at achieving energy savings with consequent reduction of carbon-dioxide emissions by adoption of cleaner and energy-efficient (EE) iron-melting furnaces was developed for the small-scale foundry units located at Rajkot in the state of Gujarat. The project location is shown on the map of India given in Fig. 2. The project proposes to change the conventionally designed melting furnaces (cupolas) of foundry units at Rajkot to EE designs with divided blast cupolas (DBC).

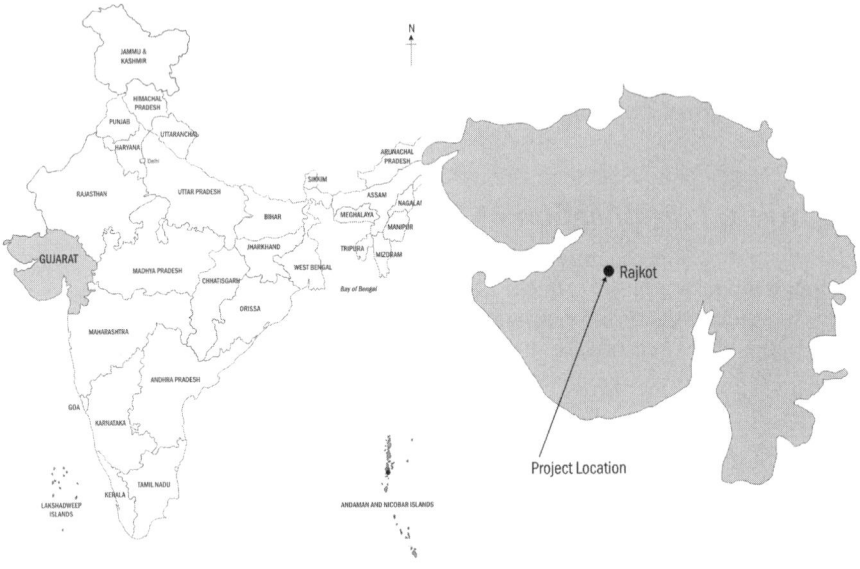

Fig. 2. Location of Gujarat in India

The baseline data used is taken from a survey of the foundry units located at Rajkot conducted by REA and TERI in the year 2002–03. The survey identified 190 foundry units which consume 17,000 tonnes of coke and emit about 46,000 tonnes of carbon dioxide (CO_2) per annum. Installation of energy-efficient DBC furnaces in these units will result in coke savings of at least 25% and will reduce CO_2 emissions by about 8,300 tonnes per year.

The sources of GHG emissions (CO_2 in this case) in a cupola furnace are as follows:

- Consumption of fossil fuel, i.e. coke – the primary source of CO_2 is from combustion of coke in the furnace.
- Consumption of limestone (calcium carbonate) which is added as a fluxing agent in iron melting. The calcium carbonate decomposes to calcium oxide (CaO) and CO_2 in the furnace.
- Consumption of electricity in auxiliary equipment, viz. cupola blowers and bucket-charging system .

Detailed formulae for estimating the baseline emissions from each of the sources are provided in the PDD which has been produced as part of the NSS project report [6]. The CO_2 emission reduction is estimated by taking the difference of the emissions in the baseline scenario and after implementation of DBC.

6 Monitoring methodology

A mix of the following approaches is possible for monitoring every variable:

a. Measurement of data of all furnaces in the cluster
b. Statistical approach for a certain cohort based on random sampling

Every variable can be monitored by either of the two approaches (a) or (b). Activities such as melt output and coke input must be monitored for 100% of the foundry units by approach (a). Variables that are more or less the same for all elements of a certain cohort may be determined with a statistical approach (b) for a certain cohort, based on suitable random sampling.

Some typical variables to be monitored are listed in Table 1 together with the monitoring approach to be applied in each case.

Table 1. Monitoring of typical variables

	Data type	Data variable	Data unit	Proportion of data to be monitored	Methodology
01	Melt output	Quantity of metal melted	tonnes	100%	(a)
02	Coke input	Quantity of coke consumed	tonnes	100%	(a)
03	Coke quality	Fixed carbon in coke	%	Random sampling	(b)
04	Operating hours	Time of cupola operation	hours	Random sampling	(b)

A database of all the records provided by individual foundry units is proposed to be maintained by the local industry association (REA). Since REA is also the nodal industry organisation at Rajkot, it has enough influence on the entities to ensure that they provide the required data. Quality assurance and quality control of the data will be the responsibility of REA.

7 Contribution to sustainable development

The project is certain to deliver significant local and national development benefits.

At the local level, the project will prevent the emission of approximately 83,000 tonnes of CO_2 over 10 years. The total emissions would amount to 465,000 tonnes of CO_2 over the next 10 years in the absence of the CDM project. In addition, the project will speed the market development of a climate-friendly technology, an essential element in accomplishing the ultimate objective of the Climate Convention. The technology will help to improve the business competitiveness of the units, helping them to survive in the face of increased market competition. In the longer term, survival of the units will lead to sustenance of a large number of jobs in several upstream and downstream sectors. The cleaner technology will lead to improvement in the local work environment that will have long-term health benefits.

The foundry industry is an important activity in the SME sector in India. There are over 5,000 foundry units employing nearly 0.5 million people. The total coke consumption by the foundry units is estimated to be about 600,000 tonnes per year. At a conservative estimate, it is possible to save about 25% of the coke by adoption of the energy-efficient technology, which translates into a reduction of 150,000 tonnes of coke or 410,000 tonnes of CO_2.

The foundry units are dispersed in about 20 well-known geographical clusters. Hence there is a good potential for replicating the project in other foundry clusters in India.

In summary, the CDM project will directly contribute to the sustainable development of the industry and local communities

8 Economic and social risks of project implementation

Although the project will greatly promote sustainable development, its implementation under CDM is constrained due to several economic and social barriers.

One of the major economic barriers with regard to implementation of small bundled projects is the high transaction cost related to monitoring, verification and certification (recurring expense cost) as compared to PDD development and validation (fixed expense). This emphasizes the need for a large volume of CERs in order to minimize the impact of transaction costs and the need for competition among DOEs (designated operational entities) in order to keep prices for such services as low as possible. For small projects ranging from 5,000 to 10,000 CERs, the up-front costs range from 11% to 18% and the monitoring and verification costs from 33% to 62% of the price on a per-CER basis. The transaction costs are in the negligible range only for large projects generating above 50,000 CERs (TERI 2005).

Another way to reduce transaction costs is to develop simplified monitoring protocols based on benchmarking or statistical approaches such as sampling. Further systems/processes impacting the emissions by less than 10% may be ignored. In the foundry case study, the major emission contribution is from combustion of coke in the cupola furnace. Hence the coke consumption must be monitored accurately before and after implementation of the CDM project. Emissions from other systems/processes, e.g. calcination of limestone and electrical consumption in auxiliaries such as blowers and charging systems, are relatively minor. Ignoring or making a fixed allowance for such minor contributors of GHG in the monitoring methodology would reduce the transaction costs significantly.

The CDM is an entirely new concept for small-scale entrepreneurs and hence considerable time and effort needs to be invested in negotiations with multiple stakeholders before the terms can be finalised. Since CDM is a commercial transaction with no provision for financing investments in project development, very few SME organisations are likely to develop small-scale bundling projects for CDM on their own. SME organisations have only limited finances for meeting up-front costs associated with project development, implementation, validation and registration; furthermore, they are not knowledgeable enough about the possibilities for bringing in equity to avail financing.

Another economic barrier to CDM implementation among SMEs is the existence of contractual clauses such as penalties for non-delivery or shortfall in CERs (certified emissions reduction) for whatever reason in the ERPA (emission reduction purchase agreement). This is particularly difficult to predict in small bundled projects, where there are multiple owners and a phased implementation.

Social or systemic barriers associated with changing existing practices, training of the workforce in new operating practices and proper documentation of data/records need to be addressed.

The furnaces use a solid fossil fuel – coke. The coke fed into the furnace is not accurately weighed and recorded: there are no on-line weighing systems installed with the furnaces. Hence the existing practice of not measuring the coke accurately will have to be changed for a CDM project.

Another barrier in the monitoring and validation process is the poor record-keeping practices in most SMEs. The reasons for this are the general reluctance of units to employ additional manpower to maintain records as well as to the widespread avoidance of paying government taxes. Since one of the governmental methods for verification of coke consumption is the checking of invoices and purchase bills, unrecorded transactions form a major barrier to implementing CDM projects in SMEs.

To gain full benefit from the DBC, it is necessary that the furnace is operated properly. Some training/orientation of the operators is required in order to ensure that they adopt the best operating procedures, thus obtaining the maximum energy savings from the DBC. However, CDM projects have limitations with respect to costs associated with training/orientation types of activities.

9 Conclusions

Clean technology is a dominant force in combating the detrimental effects of the GHG emissions. The potential of clean technology transfer to combine economic growth with the protection of the environment in developing countries was already recognized in Agenda 21, arising out of the Rio Summit in 1992.

CDM was envisaged as a tool within the Kyoto Protocol to promote technological cooperation between the developing and the developed countries and help developing countries to achieve sustainable development. The SMEs not only play an important part in developing economies: they are also significant contributors of GHG emissions. Hence, promotion of cleaner technologies among SMEs will serve the dual cause of economic growth and protection of the environment. However, the CDM and thus the Kyoto Protocol falls short of its promise of catalysing proliferation of cleaner technologies among SMEs in developing countries, as was highlighted in the case study presented in this paper.

For CDM projects to become reality in the SME sector, there is a need to revisit the CDM implementation cycle and grossly simplify the validation, registration, monitoring and verification steps. A simplified CDM project cycle for SMEs holds the promise of reducing GHG emissions by millions of tonnes in diverse manufacturing operations within developing countries, thereby fulfilling its intended objective of achieving sustainable development of their economies.

References

Dugar S et al. (1999) Energy efficiency improvement and pollution reduction in small scale foundry unit in India – results of a full-scale demonstration plant. Proceedings of the 6th Asian Foundry Congress, Calcutta, the Institute of Indian Foundrymen, Calcutta, 347 pp

Ministry of Finance (2005) Economic Survey 2003–04. Published by Ministry of Finance, Government of India

Ministry of Small-Scale Industries (1998) Second All India Census of Small-Scale Industries. Published by Development Commissioner, Ministry of Small-Scale Industries, Government of India

Ministry of Small-Scale Industries (2004) Third All India Census of Small-Scale Industries 2001–2002. Published by Development Commissioner, Ministry of Small-Scale Industries, Government of India

TERI (1995) Energy Sector Study Phase 1. Submitted to Swiss Agency for Development and Cooperation (SDC), New Delhi, 64 pp

TERI (2005) CDM implementation in India. The National Strategy Study. Summary, published by TERI Press, New Delhi, 30 pp

Part III

Adaptation: corporate responses to climate change

Developments in corporate responses to climate change within the past decade

Ans Kolk

University of Amsterdam
Amsterdam Business School
Chair of sustainable management
Roetersstraat 11, 1018 WB Amsterdam, Netherlands
akolk@uva.nl

Abstract

This chapter provides an overview of the research on business and climate change in the past decade as carried out by the author, with a particular focus on multinational companies. It covers a synopsis of policy developments, companies' political and market responses to climate change, as indicates areas for future research

Keywords: Climate change, multinationals, political responses, market responses, climate policy

Acknowledgement: Since March 2006, Professor Kolk's climate research has been embedded in the research programme Vulnerability, Adaptation and Mitigation, funded by the Netherlands Organisation for Scientific Research (NWO).

1 Introduction

On 11 June 2005, the Financial Times published an interesting cartoon that aptly represents some of the issues that have played a role in the debate on climate change all along. The comic, published well before Hurricane Katrina shook the United States, shows President George W. Bush standing on a lectern amidst a rising tide. The notes in front of him say 'climate change research', and show many crossed-out words, including 'possible' and 'not'. In the background we see melting ice caps and smoking oil refineries. The cartoon clearly illustrates the different perceptions regarding the problem of climate change and those who emphasise the dire situation due to global warming; this includes more floods and melting ice caps, as well as the impact of industrial activity within this process. We also see a representative of those who, in the face of increasing evidence, are not so convinced; they are pleading for more research. Opinions regarding the best policy responses to climate change are likewise divergent, from those who support the Kyoto Protocol, or even believe that it does not go far enough, to those who see this as undesirable and stress the negative macro- and/or micro-level economic consequences.

At roughly the same time, in the second half of 2005, British Petroleum started an advertising campaign in the Financial Times. One of the adverts, entitled "It's time to turn up the heat on climate change", read "In 1997 we became the first major energy company to publicly acknowledge the need to take steps against climate change. Since 2001, the reduction in emissions from our energy efficiency projects has now reached over 4 million tons. Over the next 4 years, we plan to implement new projects to reduce emissions by another 4 million tons." This was part of the Beyond Petroleum campaign, initially launched by the company, together with a new sunflower logo, in July 2000. Interestingly enough, at the time this new BP logo and the 'Beyond Petroleum' campaign were ridiculed within the oil industry and by NGOs. It inspired the NGO Corporate Watch to think about more appropriate phrases for the company's re-branding: '*British Petroleum: Beyond Pompous, Beyond Protest, Beyond Pretension, Beyond Preposterous, Beyond Platitudes, Beyond Posturing, Beyond Presumptuous, Beyond Propaganda, Beyond Belief...*' (Kolk and Levy 2001). Internally, inside BP, the slogan led to confusion and dissatisfaction because it threatened to hamper the company's core activities and daily operations of the various business units. At the 2001 annual meeting, management retracted the original message by emphasising that it was not meant to show the company's intention to retreat from oil. As its CEO John Browne pointed out 'Beyond Petroleum just means that we are giving up the old mindset, the old thinking that oil companies had to be dirty, secretive and arrogant'. But at this meeting he also departed from previous positive expectations about the size of future markets for renewables, and said that renewables could not even begin to substitute for oil on present conditions (Kolk and Levy 2001). So you can imagine that this author was rather surprised to see this campaign logo and slogan coming back at full speed a few years later.

Together, these two items from the Financial Times sketch the full range of interesting aspects related to climate change. It is a very fascinating topic, and one in which the dilemmas of environmental policy and corporate responses come to the fore most prominently. It is also an area where you can clearly see the importance of interaction between a variety of stakeholders, and how the development of an issue, from emergence to maturity, is accompanied by different corporate responses. So for those of us interested in what business does, and which economic factors play a role in the environment, this is a captivating ideal topic to study. This author has been intrigued by the whole complex of actors, responses and interactions since the middle of the 1990s, when policy making appeared to become more serious, and companies started to pay increasingly more attention to what was going on. In this contribution, an overview will be presented comprising the conclusions from almost ten years of research, focusing on multinationals (MNCs). In this manner, some insights into developments over the years are introduced, and some promising areas for further research are mentioned. Part of the earlier research referred to here was carried out together with David Levy, and in more recent years in cooperation with Jonatan Pinkse.

Climate change is one of the environmental issues that has increasingly attracted business attention in the course of the 1990s. Multinationals have developed different strategies over the years, initially more political, non-market in nature, but currently also market-oriented. Since 1995, the political positions of multinationals have gradually changed from opposition to climate measures to a more proactive approach or a 'wait-and-see' attitude; many have started to take steps to be prepared to deal with regulation, or to go beyond that, considering risks and opportunities. A range of aspects has played a role in companies' responses to climate change, not only at the country and sector levels, but also firm-specific and issue-specific.

2 Policy developments

Obviously, policy-making processes and outcomes, both nationally and internationally, have been very important, and have attracted much attention over the years. One of the specific issues that was discussed among colleagues and students in the 1990s was what shaped countries' positions in climate negotiations (a range of economic, geographical and political factors, see Kolk 2000 for an overview), and also how these were subject to change. An overview of policy developments since the early 1990s illustrates the scope of what has taken place (Table 1). An important milestone in the process, which set many things in motion, has been the 1997 adoption of the Kyoto Protocol.

Table 1. Overview of policy developments on climate change

Year	Policy/event	Elaboration
1992	Framework Convention on Climate Change	Adopted at the United Nations Conference on Environment and Development (Rio de Janeiro); expression of intent by industrialised countries to stabilise emissions at 1990 levels by the year 2000; no mandatory emission curbs.

1992 & 1995	EU carbon tax proposal	The European Commission proposed in 1992 a carbon tax that would raise prices of fossil and nuclear energy by 50%. The proposal was conditional on the introduction of a similar tax by the US and Japan. In 1995 a carbon tax was proposed without this condition. Both proposals failed because several EU countries refused to accept the tax.
1997	Kyoto Protocol (COP 3)	Agreement on reduction targets for greenhouse gases compared to 1990 levels, to be reached in 2008-2012. Differentiated targets per country/region, e.g. Australia +8%; Canada -6%; Japan -6%; Russia 0%; US -7%; EU -8%. EU overall target translated into specific ones for member countries, e.g. Germany -21%, France 0%, Italy -6.5%, Spain +15%, UK -12.5%.
1998	COP 4 in Buenos Aires	First Conference of Parties after Kyoto. Confirmation of the Kyoto agreement and adoption of a 'Plan of Action' to implement the Protocol.
1999	COP 5 in Bonn	A 'process meeting' which showed different views. Discussion points were: targets for developing countries (China and India refused to accept targets), the EU-US disagreement on restrictions on the use of the Flexible Mechanisms. Agreement to conclude final negotiations on global greenhouse gas emissions by November 2000.
2000	EU renewable energy proposal	Proposal of the European Commission to set 'indicative' national targets for renewable energy production with the aim to double energy consumption from renewables to 12% by 2010.
2000	COP 6 in The Hague	Failure to achieve agreement between the US and EU. Main issues concerned rules for emission trading and the Clean Development Mechanism. The issue on which the negotiations ultimately failed was the use of forests and farmlands as carbon sinks, which was favoured by the US, but contested by the EU.
2001	IPCC 3rd Assessment Report	Third report by the Intergovernmental Panel on Climate Change (IPCC), released in January. It contained expectations that the consequences of climate change would be greater than expressed in earlier assessments.
2001	US rejection of Kyoto Protocol	In March 2001 the Bush administration declared that it would not implement the Kyoto Protocol and intended to withdraw the US signature.
2001	Launch of US alternative ' science-based' climate plan	Some 'softening' of the US stance in June, shown in the proposal of an alternative 'science-based' response to climate change. Main elements were increased research expenditure for energy efficiency improvements and voluntary measures for industry.
2001	Bonn Agreement on Kyoto implementation	Agreement by the EU, Japan, Canada, Australia, Russia, and a number of developing countries on the rules for the reduction of GHG emissions as laid down in the Kyoto Protocol. Concessions of the EU included allowing emission trading, and the limited use of forests and agricultural land as carbon sinks, which enabled Japan to meet its targets.
2001	EU Emission Trading Scheme proposal	Proposal by the European Commission to set up an Emission Trading Scheme to come into effect in 2005 onwards.
2001	COP 7 in Marrakech	2001 Bonn Agreement turned into a legal text. Further concessions won by Russia and Japan on the use of carbon sinks and the ability to sell surplus emission credits.
2002	EU Kyoto ratification	EU agreement to ratify the Kyoto Protocol by the end of May 2002.
2002	Launch of UK Emission Trading Scheme	The UK government opened a national Emission Trading Scheme in April. Under the scheme, companies received a limited amount of emission allowances that served as a 'cap' on their carbon emissions, which they are allowed to trade.
2002	COP 8 in New Delhi	The eighth Conference of Parties put the position and vulnerability of developing countries central. India criticized calls for emission targets for developing countries and stressed the growing tension between the developed and developing world on climate change.
2003	McCain-Lieberman plan	Senators McCain and Lieberman propose a bipartisan plan to introduce industry-wide caps on GHG emissions and to set up an Emission Trading

		Scheme. The bill failed to pass US Congress by 12 votes, which was commonly viewed as a positive sign.
2003	Opposition of US states to federal government climate policy	Twelve US states file a lawsuit against the Environmental Protection Agency for denying responsibility for GHG emissions (reflecting their opposition to the US federal policy). US Northeast states also develop (regional and perhaps later EU-linked) 'cap-and-trade' plans.
2003	Chicago Climate Exchange (CCX)	Start of this voluntary trading scheme (which is legally binding on member organisations to meet reduction targets of 6% by 2010 compared to average 1998-2001 greenhouse gas emissions).
2003	Regional Greenhouse Gas Initiative (RGGI)	Initiative in the US by Northeast and Mid-Atlantic states to discuss a regional cap-and-trade programme that would initially cover CO_2 emissions from power plants but can be extended later.
2004	COP 10 in Buenos Aires	Disagreement about future of Kyoto Protocol after 2012 (to come up with new negotiation rules/targets by 2008); weak compromise found for a 2005 seminar to exchange information.
2005	Start of EU ETS	On 1 January 2005, the EU Emission Trading Scheme started.
2005	Kyoto Protocol entered into force	On 16 February 2005, the Kyoto Protocol entered into force with the official ratification by Russia. In 2004, President Putin had announced that Russia intended to ratify (as a 'quid pro quo' for EU's acceptance of Russian WTO admission).
2005	New South Wales Greenhouse Plan	Australian state plan to reduce greenhouse gas emissions to 2000 levels by 2025, and realise 60% reductions by 2050.
2005	Kyoto Protocol Achievement Plan	Adopted by Japanese government; implies dissimination of technology, emissions reporting and voluntary use of Kyoto Mechanisms.
2006	Asia-Pacific Partnership on Clean Development and Climate	Brings together Australia, China, India, Japan, South Korea and US in what has been labelled as an 'alternative to Kyoto' attempt that focuses on voluntary, non-binding steps relying on clean technology.
2006	California Global Warming Solutions Act	Mandates a cap of California's greenhouse gas emissions at 1990 levels by 2020.
2007	California Climate Exchange (CaCX)	Launched by the Chicago Climate Exchange to develop trading instruments related to the California Global Warming Solutions Act.
2007	Western Regional Climate Action Initiative	Initiative by Western states in the US and two Canadian provinces to realise a regional, economy-wide reduction target of 15% percent below 2005 levels by 2020, using market based systems such as a cap-and-trade programme. Builds on two earlier initiatives: the West Coast Governors' Global Warming Initiative (2003) and the Southwest Climate Change Initiative (2006).
2007	US mayors' climate protection agreement	Signed by 600 mayors in all 50 US states and Puerto Rico. Involves a commitment to cut greenhouse gas emissions by 7% in 2010 compared to 1990 (which is the US Kyoto target). Initiative was started in 2005 by the mayor of Seattle.
2007	Canadian Regulatory Framework for Air Emissions	Successor to earlier plan launched by the previous government in 2005. The 2007 plan aims to realise a 20% reduction of greenhouse gas emissions by 2030 compared to 2006.
2007	Australia Climate Exchange (ACX)	Launched Australia's first emission trading platform.
2007	Australia and New Zealand intended joint emissions trading	Announcement by Australia and New Zealand to join forces in the development of carbon-trading systems that would be compatible. Follows on earlier statement by Australia that it intends to move towards a domestic, nation-wide emissions trading system by 2012.
2007	Sydney APEC declaration on climate change	Adopted by 21 Pacific Rim countries (including Australia, US, Canada, Russia, China, Japan); includes an aspirational goal of a reduction in energy intensity of at least 25% by 2030 compared to 2005, and support for a post-2012 international climate agreement.

Source: Adapted/updated from Kolk and Hoffmann 2007; Kolk and Pinkse 2005b

3 Political responses

At the sector level, many changes have also taken place. Particularly in the period leading to the Kyoto Protocol, controversies between opponents and proponents of climate policy intensified. Before individual companies start to take positions, a main channel for expressing views was sector-wise, by trade and industry associations, or broader national or international coalitions. Sector characteristics have been important to climate issues, especially in the stage in which negotiations take place to determine the severity and specific contents of policies (Kolk 2000).

Objections to drastic or quick measures used to be raised by energy-intensive sectors such as coal, oil, steel, aluminium, chemicals, automobiles, and paper and pulp. Particularly many US MNCs joined lobby organisations, which included the Global Climate Coalition and the Coalition for Vehicle Choice. More offensive voices could be found in those sectors where this position appeared to offer new market opportunities, or where the risks of climate change predominated. These included solar and wind energy, gas, environmental technology, telecommunications, nuclear energy, insurance and banks. Their views were represented by organisations such as the Business Council for Sustainable Energy, the Pew Center on Global Climate Change and E7 (Kolk 2001).

After the adoption of the Kyoto Protocol, the opponents lost momentum, and an increasing number of MNCs left defensive organisations, sometimes even joining offensive associations. Remarkable in particular were MNCs that first broke away from more traditional sector behaviour, such as BP, Shell, General Motors and Toyota. By mid-1999, this author had compiled a list of those Fortune 500 companies that had explicitly expressed their views in favour of climate measures (around 50 companies had done this), usually underlined by the fact that they joined one of the more offensive organisations such as the ones mentioned above. At that time, an interesting change was already taking place.

It was also then that we started to analyse in more detail why and how companies change, resulting in a more detailed study that came up with the following sets of factors (see Table 2). Thus you see here that a range of aspects has played a role in companies' responses to climate change, at the country and sector levels, but there are also firm-specific and issue-specific characteristics.

Table 2. Important explanatory factors for corporate positions on climate change

Factors	Components
Home-country factors	Societal concerns about environment/climate change
	Societal views on corporate responsibilities
	Regulatory culture (litigational or consensus-oriented)
	Ability of companies to influence regulation
	National environmental policies
	National industrial promotion strategies

Firm-specific factors	Economic situation and market positioning
	History of involvement with (technological) alternatives
	Degree of (de)centralisation
	Degree of internationalization of top management
	Availability and type of internal climate expertise
	Nature of strategic planning process
	Corporate culture
Industry-specific factors	Nature and extent of threat posed by climate change
	Availability and cost of alternatives
	Degree of globalization of supply chain
	Political power of the industry
	Technological and competitive situation
Issue-specific factors	Impact of issue on various sectors, countries
	Institutional infrastructure for addressing issue
	Degree to which issue and regulation are global
	Complexity and uncertainty associated with issue

Source: Kolk and Levy 2004, p. 178

What struck us at the time were the divergent responses between multinationals in the US compared to Europe, something which you can still see to some extent if you compare, for example, Exxon Mobil and BP. We thus did an in-depth study of the oil industry to investigate this further (Levy and Kolk 2002). We posited that there were forces that would lead to convergence regarding the positions of oil multinationals across the Atlantic, regardless of their nationality, particularly their location in global industries and participation in the 'global issue arena' of climate change. At the same time, their different home-country institutional contexts as well as individual company characteristics represented pressure for divergence, for different views. Applied to the oil and automobile industries, it turned out that divergent pressures initially dominated, but that convergence increased as the issue matured. Managerial perceptions and institutional frames were important in shaping the responses of multinationals.

For companies, the issue of climate change continues to be characterised by diversity in policy developments and uncertainty as to the (potential) impact on markets, technologies and organisations. At the policy level, there has been fragmentation concerning approaches on how to implement Kyoto (if at all). The most notable regulatory development has been the introduction of the EU Emission Trading Scheme per January 2005. This is the only compulsory trading system, in addition to a number of voluntary ones (including the Chicago Climate Exchange). To influence these and other initiatives to their favour, multinationals have continued to engage in political strategies, although the specific types have changed as a result of the different context (Kolk and Pinkse 2007). It is also interesting to observe that the climate change issue has developed further, and experienced a 'secondary trigger', beyond the 'maturity' we found in our earlier work (Kolk and Levy 2004; Levy and Kolk 2002). Multinationals are also actively helping to shape the institutions that are emerging to govern climate change, i.e. the market mechanisms, particularly emission trading, which were created with the Kyoto Protocol, but have not yet been fully implemented and accepted (Pinkse and Kolk 2007).

4 Market responses

Perhaps even more exciting than the political strategies have been the corporate market responses. There is a whole range of activities that multinationals are undertaking, ranging from simply making inventories of and measuring emissions (which is most common), to process improvements, improving products or engaging in market mechanisms, especially emission trading. There is basically a distinction between innovation and compensation, which can be done either alone or in cooperation with others inside or outside the supply chain (Table 3).

Table 3. Strategic options for climate change

Organization	Main aim	
	Innovation	Compensation
Internal (company)	Process improvement (1)	Internal transfer of emission reductions (2)
Vertical (supply chain)	Product development (3)	Supply-chain measures (4)
Horizontal (beyond the supply chain)	New product/market combinations (5)	Acquisition of emission credits (6)

Source: Kolk and Pinkse 2005a, p. 8

Companies of course pursued different options simultaneously. Our analysis of multinationals showed that there were basically six groups with different characteristics (Kolk and Pinkse 2005a): cautious planners (31%, score low on all dimensions); emergent compensators (36%, internal focus, particularly box 2); comprehensive compensators (14%, which combine targets, control and production process improvements, boxes 1, 2, 4 & 6); vertical explorers (10%, supply-chain oriented, boxes 3 & 4); horizontal explorers (5%, markets beyond current scope, box 5); and emission traders (4%, boxes 2 and 6).

The sort of profile that companies have is to some extent shaped by the sector in which they operate. Automobile and oil multinationals focus, for example, mostly on developing technological capabilities (Kolk and Pinkse 2008). In the oil industry this encompasses a range of technologies, with some targeting a range of energy sources, while others explore particularly hydrogen or renewables or stick to natural gas for the time being. In the automobile industry, Toyota was a first mover with hybrid vehicles, but most other companies now are also starting to move in this direction, although they all view it as a transitional technology, and not a very profitable (even loss-making) niche market. For banks and insurance companies, organisational capabilities are more important, for example, by offering weather derivatives or facilitating/funding carbon trading or clean development/offset projects. Some oil companies are also taking steps to play a role in emission markets. General Electric, which has started a large 'Ecomagination' campaign in 2005, develops new expertise but also relies on existing ones.

5 Research agenda

In terms of a future research agenda, there are many areas that deserve further attention in this very turbulent and dynamic field. This involves not only following and tracing trends in corporate responses, both market and political, but also the way in which corporate realities help to shape policy development and the instruments that emerge to influence the behaviours of various companies. We will also assess what it is that determines which strategies companies will follow: to what extent do country of origin and location, including stakeholder pressure and local regulatory situations, sector and/or competitive pressures, geographical spread, degree of internationalisation, diversification and integration, product portfolio, perceptions of risks and/or opportunities, and other firm-specific characteristics play a role? It can again be investigated to what extent divergence or convergence is taking place, and what the performance implications of different corporate and policy approaches are.

Another important research stream is to examine how and to what extent companies implement climate approaches internally (across borders, between different subsidiaries and business units), what sorts of problems managers face in this process and whether or not climate approaches are related to 'mainstream' corporate activities. It will be interesting to see whether and how the transfer of knowledge takes place within companies, and in the case of multinationals, from which location actual innovations (technological or organisational) originate and what the role of partnerships is in addressing climate change. We also envisage studies concerning actual operations and the corporate drivers of engagement in carbon offset projects, particularly in developing countries, and the implications of policy contexts and governance characteristics for the extent and effectiveness of these market mechanisms.

References

Kolk A (2000) Economics of environmental management. Financial Times Prentice Hall, Harlow

Kolk A (2001) Multinational enterprises and international climate policy. In: Arts B, Noorman M, Reinalda B (eds) Non-state actors in international relations. Ashgate Publishing, Aldershot, pp 211–225

Kolk A, Hoffmann V (2007). Business, climate change and emissions trading: Taking stock and looking ahead. European Management Journal 25 (6): 411–414

Kolk A, Levy D (2001) Winds of change: corporate strategy, climate change and oil multinationals. European Management Journal 19 (5): 501-509

Kolk A, Levy D (2004) Multinationals and global climate change: issues for the automotive and oil industries. In: S Lundan (ed) Multinationals, environment and global competition. Elsevier, Oxford, Research in Global Strategic Management 9: 171–193

Kolk A, Pinkse J (2004) Market strategies for climate change. European Management Journal 22 (3): 304–314

Kolk A, Pinkse J (2005a) Business responses to climate change: identifying emergent strategies. California Management Review 47 (3): 6–20

Kolk A, Pinkse J (2005b) The evolution of multinationals' responses to climate change. In: Hooker J, Kolk A, Madsen P (eds) Perspectives on international corproate responsibility. Carnegie Mellon Press, pp 175–190

Kolk A, Pinkse J (2007) Multinationals' political activities on climate change. Business and Society 46 (2): 201–228

Kolk A, Pinkse J (2008) A perspective on multinational enterprises and climate change. Learning from an 'inconvenient truth'? Journal of International Business Studies, forthcoming

Levy DL, Kolk A (2002) Strategic response to global climate change: conflicting pressures in the oil industry. Business and Politics 4 (3): 275–300

Pinkse J, Kolk A (2007) Multinational corporations and emissions trading. Strategic responses to new institutional constraints. European Management Journal 25 (6): 441–452

Impacts of climate change on the electricity sector and possible adaptation measures

Benno Rothstein[I], Marion Schroedter-Homscheidt[II], Claudia Häfner[III]
Susanne Bernhardt[II], Solveig Mimler[III]

[I] University of Applied Forest Sciences
Schadenweilerhof, 72108 Rottenburg am Neckar, Germany
rothstein@hs-rottenburg.de

[II] German Aerospace Center (DLR), German Remote Sensing Data Center
P.O.box 1116, 82234 Wessling, Germany
Marion.Schroedter-Homscheidt@dlr.de
susanne.bernhardtWie@gmx.de

[III] European Institute for Energy Research (EIfER)
Electricité de France/Universität Karlsruhe
Emmy Noether Str. 11, 76131 Karlsruhe, Germany
chaefner@eifer.org
solveig.mimler@eifer.org

Abstract

Climate change is becoming an increasingly relevant factor in the planning processes of electricity companies since it affects all areas of the electricity sector, from production via distribution to consumption. It is of paramount importance for electricity companies to define adaptation measures in this new context so as to limit their risks and financial losses.

In the framework of the CLIMAGY project vulnerabilities of electricity companies in the course of climate change are assessed and possible adaptation measures are identified. In close cooperation with Electricité de France (EDF) and Energie Baden-Württemberg (EnBW), several questionnaires were designed and about 80 experts in key positions were polled.First results show that, on the one hand, the electricity sector is being affected by climate change and weather risks in almost all business units (e.g. water availability for electricity generation). On the other hand, several options for actions are already being developed to help reduce the described impacts. One of the conclusions is that it is necessary to carry out research on the way electricity companies should deal with the remaining uncertainties of future climate change.

Keywords: Adaptation, vulnerabilities, energy, electricity, climate change, extreme weather events

1 Introduction

Climate on Earth has never been constant. In fact it has continuously been changing due to both natural and human influences. The obvious increase of temperature that has been observed during the 20th century mainly results from anthropogenic emissions of greenhouse gases. Climate researchers project a temperature rise of between 2.4°C and 6.4°C by the year 2100 (IPCC 2007). Besides the temperature increase, the effects of the climate change are also manifested in the rising number of extreme weather situations all over the world. In the last few decades, global costs of extreme weather events have shown a dramatic growth. Some of the rising costs have socio-economic reasons, such as numerous settlements in high-risk zones and more expensive buildings. Nevertheless, a significant proportion of the damages and losses are ascribed to climate change. The share of climate reasons quoted for these increasing costs due to climate change is continuously growing (Munich Re 2006; Berz 2005).

During recent years important insights into the complex climate system, its natural variability and the anthropogenic interference have been acquired. In particular, as can be seen for example in the documentation of the Intergovernmental Panel on Climate Change (IPCC), the adverse role of greenhouse gas emissions has been clearly identified (IPCC 2007). There is therefore widespread agreement that there is no alternative to CO_2-emission reduction strategies (mitigation). However, rapid changes in the exploitation of fossil energy resources and successes in the reduction of their excessive use can not be expected in the short term. Irrespective of the mitigation efforts, all climate models project a continuous warming of the atmosphere for at least the next 20 years. As a result, the number of extreme weather events will probably rise as well. Extreme weather events can have severe economic consequences; e.g. EnBW lost € 18.5 million due to the heatwave in 2003.[1]

Therefore, climate change mitigation measures have to be complemented by adaptation measures. In the future, adaptation as an option for action, will become more and more important because

- on the one hand the already induced changes in the atmosphere are of an enduring nature, whatever the successes of future emission reduction measures may be;
- on the other hand it has already become clear that there will be no significant global decrease of CO_2 emissions during the next one or two decades. Thus, climate change will keep accelerating (BMBF 2004).

Many researchers all over the world are working to quantify upcoming climate changes and determine to what extent climate change could be prevented or alleviated. However, comprehensive, integrative and transdisciplinary analyses con-

---

[1] Speech of Prof. Hartkopf at a shareholders' meeting in Karlsruhe/Germany on 29th April 2004.

cerning vulnerabilities and adaptation measures to climate change in the Central European electricity sector are still lacking.

For this reason the French electricity supplier Electricité de France (EDF) funds the perennial project CLIMAGY (Climate & Energy: Adaptation Measures for Recent Climate Trends within the Energy sector). Project partners are the European Institute for Energy Research (EIFER), the German Aerospace Center (DLR) and Munich Re. The aim of CLIMAGY is to assess:

- in which ways the electricity branch is affected by climate change,
- which vulnerabilities are present and
- which adaptation measures already exist.

Even two companies in the same business area can differ significantly in their vulnerabilities to climate change. Hence, adaptation measures need to be tailored to fit the precise desires and needs of a company. Therefore, EDF and its German partner Energie Baden-Württemberg (EnBW) are taken as an example with a special focus on the state of Baden-Württemberg in south-west Germany.

The state of Baden-Württemberg is well-suited as an example for the following reasons:

- Baden-Württemberg has close alliances with its European neighbouring countries (especially France) and the cooperation between EDF and EnBW is a paradigm for German-French collaboration;
- EDF, as Europe's biggest electricity supplier, has a holding of 45% of EnBW; this constellation represents the powerful Europe-wide network in the electricity branch very well;
- Baden-Württemberg is often hit by extreme weather events (e.g. storms, floods, low water levels in rivers, heatwaves).

In this article, the methodology will be described first (Sect. 2). The authors will then point out in which way the electricity sector contributes to climate change and to what extent the electricity sector is affected by climate change (Sect. 3). Finally, some initial results of the CLIMAGY project will be presented (Sect. 4), and first conclusions will be drawn (Sect. 5).

2 Methods

Regional climate change leads to different impacts on each of the specific areas of the electricity sector (i.e. production of electricity, transport and consumption). Their sensitivities can be very different. To find out more about vulnerabilities and possible adaptation measures within the electricity sector, numerous methods were used. These methods, which led to the initial results presented in Sect. 4, are described below: they encompass expert polls and media analyses.

2.1 Expert polls

In close cooperation with Electricité de France (EDF) and Energie Baden-Württemberg (EnBW) several questionnaires were designed and altogether about 80 experts in key positions were polled, mainly in Germany but also in France.

First expert poll: detecting potential vulnerabilities

As a first step, an analysis pattern was worked out in order to detect potential vulnerabilities. To make sure that this analysis pattern entirely includes all areas of the electricity sector, a holistic view ("cradle-to-grave perspective") was taken. Coal-fired power plants and hydroelectric power stations served as examples. The former is a representative example of conventional energies whereas the latter is typical of renewable energies. Possible potential vulnerabilities were assessed and divided into seven groups: planning, purchase of resources, production and transport of electricity, storage of resources, disposal of residuals and production of power plant materials.

In a second step, approximately 50 experts were asked to assess the weight of the identified potential vulnerabilities on a scale from 1 (= "less important", "low vulnerability") to 6 (= "very important", "high vulnerability"). The whole questionnaire contained nearly 50 questions.

The specialists were from the following sectors: electricity industry, energy research and administration/insurance. The responding electricity industry and the energy research experts all work on weather-related problems. The administration and insurance sector specialists were polled in the framework of climate change conferences. Their participation in these congresses demonstrates that they are engaged in climate change issues. Nevertheless, the questionnaire remains only an initial approach towards estimating the precise impacts of climate change in the electricity sector. It should not be considered as representative.

Second expert poll: detecting potential adaptation options

Based on these detected and rated vulnerabilities within the electricity sector and on an analysis of newspaper articles, a catalogue of specific climate-related problems and adaptation options was compiled. Almost 30 experts of German and French electricity companies in different business areas (such as production of electricity, trading, facility management) were polled in the framework of intensive written and in some cases also oral interviews. They were confronted with the task of estimating the importance of specific problems due to recent climate trends and giving information about already existing and planned adaptation measures. Similarly to the questionnaire of vulnerabilities, the experts could describe the importance (from "very important" to "less important") or express the status of implementation of specific adaptation measures (from "implementation completed" to "no implementation planned"); this questionnaire also contained nearly 50 questions.

Based on these polls and on personal interviews, a ranking of existing and planned adaptation measures was made. The authors will also identify how remote sensing data could be used as a basis for adaptation measures within the electricity sector.

2.2 Media analysis

Climate change, with its occurrence of extreme weather events, also leads to a changing electricity demand. The heatwave in the summer of 2003 was taken as an example thereof. A broad analysis of about 900 media reports (esp. newspapers, but also internet and TV) allowed changed patterns of electricity consumption to be identified.

The analysis covered media reports on all parts of Germany but mainly focussed on the South-West of Germany, which proved to be the hottest region during the heatwave in the summer of 2003 (DLR 2003). The analysis targeted the time period from July to September 2003.

Due to the long duration of hot temperatures and the annual summer slump in press coverage, the extreme weather conditions were a popular topic in the media. Especially during the first half of August 2003 – extremely hot, dry and with an above-average amount of sunlight (Müller-Westermeier and Lux 2003) – the press coverage was full of reports on the effects of the heatwave and the individual reactions of the population together with stories about how people coped with it.

The results of the analysis were categorized into the topics of labour, leisure, housing and consumption so as to provide a first systematisation for deriving results. Within each of these fields, problems caused by the extreme temperatures and changed behaviour patterns were listed.

3 The electricity sector and climate change

In this section, the links between the electricity sector and climate change are briefly outlined: firstly, an overview of the ways in which the electricity sector acts as an emitter of greenhouse gases is given; this is followed by a summary of how climate change affects the electricity sector.

3.1 Electricity sector – an emitter of greenhouse gases

An analysis of climate change literature shows that the electricity sector has an impact on global climate change, especially because of its contributions to the atmospheric CO_2 concentration resulting from the use of fossil energies. Approximately one third of the CO_2 emissions in the EU are produced because of electric power supply (Mathes 2001). In Germany the proportion of the electricity sector in the CO_2 emissions is higher than in many other European countries, because of the fact that in Germany 62% of the electricity generation is based on fossil energy (Trittin 2004). Nevertheless, during the 1990 to 2001 period the German CO_2 emissions in this branch were reduced by 15% (BMWi 2002).

Besides the CO_2 emissions by the electricity sector, the use of sulphur hexafluoride (SF_6) also contributes to the anthropogenic greenhouse effect. On a global level the SF_6-emissions by the electricity sector continuously increased during the period from 1970 to 1998 (Olivier and Baker 2005, p. 9).

The emissions of SF_6 are relatively low compared to CO_2, but on the other hand SF_6 is one of the strongest greenhouse gases. Its global warming potential is 23900 times larger than that of CO_2 over 100 years. In the electricity sector SF_6 is used for insulation of electrical equipment (Ochmann and Braun 2005). Since about 1990 the electricity sector has become the largest source of fluorinated gases at the global level (Olivier and Baker 2005).

3.2 Electricity sector – affected by climate change

According to the interview partners' opinions, it is assumed that an electricity company having strategies for adaptation to climate change will experience advantages in comparison with competitors which are less adjusted to the risks of climate change. Current company strategies often take the climate change aspect into account.

The measured trend of higher surface average temperatures implies risks as well as opportunities for the electricity sector. Opportunities for the electricity branch arise from the expansion of the market due to increasing electricity needs, e.g. because of cooling systems and delivery rates of water pumps. Higher temperature averages flatten the seasonal differences in electricity demand and thus improve the utilisation of power plants. A slightly declining demand in winter and an increasing demand in summer due to cooling and climate technology can be expected. Although these arguments seem plausible, they are not proven and quantified yet (European Commission 2005; UVM 2003).

Climate-change-induced risks in the electricity branch primarily involve short-term variations of electricity demand as a result of extreme weather situations. In addition, problems such as power plant failures due to lack of cooling water (or lack of operating water in the case of hydro-electric power) have to be dealt with (e.g. Luhmann and Fischedick 2004). Moreover, higher costs have to be taken into account for building and plant safety due to storms, hail, etc.

4 Impacts of climate change and adaptation measures

In this section some initial results of the CLIMAGY project are presented. Firstly, some impacts of climate change for the electricity sector are highlighted with a focus on electricity generation and consumption, then possible adaptation measures are described.

4.1 Impacts of climate change on the electricity sector

Vulnerabilities of electricity generation

Until now, vulnerabilities of power plants in Central Europe have rarely been investigated. Only a few experts are available in the electricity sector for the identification and rating of vulnerabilities in their daily business. Additionally, there are no commonly accepted measurement scales available for vulnerabilities in the electricity sector. For this study, a subjective scale based on numerical values was selected. These numerical values can be statistically analysed, although of course the ratings given by the specialists must be interpreted carefully due to the inherent individuality and subjectivity.

Experts from three different sectors (electricity industry, energy research and administration/insurance; see Sect. 2.1) were asked to estimate the vulnerability. All experts interviewed also answered the questionnaire. The answers given by the experts of each sector perfectly match those from the other sectors. Therefore, the results given here in brackets are the mean values of the three sectors that were subjected to the questionnaire. The authors have decided not to indicate the variance.

Concerning vulnerabilities of the various electricity generation methods, wind (5.1) and hydro-electric power (4.1) are considered as the most vulnerable. Solar (3.3), nuclear (3.2) and coal-fired power stations (3.2) are regarded as having medium vulnerability, while gas-fired power stations (2.6), fuel cells (1.6) and geothermal energy (1.6) are estimated as least vulnerable.

Vulnerabilities of hydro-electric and coal-fired power stations in various activity partitions were also assessed. With the help of a holistic approach it was possible to examine the impacts of climate change on the electricity sector in detail. Furthermore, thanks to this method no aspect is overseen. During the study the following activity partitions were analysed: planning, purchase of resources, production and transport of electricity, storage of resources, disposal of residuals and production of power-plant materials. This closer examination of hydro-electric and coal-fired power stations showed that their vulnerabilities for the seven activity partitions can be very different (the numbers are the mean value of the estimated vulnerability):

- Transport of electricity (via transmission line) is considered as being the activity which is the most vulnerable to climate change (4.0) independently of the type of electricity generation.
- Furthermore, production of electricity, which includes the water availability, plays an important role for both types. Coal fired power plants need water especially for cooling (for direct cooling as well as for cooling towers) (4.0), but also for the transport of coal by ship (3.8). Hydro-electric power stations depend on the water level of rivers or lakes in order to produce electricity with water turbines (4.3).

- Taking into account the climate change aspect in the future planning of power stations is considered as quite important for hydro-electric power (4.6) and less important for coal-fired power stations (2.4).
- Purchase and storage of resources were also rated as quite important (4.2) for hydro-electric power. For coal-fired power stations, the vulnerability of these aspects was evaluated as only moderately important (2.7).
- The disposal of residuals and production of power-plant materials were regarded as less vulnerable (2.1; 2.2).

Of course, the assessment of the vulnerabilities in the various activity partitions also strongly depends on the choice of the experts being interviewed. Nevertheless, this study showed that these different groups of experts estimated the seven activity partitions very similarly.

Changing electricity consumption patterns

Regarding changing patterns of electricity consumption during the heatwave in 2003, the analysis of media reports from the period of July to September 2003 (see Sect. 2.2) showed two major qualitative results:

1) A higher total demand for electricity due to the following reasons:

- More frequent use of air-conditioning systems:
  On internet forums people spoke about their increased use of air-conditioning even during the night. At the same time providers registered increased sales of air-conditioning systems to private households. The electricity company EnBW called for electricity savings by either switching off or raising the temperature of air-conditioning.

- More frequent use of fans:
  There was a huge demand for fans that could not be satisfied any more by providers. The use of a fan to cope with high temperatures, especially during working time, was one of the most often-mentioned strategies on internet forums.

- Higher utilization of cooling devices such as refrigerators and freezers:
  In general, there were increased sales of refrigerators and freezers, but also of small fridges as additional devices in households. Furthermore, providers reported a higher number of repairs and service calls concerning damaged and overloaded cooling devices.

2) Displacements in electricity consumption due to altered daily routines:

- Earlier beginning and ending of working days in order to take advantage of cooler temperatures in the morning hours:
  According to flexitime regulations or prescriptions concerning noise control, employers encouraged their employees to start earlier in the morning and therefore end earlier in the afternoon. They also gave the possibility to reduce overtime.

- Leisure activities either in the early morning or in the late evening to avoid the highest daily temperatures especially between 11 a.m. and 4 p.m.:
    Festivals such as parish fairs and other outdoor events were highly attended in the evenings but sparsely in the afternoons. In shops, supermarkets and gastronomical establishments such as ice-cream parlours more customers were registered in the early morning and evening hours, but almost no customers in the afternoon. In media reports medical doctors regularly advised the population to avoid any effort between 11 a.m. and 4 p.m. in particular.
- Open-air leisure activities:
    All sorts of open-air events were highly attended, especially during late evening hours. In contrast, TV stations suffered from very low audience ratings especially during prime time. Furthermore, there were a higher number of police operations due to breach of the peace at night.

Climate change is becoming a more and more relevant factor in the planning processes of electricity companies since it affects all areas of the electricity sector from production via distribution up to consumption. It is of paramount importance for electricity companies to define adaptation measures in this new context so as to limit their risks and financial losses.

4.2 Adaptation measures within the electricity sector

In order to detect existing adaptation measures a second poll was conducted (see Sect. 2.1). The first results concerning the written questionnaire and the telephone interviews among the German experts are as follows:

- About 50% of the people interviewed emphasized that "Adaptation measures for recent climate change" are an important issue.
- There are a large number of different problems due to climate change, which the electricity sector must cope with (e.g. low water level, high water level, increasing water temperature and melting of alpine glaciers).
- Problems rated as most important by the polled departments or business areas are very different and individual.
- The solutions or adaptation measures corresponding to these diverse problem descriptions are also very individual and need to be discussed before they can be implemented (e.g. adaptation emergency plans, alternative cooling systems, changing of power-plants portfolio).
- As an outcome, individual departments and business areas have a very specific and variable interest in environmental parameters (e.g. weather forecast, water temperature of rivers, wind speed).

Furthermore, the analysis showed that experience with remote sensing information products is still in its infancy within the electricity sector. The main reasons for the low degree of utilisation are:

- lack of information about remote sensing product contents;
- lack of awareness of potential advantages that remote sensing information products might offer to the electricity sectors.

A study on the usage of remote sensing information revealed that less than half (43%) of the respondents use such products. This motivated the analysis which is currently being performed on linkages between the electricity sector and remote sensing services. Additionally, a study is needed in order to assess which current requirements in the electricity sector can be met by remote sensing services and which cannot.

One adaptation strategy for coping with changing demand patterns is load management. Load management is important for the electricity industry because it allows loads to be adjusted for an optimised generation of electricity: power plants work according to their capacities. Furthermore, over- and under-voltage levels are avoided. Suitable tools for load management include e.g. controllable consumers, calls for energy savings, as well as special tariffs and contracts.

5 Conclusions

One can conclude that, on the one hand, the electricity sector is affected by climate change and weather risks in almost all business units. On the other hand, several options for actions have already been developed to help reduce the described impacts.

Electricity companies have to accept unclear projections of the future climate in the foreseeable future. It is thus necessary to decide on investments despite incomplete information. The electricity industry considers that further research is needed for the improvement of regional climate projections. The analysis of the occurrence probability of extreme events (e.g. low water periods in summer, intense snow fall at higher winter temperatures) is especially important, but accurate forecasts regarding the further shift of mean values are also required.

Moreover, research on how electricity companies should deal with the remaining uncertainties of future climate change is needed, because there are circumstances in which incomplete information can not be accepted, especially when adaptation measures require long-term decisions (e.g. general issues of choice of power-plant location, choice of power-plant type). Therefore, more research on impacts (and their interplay) and adaptation measures will be necessary in the future, in order to provide a well-founded information basis for rational action. Initially, the problem of the present absence of data has to be solved so that research can function efficiently. The large amount of information required (from planning offices, catastrophe service and as far as possible from companies) therefore needs to be acquired and systematised. Subsequently, a GIS-based company-specific risk mapping will for instance be possible. Based on this GIS, detailed information, such as about the specific design of all facilities (power plants and transformer stations, high-voltage pylons etc.) of the company in relation to climate parameters (e.g. ice loads) will be able to be accessed and connected with information of various regional climate models.

It is not important to know the future but rather to be prepared for it (Perikles, 490–429 B.C.). In order to be prepared for the future, an exact analysis of the influencing parameters and their cross-linking is required. The CLIMAGY project offers an initial contribution towards meeting this requirement.

References

Berz G (2005) Klimawandel – Kleine Erwärmung, dramatische Folgen. In: Munich Re (ed) Wetterkatastrophen und Klimawandel – Sind wir noch zu retten? Edition Wissen, Munich, pp 98–105

Bundesministerium für Bildung und Forschung (BMBF) (2004) Forschung für den Klimaschutz und Schutz vor Klimawirkungen. Bonn

Bundesministerium für Wirtschaft (BMWi) (2002) Energiedaten 2002. Berlin

Deutsches Zentrum für Luft- und Raumfahrt (DLR) (2003) Weekly average surface temperature [°C] from 4th to 10th August 2003. (not published)

European Commission (2005) Winning the battle against global climate change. COM(2005) 35 – Official Journal C 125 of 21st May 2005

Intergovernmental Panel on Climate Change (IPCC) (2007) Climate Change 2007: The Physical Science Basis. Summary for Policymakers. Geneva

Luhmann HJ, Fischedick M (2004) Renewables, adaptionspolitisch betrachtet. Aus Politik und Zeitgeschichte. Beilage zur Wochenzeitung Das Parlament 37, 18–24

Mathes F (2001) Vor neuen Herausforderungen. Ökologisches Wirtschaften 1: 5

Ministerium für Umwelt und Verkehr Baden-Württemberg (UVM) (2003) KLARA. Klimawandel, Auswirkungen, Risiken, Anpassung. Analyseraster. Anhang 1a. In: Informationssystem KLARA 1.0. CD, Karlsruhe

Müller-Westermeier G, Lux G (2003) Die extreme Witterung in der ersten Hälfte des August 2003. http://www.dwd.de/de/FundE/Klima/KLIS/prod/spezial/temp/Die_extreme_Witterung_in_der_ersten_H%e4lfe_des_August_2003_4GMW.pdf

Munich Re (2006) Topics Geo – Jahresrückblick Naturkatastrophen 2005. Edition Wissen, Munich

Ochmann F, Braun R (2005) Alles ganz normal? Stern 7: 26

Olivier J, Baker J (2005) Historical global emission of the Kyoto gases HfCs, PFCs, and SF_6. http://www.epa.gov/highgwp/electricpower-sf6/pdf/olivier1doc.pdf

Trittin J (2004) Energiesparen ist auch eine Innovation. Frankfurter Rundschau, 09.01.2004: 9

Modelling impacts of climate change policy uncertainty on power investment

Ming Yang[I], William Blyth[II]

[I] Energy and Environment Division
International Energy Agency
9 rue de la Fédération, 75395 Paris Cedex 15, France
ming.yang@iea.org

[II] Oxford Energy Associates
28 Stile Road, Oxford OX3 8AQ, UK
william.blyth@oxfordenergy.com

Abstract

The changing electricity prices in competitive electricity markets, the uncertain carbon prices, and the increasing energy prices have forced power investors and government policy-makers to search for, and use, more sophisticated methods for project evaluations. The objective of this paper is to present a computer model currently developed by the International Energy Agency (IEA) to quantify the impacts of climate change policy uncertainties on power investment. The methodologies used include the traditional discounted cash flow approach to calculating project net present value, stochastic simulation to capture the characteristics of uncertain variables, and real option valuation to capture investors' flexibility to optimize the timing of their investment. We applied these modelling methodologies in a case study to evaluate the effects of the changing carbon prices on firms' decisions to invest in more energy efficiency power technologies. The study concludes that (1) the simulation of stochastic processes and real options could be very useful tools for power investors when dealing with uncertainties about future carbon prices; and (2) the more uncertain the primary energy prices and carbon trading prices, the more the economic case for lower emitting technologies deviates from the traditional discounted cash flow.

Keywords: Stochastic analysis, real option modelling, regulatory uncertainty, investment in energy efficiency technology

Acknowledgement: Acknowledgement is due to the Electric Power Research Institute in the United States for providing the first version of GHG-CAM model, continuous technical assistance and the financial sources for this project. The authors also wish to thank the governments of Canada, Italy and the Netherlands and the power companies Enel, EON-UK and RWEnpower for their financial contributions and assistance in data collection for the study. The authors are indebted to Richard Bradley, Richard Baron, Julia Reinaud and Nicolas Lefevre of the IEA for their guidance on the project, data collection for the study, and their comments on the paper. The viewpoints expressed in this article are solely those of the authors. They do not represent those of any other organization, nor do they reflect those of the IEA or its member countries.

1 Introduction

The power industry worldwide has encountered three major challenges: reforms of power markets for competition, greenhouse gas (GHG) mitigation and rising energy prices. In a competitive market, the risks facing decision makers in the power sector are significantly different from those facing monopolies in integrated power markets. In addition, GHG mitigation policies have put monetary values on carbon dioxide (CO_2) emissions. Power producers in Europe are now subject to an Emission Trading Scheme - arguably the most efficient form of regulation as far as companies are concerned - and face a very uncertain outlook for carbon prices. Furthermore, current high oil and gas prices have considerably increased production costs and affected energy demand. In today's liberalised market environment, the key risks to recovering investment are caused by uncertainties about energy and CO_2 prices, and other regulatory uncertainties.

Risks and uncertainties tend to lead investors toward flexible technologies with a short investment return period, short construction time and the ability to switch fuels. On the other hand, economies of scale encourage investors to develop large-scale power facilities. In addition to the traditional deterministic discounted cash flow approach, power investors are adopting new financial assessment approaches and methodologies to address the above-mentioned uncertainties and risks. Policy makers are also trying to understand what actions they should take to alleviate the impacts of these uncertainties on the sustainability and security of the domestic power industry.

A stochastic simulation imitates or evaluates the desired true characteristics of a variable with a random component. Thus, modelling uncertainties with stochastic approaches has been widely used in the energy sector (Ruszczynski and Shapiro 2003). In our study we developed a cash flow model with Monte Carlo to simulate multiple uncertain variables: the prices of CO_2, primary energy and electricity. We also modelled the correlations among CO_2 price, gas price and electricity price.

Real option analysis (ROA) is the modern approach to capital budgeting. This approach considers the value of the opened options for the decision makers. The origin of the term "real options" can be attributed to Myers (1977) who first identified the fact that many investments in real assets can be viewed as options. The real option problem can be viewed as a problem of optimization under uncertainty of a real asset such as power project investments given the available options.

The theory and application of option pricing in the area of quantifying climate change policy impacts on power sector investment are still in their infant stage. Laughton et al. (2003) applied ROA to the evaluation of a geological GHG sequestration option. They used a simplified model of an option to sequester a part of the CO_2, to show how the risks are valued. Their analysis showed that the net present values of the project can be quite different under the traditional deterministic discounted cash flow (DCF) approach and the real option analysis approach. It concludes that the traditional DCF approach can give misleading results because it does not take into account the complex effects of risk and uncertainty on values.

As acknowledged by the author, Laughton's study was quite preliminary, since some key elements such as price variables were not modelled.

Sekar (2005) evaluated the investments in three coal-fired power generation technologies using real option valuation methodology in an uncertain CO_2 price environment. The technologies evaluated include pulverized coal, integrated coal gasification combined cycle (baseline IGCC) and IGCC with pre-investments that make future retrofit for CO_2 capture less expensive (pre-investment IGCC). Cash flow models for specific cases of these three technologies were developed, but the problem was formulated with CO_2 price as the only uncertain cash flow variable. Sekar's approach combines two elements: market based valuation to evaluate cash flow uncertainty and dynamic quantitative modelling of uncertainty. Monte Carlo cash flow simulation methods have been applied to simple scenarios to incorporate cash flow uncertainty. However, energy prices are not modelled as stochastic variables in the study.

EPRI (2005) developed the Greenhouse Gas Emission Reduction - Cost Analysis Model (GHG-CAM). GHG-CAM uses a discounted cash flow (DCF) analysis methodology to evaluate the revenues, costs and expected after-tax gross margin expected to flow from investments in GHG reductions. Furthermore, the model incorporates sophisticated statistical and economic tools, including Monte Carlo simulation, real options analysis and decision analysis methods, which make it possible to include risks, uncertainties and real options directly into the evaluation of specific GHG reduction strategies. Energy prices and CO_2 trading prices can be modelled with correlations. The model is developed in MS-Excel and is supported by another commercial software programme called Real Option Calculator or ROC.

The IEA is undertaking a study to quantify climate change policy uncertainty or risk on power sector investments. Specifically, the goals of the study are (1) to analyze the effect that climate change policy uncertainty may have on individual investments in the power sector; (2) to explore the consequences of policy uncertainty for the evolution of the power sector and associated risk for policy objectives such as greenhouse gas emission mitigation and energy security; and (3) to examine the reduction of impacts of climate change policy uncertainty through improved policy design.

As a part of the study, the IEA is developing a model called **M**odelling **IN**vestment with **U**ncertainty **I**mpac**T**s (MINUIT). The IEA model is developed along similar lines to EPRI's first version of GHG-CAM model. It keeps all the merits of EPRI's model, but extends the number of different options available; it also adds some new features such as accounting for plant build-time and the ability to model the development of a new power plant.

We applied the model in a case study which covers new work. In contrast to Laughton et al. (2003), we incorporated and modelled all the energy prices with stochastic variables. In addition to treating CO_2 as a random variable as in Sekar (2005), we also modelled the electricity price and primary energy prices as random variables.

The objective of this paper is to illustrate how this type of analytical tool can be used to improve the understanding of policy-makers and investors regarding the impacts of regulatory uncertainty. In order to test the model, we examined under which market conditions a relatively inefficient coal-fired power plant will be closed and a new efficient combined cycle gas turbine (CCGT) built. We also collected intensive data such as forecast energy prices, capital costs for the technologies, and CO_2 emission trading prices. A database was also developed to support the model analysis.

This paper is comprised of four sections. Following the introductory section, Section 2 presents the methodologies used in the study. In Section 3, we show a case study with data collection, parameter assumptions and result analysis. Section 4 concludes this paper.

2 Modelling methodology

There are three key elements in our methodology and model: deterministic analysis based on discounted cash flow (DCF), stochastic simulation (which provides an explicit representation of the most relevant uncertainties), and real option optimization. In the following, we briefly introduce these elements. They represent progressively more sophisticated ways of incorporating uncertainty and risk into the analysis of a project's viability. By comparing the different valuations and investment criteria offered by the different approaches, we identify the potential effects of uncertainty on actual investment behaviour.

2.1 Deterministic method with discounted future cash flow

In traditional project analysis, the deterministic analysis, which fixes and discounts future cash flow to calculate the present values, has been widely used to evaluate the costs and expected gross margin or profits. Net present values (NPVs) or levelized costs per unit of output are the key criteria to evaluate the financial viability of the project. This approach assumes that the future energy prices and carbon trading prices are fully known. The future cash flow is often discounted by the weighted average cost of capital (WACC) of a firm. In DCF analysis, if the project revenue value is higher than the costs of the project investment and operations, the project may be good for investment. A detailed description on how to calculate DCF and WACC can be found in ADB (2002).

Project risk evaluation can be incorporated into DCF analysis in a simplistic way by setting higher discount rates, although this has several shortcomings. DCF is best suited to a situation when a company operates in a stable environment and is in the maturity of its life cycle. Cash flows for the next year can be forecast more or less exactly, and risk premiums for the type of project being developed are well established. This might not be true in a dynamic environment or in the period of launching a new power generation technology, where it is very difficult or impossible to determine the potential revenue because of uncertainties about the raw materials, primary energy prices and CO_2 trading prices. In any case, using a single parameter (discount rate) to represent many different sources of risk can be a blunt instrument, and raises the difficulty of choosing an appropriate discount rate, particularly in novel situations where risk premiums are not well established. Perhaps most importantly, the DCF does not account for the flexibility that investors often have to choose when making their investments. For example, some projects may not be cost-effective at current economic and technical conditions. However, these projects may become cost-effective in the near future when these conditions are changed. The DCF method may give up these project opportunities, but the ROA method may suggest that the project developer postpone the investment. Project valuations by DCF analysis may therefore underestimate project value by not accounting for optimal timing or by not evaluating the strategic benefits of the advanced technologies. It may also overestimate project value by not adequately accounting for project risk.

In order to assess the extent of these shortcomings in the DCF method, we apply stochastic simulation which adds explicit analysis of uncertainty in the cash-flow variables, and real option analyses which further adds optimality to the investment decision.

2.2 Dynamic stochastic analysis method

A stochastic analysis method uses Monte Carlo simulation to incorporate the estimations of uncertainties into the model variables. In our model, we present the input data such as primary randomized energy prices, electricity prices and carbon trading prices, with some statistical distributions rather than fixed and known price points. For example, we treat future energy prices subject to beta distributions with minimum, maximum values and mean values (these prices represent the long-run marginal cost of production in that period).

The results from this stochastic analysis are in the form of stochastic distributions of NPV. In contrast to the deterministic modelling results, the stochastic method results provide decision makers with information such as: "in a confidence interval of 95%, the NPV of this project will be less than $ 200 million". Choosing the stochastic process for modelling energy and carbon prices is likely to have an important influence on the results, but a detailed discussion of the merits of different approaches is beyond the scope of this paper.

The MINUIT model is capable of modelling almost any stochastic process. In the current context, we have chosen to model energy and carbon prices using the Exponential Mean Reverting (EMR) process. EMR is a mean reverting model in which the value of the logarithm of the process fluctuates randomly around the logarithm of the users' forecast values. The speed of reversion and the variance of the process are determined by the long-term volatility (LTV) and half-life (H) parameters. LTV describes the constant long-term uncertainty level exhibited by the process, and the H parameter describes the expected time it takes the logarithm of the process to travel half the distance towards the reversion level.

Mathematically, in the EMR process, the change of the price at time t, $t = t_0, t_1, t_2, \ldots t_n$ $[P(t_i)]$ is controlled by the following equation:

$$dP(t) = \kappa(\mu_i - \ln P(t))P(t)dt + \sigma P(t)dB(t) \tag{1}$$

Where: The first term in the equation describes the price change due to long-term trend with reverting capability to the assumed price. The second term captures the volatility of the prices. In the equation,

$$\mu_i = \ln y(t_i) + \frac{\sigma^2}{2\kappa} \tag{2}$$

is a linear shift of $\frac{\sigma^2}{2\kappa}$ of the logarithm of the reversion level $y(t_i)$, and $y(t_i)$ is the assumed price function that revolves around certain level in t_i. (long-run marginal cost of production or reduction).

$$\sigma = LTV \times \sqrt{2\kappa} \tag{3}$$

is the instantaneous volatility of the price change, and LTV is long-term volatility, or standard deviation of the logarithm of the price.

$$\kappa = \frac{\ln 2}{H} \tag{4}$$

is the speed of the reversion, and H is the half-life, the expected time it takes for the price to return halfway back to the reverting level after a deviation. B(t) is a stochastic (randomized) function generated by the computer.

The above mean reverting model is dynamic, stochastic, and log-normal. It will ensure the simulated distribution of the future forecast energy prices and carbon prices to be greater than zero and generally stick to the long-run marginal production costs for energy prices or long-run marginal reduction costs for CO_2 prices.

2.4 Real option optimization method in this study

The final element introduced in our model is the ability to decide when to invest. Flexibility in the timing of investment is one of the key strategic tools available to management in response to uncertainty. In general, the combination of increased volatility of prices and the flexibility to invest when conditions are optimal, leads to opportunities for making greater profits than under conditions of complete price certainty, since investors can time their investments to avoid the down-side risks, and maximize their up-side gains. The extent to which they can make these optimal timing decisions depends on the type of price uncertainty they face, and investors may have different attitudes toward price uncertainties caused by regulatory interventions as opposed to price effects caused by market fluctuations (resulting from non-policy effects such as economic growth, technical progress, etc.).

Optimal timing under uncertainty can manifest itself in two ways. First consider projects that appear not to be cost-effective (i.e. have a negative NPV) when measured using a standard DCF approach. If no account is taken of price uncertainty, then the decision would be never to invest in the technology. However, this same investment opportunity may actually be cost-effective under some conditions if prices are uncertain or volatile. In this case a real options analysis would indicate that some of the time the investment should be made; the effect of optimal timing is to accelerate the investment decision. This is the example we illustrate in our case study below.

Conversely, if an investor considers a power project that has a positive NPV under a DCF approach, the standard investment rule would be to invest immediately. However, a real options analysis that takes account of price uncertainty might indicate that greater profits could be made by waiting for more optimal price conditions. In this case, the effect of optimal timing would be to delay the investment relative to the case where no account is taken of uncertain prices.

Timing of investment (and choice of technology) is also of key interest to policy-makers, particularly in the power sector where investment is crucial to balancing supply and demand for electricity. Changes in the timing of investment could therefore have important effects on power prices. A key feature of the MINUIT model is its ability to derive an optimal investment rule which takes account of the uncertainties in costs and revenues, as well as the flexibility of investment timing.

In our case study, we started with a relatively inefficient coal-fired power plant of 472 MW capacity and 33% generation efficiency. We denoted the current power plant status as a Base Strategy (BS). One of the options available to BS is to replace the coal-fired power plant by a CCGT, which we call the target strategy (TS)[1]. We assumed that the project investment would take place within a period of t years $\{t=1, \ldots, T\}$. The capital investment in CCGT is a constant K. We denoted C_{BK} and C_{TK} cash-flows corresponding to BS and TS at year t respectively. We also have a discount curve where $d(t1, t2)$ denotes the discount factor applied at time t_1 to cash-flows occurring at time t_2. By definition, we have $d(t1, t1) = 1$. In any year t the random value of exercising the target option is the present value of the target scenario's cash-flow, that is:

$$V_t^{ex} = \sum_{k=t}^{T} d(t,k)\, C_{TK(k)} \tag{5}$$

The continuation value which is the value of the project if one chooses not to exercise the option in period t is given by:

$$V_t^{cont} = C_{BK} + d(t, t+1)\, V_{t+1}^{*} \tag{6}$$

---

[1] Other options include (1) switching to bio-mass coal co-firing; (2) improving the heat rate; (3) expanding the life time of the power plant; (4) early retiring of the power plant; (5) completely re-building a new coal-fired power plant or gas power plant, etc.

Where V^*_{t+1} is the summed cash flows of the base scenario project during year $t + 1, t + 2, ... T$, discounted back to year $t + 1$.

The optimal exercise policy is derived by estimating the value of exercise versus the continuation value using dynamic programming techniques (Dixit and Pindyck 1994). At any time t, we derive the optimal exercise policy according to the following rules:

$$\text{Excercise, if } V_t^{ex} - V_t^{cont} - K_{x(t)} > 0 \tag{7}$$

$$\text{Do not excercise, if } V_t^{ex} - V_t^{cont} - K_{x(t)} \leq 0$$

2.5 Computer software programming

MINUIT model uses a Real Option Calculator (ROC) Excel Add-in software programme also used by the Electric Power Research Institute (EPRI) in the late 1990s to support EPRI's GHG-CAM. The ROC uses the Monte Carlo simulation tool and runs the optimization routines. When this project started, the software was commercially available from Onward Inc., but is now no longer available. Other commercially available decision-support software could be used to carry out similar analysis.

3 Case study

We applied the model in a case study to examine the cost-benefits of switching an existing coal-fired power plant to a combined cycle gas turbine (CCGT) under uncertainties. The capacity of the power plant is 472 MW. It is a base-load power plant with a load factor of 85%. To simplify the analysis, we assumed the capacity and load factor would remain unchanged after the option is exercised (i.e. investment occurs).

3.1 Data collection

We collected techno-economic data from various sources such as capital investment costs (or option exercise costs), operation and maintenance (O&M) costs, fuel prices and CO_2 emissions for the two different technologies. The efficiencies of the coal-fired power plant and the CCGT plant are 33% and 54% respectively. CO_2 emission factors from the coal power plant and the gas power plant are 95 tCO_2/TJ (1.03 tCO_2/MWh) and 56 tCO_2/TJ (or 0.37 tCO_2/MWh) respectively. Table 1 shows some of the data.

Table 1. Techno-economic data for different technologies

Project Specific Assumptions	Base case (coal)	Option 1 - CCGT
Plant duration based on existing plant or new build?	Existing	New
Project Lifetime (Years)	20	25
Local Property Tax Rate	0.00%	0.00%
Production Tax Credit ($/MWh)	0.00	0.00
Capacity Retrofitted (MWe)	472	472
Capital Cost ($/kW)	0	630.84
Construction Period (No. years of capital of payment)	0	3
Construction Period (No. years deferral of revenue)	0	3
Capacity/Load Factor	85%	85%
Average annual efficiency of generation	33%	54%
% of coal in fuel mix	100%	0%
% of oil in fuel mix		
% of gas in fuel mix		100%
CO_2 Emissions Factor for fuel (tCO_2/TJ)	95	56
Environmental cost multiplier ($/MWh)		
Fixed Op&Maint ($/kW-Yr)	51.89	24.14
Unit Variable Op&Maint (excluding fuel cost) ($/MWh)	2.00	2.00

3.2 Assumptions in the scenarios

In our case study, we designed and ran three scenarios. We made the following assumptions in our basic scenario (or Scenario 1):

1. The discount rate is 10%. The project discount rate was the same as a firm's Weighted Average Capital Cost (WACC).
2. Coal price is $ 2,725/TJ at year 2005, with an annual growth rate of 0.5%.
3. Gas price is $ 6,521/TJ at year 2005, with an annual growth rate of 1.1%.
4. Underlying electricity price (excluding carbon price) is $ 47.9/MWh in year 2005, with an annual growth rate of 1.1%.
5. CO_2 trading price in year 2005 is $ 15/ton. The price will increase by 5% each year.
6. Carbon prices will be fully passed through to electricity prices at a rate determined by the marginal plant emissions.
7. The firm's combined corporate tax rate is 31%.
8. The project planning period is 20 years between 2005 and 2025.

In addition, we assumed that there is some degree of correlation between gas, electricity and CO_2 prices. The correlation coefficients are likely to be an important parameter in affecting the result, but a detailed analysis of the sensitivity of the model to these parameters is beyond the scope of this paper. For illustrative purposes, we present the correlation coefficients in Table 2.

Table 2. Correlation coefficients used in the study.

	Electricity price	Natural gas price	CO_2 price
Electricity price	1	0.5	0.5
Natural gas price		1	0.5
CO_2 price			1

In Scenario 2, we keep all of the above assumptions and add volatilities of CO_2 prices. Specifically, we set the volatility of CO_2 prices at 50% and reversion half-life of five years[2]. In Scenario 3, we keep all the assumptions in the first scenario and the second scenario, and add volatilities of energy prices. Specifically, we set volatilities at 10% and 20% respectively for coal and gas prices and reversion half-life at one year for both coal and gas volatilities. The purpose of the scenario analyses is to quantify the impacts of uncertainties (volatilities of CO_2 trading prices and energy prices) on the revenues or profits of new power technology investments and operations, and the consequences on the overall investment decision.

3.3 Results of the analysis

3.3.1 The higher the price volatility, the higher the probability to shift investment

The particular case study illustrated here is an investment that would not be considered cost-effective under DCF analysis with the price assumptions we have made. The simple NPV for the investment is -$ 120 million under deterministic prices. Consequently, the project would not be undertaken if no account were taken of uncertainty.

The effect of introducing optimal timing can be seen by allowing the model to run as a real option decision, but with volatility values set very low (Scenario 1)[3]. Even with optimal timing, with the new CCGT plant being allowed to be built at any time during the existing plant's 20 year remaining lifetime, the new CCGT would not be built in the absence of price volatility. This can be seen from the results in Table 3 where the recommended investment policies are illustrated. In Scenario 1, where we set all price volatilities to near zero, meaning that the energy prices and CO_2 trading prices are known or there is no price uncertainty, it is more economic for the project owner to keep running his current coal-fired power plant without any investment.

---

[2] The reversion speeds chosen for this case study are largely arbitrary – at this stage, a detailed analysis of suitable values has not been carried out. The slower reversion speed (longer half-life) of five years (compared to the rate for the other stochastic variables of one year) was chosen for CO_2 to reflect the possibility for longer-term swings away from expected price forecasts due to the relative lack of fundamentals What do you mean here? driving the price compared to the other energy commodities.

[3] The model does not allow zero values for volatility, so for illustration we set values for this scenario to 0.1%.

In Scenario 2, where the volatility of the CO_2 price is set at 50%, the project owner may consider the investment in a new CCGT. In 2009, should the project owner invest in CCGT, he would have 0.4% chance of achieving a NPV of up to $ 90 million. The probability to make profit by shifting from coal-fired power plant to CCGT during the whole period of planning is not high, only 14%. However, if all the prices (i.e. including gas, coal and electricity) are volatile during the planning period as in Scenario 3, CCGT will become much more attractive. The total probability of making a profit by shifting coal-fired power plant to CCGT is over 30%, much higher than the probability values in the previous two scenarios.

Table 3. Investment (option exercise) policies

Years	Scenario 1			Scenario 2			Scenario 3		
	exercise probability	exercise threshold	exercise relationship	exercise probability	exercise threshold	exercise relationship	exercise probability	exercise threshold	exercise relationship
2005	0.0%		Don't Exercise	0.0%		Don't Exercise	0.0%		Don't Exercise
2006	0.0%		Don't Exercise	0.0%		Don't Exercise	0.0%		Don't Exercise
2007	0.0%		Don't Exercise	0.0%		Don't Exercise	0.0%		Don't Exercise
2008	0.0%		Don't Exercise	0.0%		Don't Exercise	0.0%		Don't Exercise
2009	0.0%		Don't Exercise	0.4%	90,406,599	Greater than	0.0%		Don't Exercise
2010	0.0%		Don't Exercise	0.6%	93,490,322	Greater than	0.0%		Don't Exercise
2011	0.0%		Don't Exercise	0.6%	96,616,294	Greater than	0.0%		Don't Exercise
2012	0.0%		Don't Exercise	0.6%	100,316,608	Greater than	0.0%		Don't Exercise
2013	0.0%		Don't Exercise	0.4%	104,800,621	Greater than	0.0%		Don't Exercise
2014	0.0%		Don't Exercise	0.6%	109,441,596	Greater than	0.0%		Don't Exercise
2015	0.0%		Don't Exercise	0.5%	114,701,049	Greater than	0.0%		Don't Exercise
2016	0.0%		Don't Exercise	0.6%	119,969,911	Greater than	0.0%		Don't Exercise
2017	0.0%		Don't Exercise	0.4%	126,165,557	Greater than	0.0%		Don't Exercise
2018	0.0%		Don't Exercise	0.5%	132,422,628	Greater than	0.0%		Don't Exercise
2019	0.0%		Don't Exercise	0.4%	139,027,663	Greater than	0.0%		Don't Exercise
2020	0.0%		Don't Exercise	0.5%	146,235,927	Greater than	0.0%		Don't Exercise
2021	0.0%		Don't Exercise	0.4%	153,720,357	Greater than	1.4%	236,908,792	Greater than
2022	0.0%		Don't Exercise	0.4%	161,315,711	Greater than	1.9%	228,379,169	Greater than
2023	0.0%		Don't Exercise	0.8%	166,988,118	Greater than	2.7%	218,695,416	Greater than
2024	0.0%		Don't Exercise	0.9%	171,948,532	Greater than	4.1%	208,709,667	Greater than
2025	0.0%		Don't Exercise	5.4%	159,191,779	Greater than	20.0%	159,191,779	Greater than

More volatility increases uncertainty in project NPVs

Energy prices and CO_2 price uncertainties will considerably increase the uncertainties of the project NPVs. Figure 1 shows the histograms of the project NPVs under three scenarios. The range of NPV ($ 45.6 million ~ $ 48.2 million) under stochastic analysis in Scenario 1 where all the future prices are known and stable, is much smaller than the range of NPV in Scenario 3 where all the prices are unknown (-$ 116 million ~ +$ 245 million). It means that the higher the uncertainties of the project, the wider the range of the project NPVs. You may want to include the NPV distributions of both BS and TS, to illustrate the larger volatility of the latter (if that exists).

This characteristic can also be seen in Figure 2, which plots the results of Figure 1 as a cumulative function. This phenomenon of real options in the power investment is quite similar to that in the stock market.

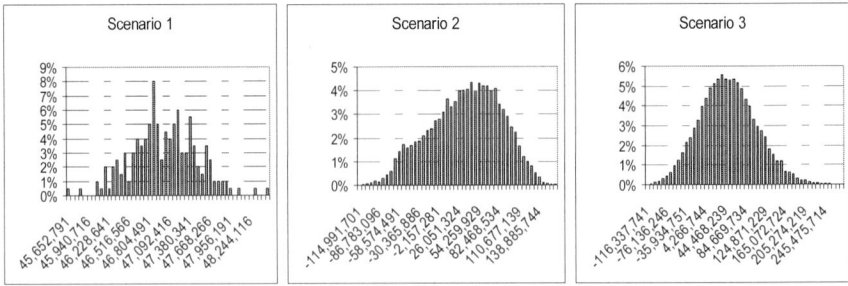

Fig. 1. Histograms of project NPVs ($)

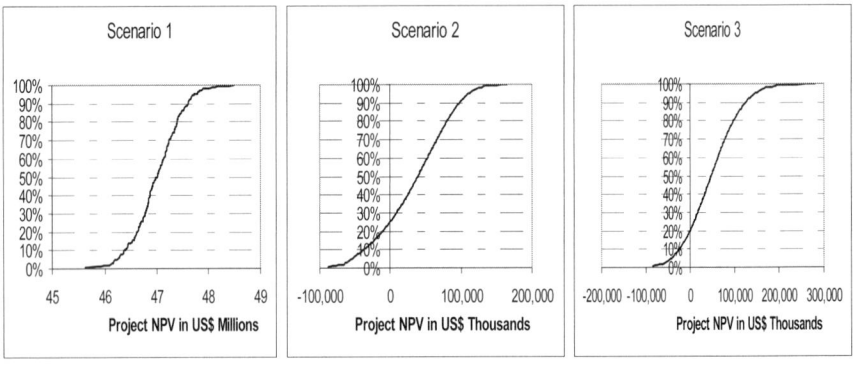

Fig. 2. Probability cumulative distribution functions

To summarize, this case study shows the following:

1. **Scenario 1**: If all the energy and CO_2 prices are stable, under the price assumptions used in the case study, investment in CCGT would not take place. The current coal-fired power plant will have an NPV of at least

$ 45.6 million during its lifetime. In a confidence interval of 95%, the NPV of the existing coal-fired plant will not be more than $ 48 million.
2. **Scenario 2**: If all the energy prices are stable but CO_2 trading prices are volatile, investment in CCGT may occur after 2009. Taking into account the investment cost ($ 300 million), the whole project has a loss probability of about 25%. The maximum value of loss can reach $ 110 million. On the other hand, however, the project NPV still has a 75% probability of being greater than zero, and 43% of being greater than the maximum value in Scenario 1 ($ 48 million);
3. **Scenario 3:** If all the energy prices and CO_2 trading prices are volatile, CCGT may be chosen after 2020. The loss probability of the project is about 20%, and the maximum value of the loss could reach $ 116 million. In a confidence interval of 95%, the project will not make more than $ 147 million. However, the project NPV will have a 49% probability of being greater than the maximum NPV ($ 48 million) in Scenario 1.

4 Conclusions

Different tools used for project finance analysis may lead to different results and investment choices. Real options analysis can introduce two types of deviation from the traditional DCF analysis. Firstly, consideration of risk and uncertainty can introduce a strategic value to keeping options open that would otherwise appear not to be cost-effective under a DCF approach. This is the effect that we have illustrated in our case study. Under the price assumptions used in the study, and with the DCF analysis methodology which cannot capture the uncertainties of climate change policy regulation and the primary energy prices, investment in new CCGT would not be attractive compared with continued operation of the existing coal-fired power plant. According to the results of the DCF method, the project developer would not consider CCGT as a replacement for the existing coal-fired plant in 2005–2025. However, by using the stochastic and real option approaches which capture the investment uncertainties, CCGT could be competitive after 10 or 15 years under some conditions which are met up to 30% of the time.

Secondly, projects that look attractive under a DCF type analysis can become less attractive or even not cost-effective at all under some conditions if volatility is introduced. However, having flexibility on timing of investments allows decision makers to avoid down-side risk and maximize up-side benefits so that increased volatility can often increase the value of the option to invest. However, this optimal timing of investment can often mean delaying investment relative to the case without uncertainty, and this could be a cause for concern for policy-makers particularly in the context of investment in the power sector. Our model is also capable of analyzing these effects, although we do not present this type of analysis in this paper.

Uncertainties in energy prices and CO_2 prices will significantly affect decision-making. Fortunately, to some extent we can now model and quantify the impacts of these uncertainties.

References

ADB (Asian Development Bank) (2002) Guidelines for the Financial Governance and Management of Investment Projects Financed by the Asian Development Bank, Asian Development Bank, January, Manila, Philippines

EPRI (Electric Power Research Institute) (2005) Green Gas Emission Reduction Cost Analysis Model (GHG-CAM), Version 1.1, Software Users Manual, Palo Alto, California, US

Laughton D, Hyndman R, Weaver A, Gillett N, Webster M, Allen M, Koehler J (2003) A Real Options Analysis of a GHG Sequestration Project. The paper is available from David Laughton at: david.laughton@ualberta.ca)

Sekar C (2005) Carbon Dioxide Capture from Coal-Fired Power Plants: A Real Options Analysis, Massachusetts Institute of Technology, Laboratory for Energy and the Environment, 77 Massachusetts Avenue, Cambridge, MA 02139-4307, US

Dixit AK, Pindyck RS (1994) Investment under Uncertainty, Princeton University Press, New Jersey 08540, US

Onward (2004) Real Options Calculator Excel Add-in Version 1.1, User's Manual, Onward Incorporated 888 Villa St. Suite 300, Mountain View, CA 94041, US

Myers S (1977) Determinants of Capital Borrowing, Journal of Financial Economics, vol 5, Simon School of Business, University of Rochester, New York 14627, US.

IEA – International Energy Agency (2005) Inquire 3 (IEA Energy Database), Paris, France

Ruszczynski A, Shapiro A (ed) (2003) Stochastic Programming – Handbooks in Operations Research and Management Science, ISBN: 0-444-50854-6, 700 pages, NORTH-HOLLAND. http://books.elsevier.com/elsevier/?isbn=0444508546

Reputational impact of businesses' compliance strategies under the EU Emissions Trading Scheme

Corinne Faure[I,II], Arne Hildebrandt[III], Karoline Rogge[IV,V], Joachim Schleich[VI,VII]

[I] European Business School
Schloß Reichartshausen, 65375 Oestrich-Winkel, Germany
corinne.faure@ebs.de

[II] Johann Wolfgang Goethe-University, 60054 Frankfurt, Germany
faure@wiwi.uni-frankfurt.de.

[III] KPMG Deutsche Treuhandgesellschaft
60439 Frankfurt, Germany
ahildebrandt@kpmg.com.

[IV] Fraunhofer Institute Systems and Innovation Research
Breslauer Str. 48, 76139 Karlsruhe, Germany
karoline.rogge@isi.fraunhofer.de

[V] ETH Zurich, Group for Sustainability and Technology
Department for Management, Technology, and Economics
8092 Zurich, Switzerland

[VI] Fraunhofer Institute Systems and Innovation Research
Breslauer Str. 48, 76139 Karlsruhe, Germany
joachim.schleich@isi.fraunhofer.de

[VII] Virginia Polytechnic Institute and State University
Blacksburg, Virginia 24060-0401, USA

Abstract

Since January 2005, more than 11,000 installations in the EU are participating in the CO_2 Emission Trading Scheme. Within this system companies may choose from a variety of compliance strategy options. To assess the relevance of reputational effects on companies' climate strategies, a survey of the 300 largest emitting companies in Germany was conducted. Results indicate that reputational effects matter, but are dominated by other factors such as compliance costs, technical and economic risks, or practicabilty. Long-run reputational risks are most relevant for the Clean Development Mechanism and Joint Implementation.

Keywords: Emission Trading Scheme, Clean Development Mechanism, Joint Implementation

1 Introduction

Since January 2005, about 11,400 large installations from the energy sector and from most other carbon-intensive industries in the European Union have begun participating in the EU-wide CO_2 Emission Trading Scheme (EU ETS). To comply with the system, companies may choose from a variety of compliance strategy options.

From a purely economic perspective, companies are expected to choose the most cost-efficient compliance option. However, other decision factors may also be relevant. Among these factors, reputational risk appears particularly interesting, as the various options available have different image impacts. In this paper, we empirically compare the importance of reputational aspects relative to other decision factors and explore the perceived impact of diverse compliance strategy options on reputation.[1]

2 Overview of EU ETS and project-based Kyoto mechanisms

The EU ETS is the world's largest emissions trading program and is expected to help the EU and its Member States (MS) fulfil their obligations under the United Nations Framework Convention on Climate Change, the Kyoto Protocol and the Burden-Sharing Agreement in a cost-efficient way. The types of installations to participate in the EU ETS are listed in Annex I of the Directive and include combustion installations with a rated thermal input capacity of at least 20 Megawatts, refineries, coke ovens, steel plants, and all installations producing cement clinker, lime, bricks, glass, pulp, and paper which exceed certain output thresholds. As a result, the EU ETS covers about 50% of Europe's CO_2 emissions.

There are several compliance options available to companies participating in the EU ETS. These options fall into two main categories: internal and external reduction of emissions. The *internal* options include the adoption of new technologies and measures to improve *energy efficiency* such as heat recovery, thermal insulation, improved control systems, cogeneration and combined heat and power (CHP), or investments in more energy-efficient installations. Likewise companies pursue *fuel substitution* of carbon-intensive fuels such as coal and oil by gas or bio-fuels, or they may *reduce production*.

A number of compliance options are available for an *external* reduction of greenhouse gas emissions. First, companies can buy (and sell) emission allowances on the market (*emission trading*). In addition, the so-called EU Linking Directive (2004/101/EG) allows companies to use credits from the two project-based Kyoto mechanisms *Clean Development Mechanism* (CDM) and *Joint Implementation* (JI). This can be done either through direct investment in such climate projects or through indirect investment in *climate funds*, such as the World Bank Pro-

---

[1] To a large extent, this paper relies on Hildebrandt (2007).

totype Carbon Fund or the German Climate Fund operated by the German Reconstruction Bank (KfW).[2] Through this linkage of EU ETS with CDM and JI, companies can therefore also benefit from emission reductions achieved outside the EU ETS, thereby lowering companies' compliance costs. For the second phase of the EU ETS (2008–2012) the European Commission has limited the amount of credits from JI and CDM projects that companies may use to cover their emissions. First analyses suggest however that – at least theoretically – the maximum total amount of these credits allowed, would almost be sufficient to make up for the projected shortage of allowances (Betz et al. 2006; Schleich et al. 2007).

According to Article 12 of the Kyoto Protocol, the purpose of CDM is to assist developing countries in achieving sustainable development and to contribute to the ultimate objective of the UNFCCC – the stabilization of the world's GHG emissions at a level that would prevent dangerous anthropogenic interference with the climate system. The CDM also aims at assisting Annex-I-countries (i.e. those with a binding commitment which are mainly industrialized and transformational economies) in complying with their quantified emission limitation and reduction commitments. In contrast to this, Article 6 of the Kyoto Protocol specifies JI as a mechanism which allows any Annex-I-country to transfer to, or acquire from, any other Annex-I-country emission reduction units resulting from projects aimed at reducing anthropogenic emissions by sources or enhancing anthropogenic removals by sinks of greenhouse gases in any sector of the economy. Both baseline-and-credit mechanisms require that certificates are only granted if emission reductions are additional to those that would have occurred without the incentives of these two policy instruments. Before certificates from CDM- and JI-projects are issued, they have to be generated through a fairly complex and formalized project cycle. For the CDM, this cycle includes the development of a baseline (emissions which would have occurred in the absence of the project) based on approved methodologies, validation of the project through specific organizations (operational entities), the monitoring of actual emissions as well as the verification and certification of actual emission reductions. The process for JI projects may be simpler if the host country fulfils certain requirements for national emission monitoring and emission inventories (JI First Track).

In summary, the following options are available for companies involved in the EU ETS: (1) internal reduction through efficiency gains or innovation; (2) internal reduction through fuel substitution; (3) internal reduction through production reduction; (4) external reduction through emissions trading; (5) external reduction through CDM; (6) external reduction through JI, and (7) indirect external reduction through investment in climate funds. These options will be compared to one another in the rest of the paper.

---

[2] For a recent overview of these funds see Betz et al. 2005, pp. 41.

3 Determinants of firms' compliance strategy choices

Our first objective was to test the relative impact of reputation factors compared to other decision factors when choosing a compliance strategy. A number of factors are mentioned in the literature (Betz et al. 2005). The most important are cost-related factors, including direct (monetary) costs and transaction costs (i.e. additional expenditure of time and effort required in order to find a suitable measure/project and to acquire the allowances/credits). Next in order of importance are risk factors of both technical and economic natures. Finally, we consider short-term and long-term reputation effects. We expect the reputation effects to be perceived as less important than the economic or technical factors typically considered, but nevertheless to have some impact on the choice of a compliance option.

Reputational risks are linked to the potential effects (positive or negative) that a given action might have on an image. We understand image to be the sum of deliberate and unconscious feelings, attitudes, experiences and opinions which a person or group of persons holds towards a subject matter. According to Essig et al. (2003, p. 21) image represents a stereotypical simplification of an objective issue. The literature distinguishes between six types of images in the business world: product, product category, brand, company (or corporation), sector, and country image. A firm can only influence its product, brand, and company images (Möhlenbruch et al. 2000, pp. 29–33). The choice of compliance strategies is a corporate decision, and thus has an influence on image at the corporate level. We will therefore focus on reputational risks at the corporate level. Corporate image is determined by a firm's corporate identity, communications, design and behaviour (Meffert 1998, p. 670f; Birkigt, Stadler 1986, p. 28). The corporate climate strategies chosen, when communicated, have therefore the potential to affect corporate image.

Image can vary across persons or groups. It is therefore important to consider the various recipients of corporate actions. For internal stakeholders (such as employees, owners and partners), high returns and financial strength, good management quality and the high satisfaction of employees are indicators of a positive corporate image. For external stakeholders (such as suppliers, customers, competitors, shareholders, financial institutions, public authorities, NGOs, the general public and opinion makers), the most important factors to positively influence a firm's image include a strong customer-orientation, good product quality and price-performance-ratio, and good communication (Manager-Magazin 2004; Weis 2005, p. 525).

It is interesting at this point to consider the characteristics of the companies covered by the EU ETS. First, most EU ETS firms manufacture commodity products and can hardly set themselves apart from competitors through product differentiation[3]. Second, products from EU ETS firms are mainly consumed by indus-

---

[3] Certainly, producers of paper and – to a lesser extent – even cement are able to differentiate a few of their products. Similarly, some power companies have established successful product brands after liberalization and some utilities companies are able to differenti-

trial customers. For these customers, the most important decision criteria are the price-performance ratio of products and a strong customer-orientation; the image of the supplier is of secondary concern. Third, and possibly as a consequence of being industrial companies, EU ETS firms do not have a strong corporate image. In a recent image ranking study of 171 German companies from 16 sectors, the large power producers E.ON and RWE achieved ranks 48 and 50 respectively, while the cement companies HeidelbergCement and Dyckerhoff were ranked 138 and 155 (Manager-Magazin 2004). These low ranks partly reflect the fact that many of these companies have a comparatively negative image due to their high production focus and its associated environmental side effects. These companies therefore have few possibilities to distinguish themselves positively from others and generally suffer from a poor environmental image. It is therefore particularly important for these companies to manage their environmental image effectively. The Responsible Care Program ® (International Council of Chemical Associations 2002) in the chemical industry or the diverse "green power" products in the utilities industry are good examples of the attempts made by some of these companies to actively manage their environmental image.

Compulsory emissions trading causes two fundamental problems for the communication of a firm's environmental performance: (1) Firms lose the marketing potential of *voluntary* reductions of CO_2 emissions and cannot set themselves apart from competitors that are also covered by EU ETS. (2) Future CO_2 emission reductions cannot as easily be marketed as a company's green involvement because motives of cost savings and profit maximization may be suspected to be the main drivers for CO_2 reductions. On the other hand, a study conducted in 2004 indicates that there is little general public knowledge about emissions trading (only 30% of the surveyed Germans knew about the new instrument (TND Emnid 2004, p. 14)), so that activities undertaken under the EU ETS might still be communicated independently of this program. Thus, the choice of compliance strategies might not only be influenced by cost and efficiency factors, but also by the potential effects of these strategies on corporate image.

The different strategies available can be distinguished in a number of dimensions. One dimension is the level of involvement of the company in the CO_2 reduction activities. While internal strategies require a direct involvement of the company, external strategies signify that the company obtains CO_2 credits through investments in external projects. We would expect such projects to be perceived as more sensitive to reputational concerns, as these external strategies could be interpreted as the company trying to escape its responsibilities and especially its local, regional and national responsibilities. These strategies are therefore likely to be perceived more negatively by the domestic general public. This should be particularly true for climate fund investments, which represent the lowest level of involvement of all the strategies. On the other hand, external strategies have the potential to be communicated positively as responsible endeavours to support development in less favoured countries; they exhibit therefore an additional di-

---

ate themselves by emphasizing their regional involvement. In general, however, EU ETS firms only differentiate to a small extent between different brands.

mension, their international involvement, which can be interpreted positively, especially by NGOs and political institutions, but also to some extent by the general public. This positive effect should be particularly pronounced for CDM projects, which, under the rules set out in the Kyoto Protocol, have to contribute to sustainable development in the host country. Both effects together suggest that external strategies should be viewed as more reputation-sensitive than the other options. Since the direction of the reputation effects appears ambiguous, reputation concerns are likely to play a bigger role in the choice of these strategies.

Another dimension used to differentiate strategies is the extent to which they can be interpreted as being voluntary. For instance, fuel switching, efficiency gains or interesting CDM or JI projects have the potential to be communicated as voluntary proactive company initiatives. On the other hand, strategies such as production reduction (likely to be interpreted negatively because of potential job losses), climate funds, or the purchase of emissions certificates through the trading system appear more reactive, simply in order to comply with the system. We therefore expect that the more proactive strategies will be associated with a higher reputational impact.

Finally, the various strategies available have different time horizons. Some have a short-term character (fuel switching or emissions trading for instance), whereas others imply long-term projects (CDM or JI, but also climate funds, which are typically long-term investments). We therefore expect that long-term reputational concerns will play a comparatively greater role for strategies with a longer time orientation, whereas short-term reputational concerns should dominate for strategies with a shorter time orientation.

In summary, we expect that reputational aspects will play a minor role in decisions concerning compliance strategies. We expect that their role will be largest for CDM and (possibly) JI projects, particularly in the long term. Because of the low awareness of emissions trading among the general public, we expect that reputational effects will become more important in the long run as the general public becomes better acquainted with the EU ETS and reduction goals become more stringent.

4 Method

In order to test the reputational impact of the diverse strategies, we conducted expert interviews and administered a standardized survey to companies involved in the EU ETS. Because of differences in the timing and scope of the national implementation of the EU ETS across EU Member States (see Betz et al. 2004), we decided to focus on a single country. With an emission budget corresponding to 499 million tons of CO_2 p.a. for the installations covered by the EU ETS in the first phase (2005–2007) of the EU ETS, Germany holds about 25% and thus the largest share of the EU-25 cap, and also contributes the largest number of installations covered by the EU ETS (1,849 installations). Germany was also among the first to transpose the EU linking directive into national law, thereby allowing

German EU ETS firms to use credits from JI and CDM to fulfil their obligations under the scheme (ProMechG 2005). Germany can therefore be seen as one of the forerunners in the EU ETS, where all the compliance options of the ETS were available at an early stage.

In a first step, we conducted 15 telephone interviews with leading experts from 9 energy-intensive companies, 2 industry associations, 2 non-governmental organisations (NGOs), and 2 specialized consulting companies.

In a second step, we conducted a survey among large industrial emitters covered by the EU ETS in Germany in order to assess the relevance of various decision factors for companies' compliance strategies empirically. For the survey, we contacted 303 companies, each with at least 100,000 allocated allowances. We identified the person responsible for EU ETS implementation and sent them an email asking to participate in our survey. The effective response rate of our online survey was 16.5%, with approximately half of the 50 respondents coming from the energy sector (Figure 1 provides descriptive statistics about the sample).[4]

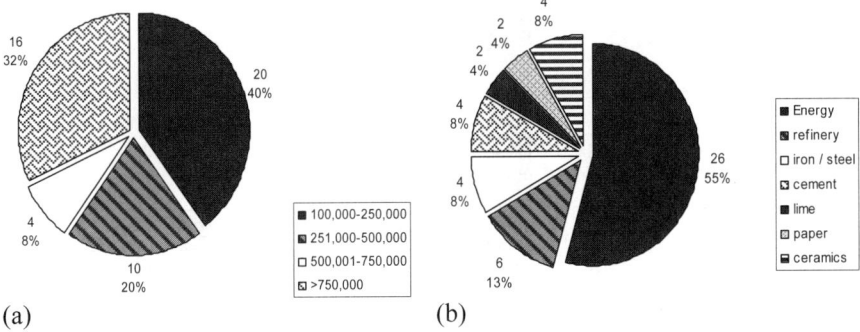

Fig. 1. Number and percentage of respondents according to (a) allocated allowances and (b) sector

Interviews and survey took place in early 2005, i.e. shortly after the launch of the EU ETS. Both included general questions about the EU ETS and its consequences for the respondent's organisation, as well as information about the compliance strategies chosen and up-to-date experience. The core of the survey was a ranking of the diverse decision factors in the process of choosing a compliance strategy (on a scale from 1 to 6, with 1 being not important and 6 being important), and a specific evaluation of the decision criteria: costs, transaction costs (effort and practicability), risk (technical and economic), and short-term and long-term reputation for specific strategies. Therefore, the relevance of each decision factor was first evaluated globally, before being evaluated specifically for each strategy available (internal through efficiency gains, fuel substitution or production reduction, or external through emissions trading, CDM, JI or climate funds).[5]

---

[4] Because the sample size is relatively small, statistical tests were not conducted.
[5] Note that the production reduction strategy was included in the interviews but not in the survey. At the time the survey was conducted, the EU ETS was (and still is) subject to

5 Results

Descriptive results. Among the 50 companies surveyed, the compliance strategy option used most at present is the trading of allowances. This is followed (but a lot less frequently) by internal measures – predominantly efficiency improvements – and investment in climate funds (see Figure 2). JI is not used within the sample. Interestingly, most companies seem to use only one strategy (even though multiple responses were allowed, only 2 companies mentioned multiple strategies).

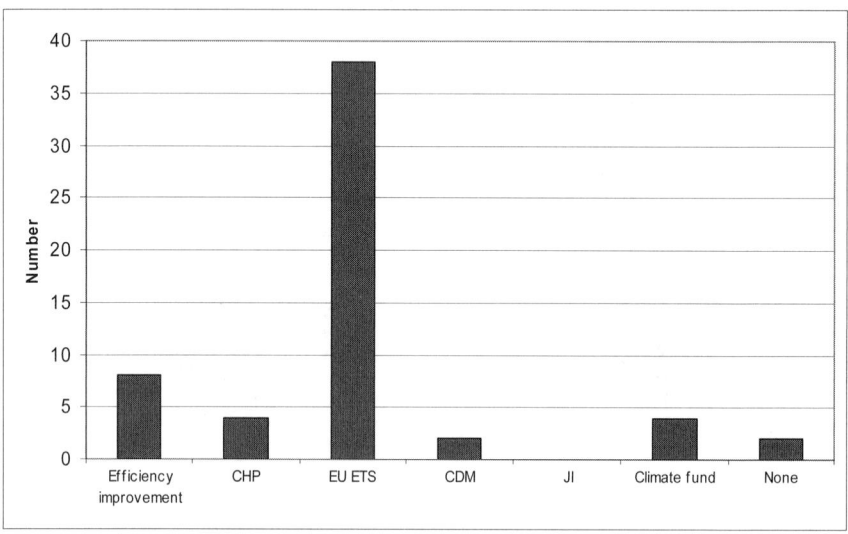

Fig. 2. Compliance strategy options currently in use (sample size = 50)

Importance of decision factors. The global assessment of each decision factor led to the following results (Figure 3): direct abatement costs appear to be the most important decision factor (mean = 5.16), followed by economic risk and practicability (mean = 4.32), technical risk (mean = 3.96), and time and effort (mean = 3.90). Reputational aspects were found to be least important when considering compliance strategy options. In line with our hypothesis, respondents put a stronger emphasis on long-term reputational effects (3.00) compared to short-term reputational effects (2.88), although the difference is only marginal. A similar pattern of results was obtained for the expert interviews.

---

heavy criticism by companies and industry lobbies claiming that the EU ETS would result in production reduction and job losses. To avoid receiving "strategic" responses from the companies, the category of production reduction was not included in the survey.

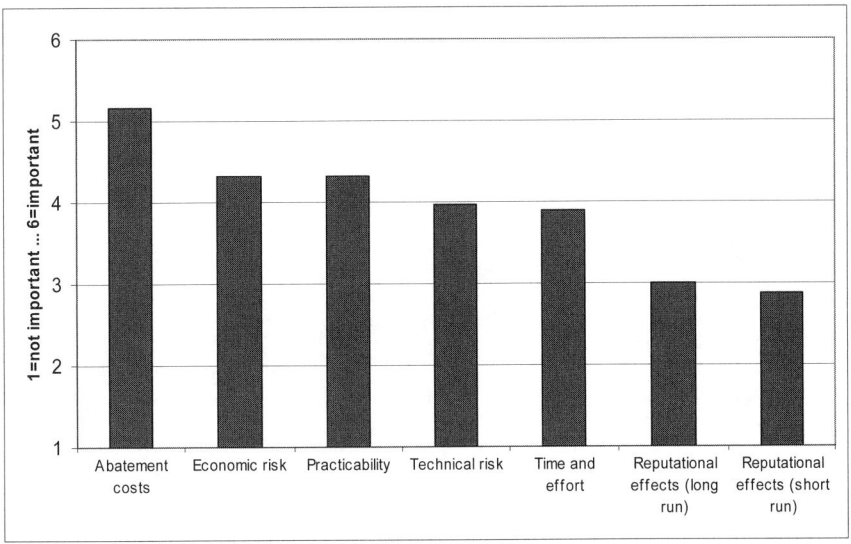

Fig. 3. Importance of decision factors for determining compliance strategy options (sample size = 50)

Reputational impact of compliance strategies. After respondents had evaluated the importance of the suggested decision factors for selecting a compliance strategy option, they were asked to assess selected compliance strategy options with regard to our decision factors. Respondents could rate the relevance of given concerns as a decision factor for a particular strategy on a scale from 1 (low) to 6 (high). Figure 4 shows the mean values for the seven compliance strategy options under consideration. Similarly, Figure 5 displays the relevance of each decision factor for the various compliance strategies. Abatement costs are estimated to be highest for the two internal measures of fuel switching and efficiency improvement as well as EU ETS and lowest for JI projects and investment in climate funds. Costs for project-based certificates are lower than for internal measures or EU ETS; this can be explained by the fact that production processes are less efficient in transforming and developing countries, partly as a result of less stringent environmental regulation. Thus, it is cheaper to reduce emissions in those countries than it is to do so domestically or through the EU ETS. That JI projects are assessed to exhibit (slightly) lower costs than CDM projects may be somewhat surprising at first, but corresponds with empirical findings in the literature at that time (e.g. Lückge, Petersons 2004; Betz et al. 2005, p. 58). Possible reasons include higher exchange rate risks, buyer liability, higher political risks, higher project risks or a premium for the additional environmental and social benefits (reputational benefits!) associated with CDM projects.

Technical risk is judged to be highest for both internal measures (efficiency improvement and fuel switching), and lowest for investments in climate funds. In the sectors covered by the EU ETS, improving energy efficiency and changing the fuel mix typically affect the core production processes, possibly exposing the company to considerable risks. By contrast, buying allowances on the market for EUAs is associated with virtually no technical risk.

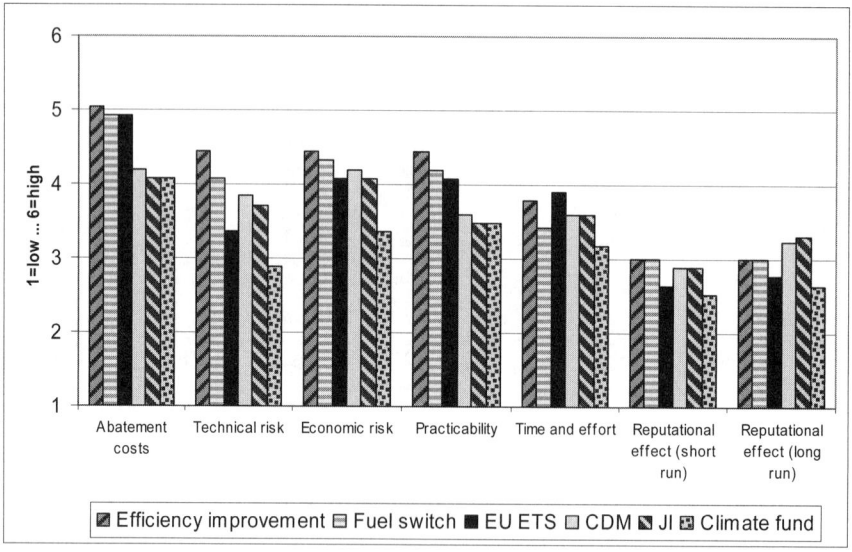

Fig. 4. Assessment of decision factors for compliance strategy options (sample size = 50)

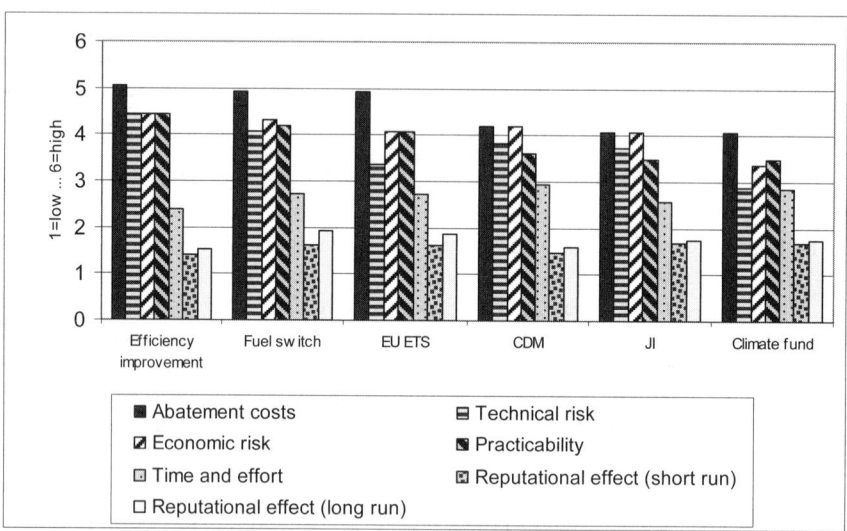

Fig. 5. Importance of decision factors for compliance strategy (sample size = 50)

Economic risk appears to be highest for energy efficiency improvements followed by fuel switching and lowest for climate funds. Internal measures involve investment costs and could result in technological lock-in effects, which may turn out to be costly if market prices for EUAs fall after the investment is made.[6] In this case, it would have been cheaper and less risky to use external measures. Likewise, companies face uncertainty about future allocation rules. In particular, if the quantity of allowances allocated for future phases depends on emissions in earlier phases, using internal measures would lower the quantity of allowances allocated (updating problem).[7] By contrast, using external measures does not affect the emissions of an installation covered by the EU ETS.

The two internal measures, followed by allowance trading, are considered to be the most practical compliance strategy options, while JI, CDM and climate funds are seen as less practical solutions. This result may be explained by the complex, formalized approval procedures for the new instruments JI and CDM and the respondents' potential lack of information about them. Similarly, investing in JI- and CDM-projects as well as participating in a climate fund requires large financial outlay and additional know-how which can probably only be afforded by large companies. In comparison, companies are more familiar with internal measures, and outside expertise may not be required at all.

Time and effort are estimated to be highest for allowance trading, followed by efficiency improvements and lowest for climate funds. This latter result is surprising as CDM and JI are usually mentioned in the literature as the options with the highest transaction costs (search for project partners, monitoring and project registration or the issue of certificates) (see Michaelowa et al. 2002). However, at the time the survey was conducted, respondents may have been very aware of the transaction costs associated with EU emission trading for using brokers, exchanges, and various consultants for monitoring, reporting and validation of emissions etc. (see Schleich, Betz 2004), whereas the time and effort required for other external options may not have been as well known to many respondents due to a lack of experience.

In contrast to our interview results, where CDM had the highest reputational effect (both short- and long-term), our survey results show that internal measures (efficiency improvements and fuel switch) have the highest reputational effect in the short run, but the differences when compared with JI and CDM are rather small. The highest long-term reputational effect is attributed to JI followed by CDM, thereby confirming our expectations. For both short-term and long-term reputation, investment in climate funds has the lowest effect.

In order to draw conclusions about the importance of the decision factors in the decision making process, we weighed the assessment of the factors by their overall importance for each strategy. We therefore converted the importance of decision factors into weights. The weights are normalized such that the decision factor

---

[6] In fact, when the European Commission first published verified emissions data for 2005 for the companies covered by the EU ETS (CEC 2006) in May 2006 prices for EUAs plummeted from around € 26/EUA to around € 10/EUA because the data suggested that the market is fairly long in the first phase. For the second phase allocation appears to be more stringent (e.g. Betz et al. 2006) and future prices range around € 20/EUA.

[7] In Germany, allocation rules for phase 2 were not known until mid 2006. According to these rules, allocation for existing installations for phase 2 also depends on verified emissions in 2005.

with the highest value, i.e. 5.16 for abatement costs, was assigned a weight of 1.0. Figure 6 shows the resulting relevance of the decision factors.

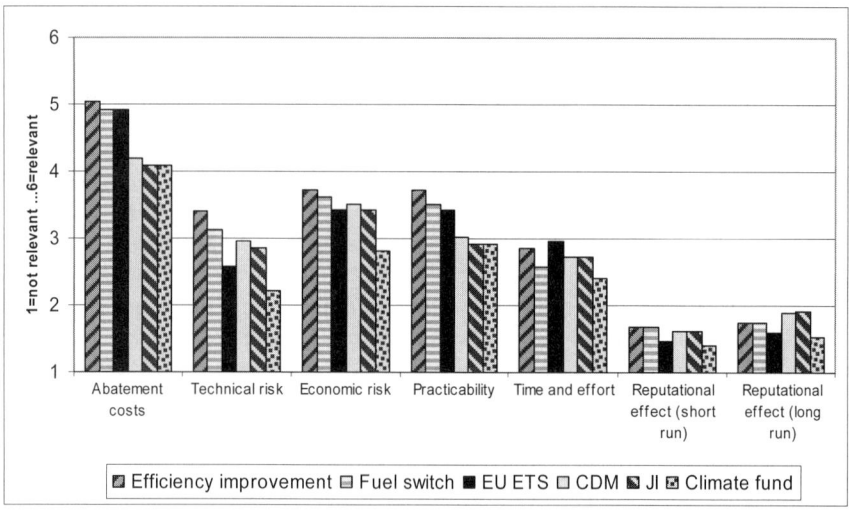

Fig. 6. Relevance of decision factors for compliance strategy options (sample size = 50)

In line with the results of our interviews, reputational aspects play the least important role in deciding on compliance strategy options, with a factor of 2 for technical risk, time and effort, and a factor of 3 for abatement costs. As expected, the reputational effects of all options, and especially of JI and CDM, grow in importance in the long run. These two strategies clearly appear to be the ones where long-term reputational concerns have the biggest impact.

While the space limitations of the online survey did not allow the stakeholder issue to be covered in detail, the interviews yielded additional information about the impact of reputational concerns on the choice of the different strategies. Within the interviews, we investigated the importance of a company's image with regard to its various stakeholders in determining a compliance strategy. The political domain appears the most important stakeholder for EU ETS firms, closely followed by regional stakeholders and NGOs. Within the EU ETS the political domain is of particular importance to companies, since political decision-makers decide on the quantity of EUAs allocated for free. Depending on the market price of EUAs and the companies' ability to pass on the additional (opportunity) costs, the primary allocation results in a huge income transfer to those companies. In any case, company profits depend on primary allocation and organised industries, like the power industry, may lobby the governments for additional cost-free allowances at the expense of other industry groups, or other domestic sectors outside the EU ETS which are less well organised (Olson 1965; Markussen and Svendsen 2005).

Within the interviews, respondents also discussed the increased importance of environmental concerns when selling products to private consumers. Indeed, a firm's environmental orientation appears particularly important for private consumers when products are relatively homogeneous and have the same price – as is the case for the majority of products of EU ETS firms (KLD 2005). This suggests that EU ETS firms with private customers might put more emphasis on their environmental image when selecting a compliance strategy than EU ETS firms with only industrial customers.

6 Summary and final remarks

In this project we tested the impact of reputational concerns on the choice of corporate compliance strategies in response to EU ETS, using expert interviews and a standardized survey. Our results show that reputational concerns play a minor role when selecting a compliance strategy option. This might be due to the fact that the companies covered by EU ETS – with a few exceptions – serve industrial consumers, not private ones, and therefore show little concern for risks to their reputation. These results might also be due to the fact that the individuals responsible for implementing the EU ETS are typically from a technical or financial background and do not regard image management as a major concern. However, since the different climate options clearly exhibit different characteristics as far as their reputational impact is concerned, we expect this situation to change and that the image effects of the diverse strategies will play an increasingly important role in the future. This is particularly likely to happen as the EU ETS matures and receives more public attention.

Long-term reputational effects were found to be most relevant for the two project-based mechanisms, CDM and JI. This can be interpreted both as an opportunity and as a threat, as these effects could be either negative (escape from national obligations) or positive (contribution to sustainable development abroad). Our survey does not allow for direct conclusions on this issue, but from our interviews it appeared that companies view the direct involvement in climate protection projects as an opportunity. This would mean that companies will be closely examining the quality of their projects, thereby showing their commitment to sustainable development in transforming and developing countries. Future studies should investigate the ambivalent influence of CDM and JI projects on a company's reputation in more detail.

References

Betz R, Eichhammer W, Schleich J (2004) Designing National Allocation Plans for EU emission trading – A First Analysis of the Outcome. In: Energy & Environment 15 (3): pp 375–426

Betz R, Rogge K, Schleich J (2005) Flexible Instrumente im Klimaschutz. Emissionsrechtehandel, Joint Implementation, Clean Development Mechanism. Eine Anleitung für Unternehmen. Umweltministerium Baden-Württemberg (ed), Stuttgart: Umweltministerium Baden-Württemberg

Betz R, Rogge K, Schleich J (2006) EU Emissions Trading: An Early Analysis of National Allocation Plans for 2008–2012. In: Climate Policy 6 (4): 361–394

CEC (2006) EU emissions trading scheme delivers first verified emissions data for installations, IP/06/XXX, Brussels: CEC

Birkigt K, Stadler M (1986) Corporate Identity – Grundlagen, Funktionen, Fallbeispiele . 3.Auflage, Landsberg am Lech

Essig C, Soulas De Russel D, Semanakova M (2003) Das Image von Produkten, Marken und Unternehmen. Sternfels

Hildebrandt Arne (2007) Strategien und Unternehmensimage beim CO_2-Emissionhandel. Vdm Verlag Dr. Müller

International Council of Chemical Associations (2002) Responsible Care Status Report 2002

KLD (2005) Consumers Respond to Brands Affiliated with Social Issues. KLD Research & Analytics, Inc., Boston

Lückge H, Petersons S (2004) The role of CDM and JI for fulfilling the European Kyoto commitments, 1232. Kiel Working Paper, Kiel: IfW

Manager-Magazin (2004) Imageprofile 2004 – Ranking nach Gesamtbewertung. In: Manager-Magazin, 2004

Markussen P, Svendsen GT (2005) Industry lobbying and the political economy of GHG trade in the European Union. In: Energy Policy 33: 245–255

Meffert H (1998) Marketing – Grundlagen marktorientierter Unternehmensführung. Wiesbaden

Michaelowa A, Stronzik M, Eckerman (2002) Transaction Costs of the Kyoto-Mechanisms. HWWA-Disscussion Paper 175, Hamburg

Möhlenbruch D, Claus B, Schmieder UM (2000) Corporate Image und integrierte Kommunikation als Problembereiche des Marketing. Halle/Saale

Olson M (1965) The Logic of Collective Action. In: Cambridge University Press

ProMechG (2005) Projekt-Mechanismen-Gesetz

Schleich J, Betz R (2004) EU-emissions trading and transaction costs for small companies. In: Intereconomics: pp 121–123

Schleich J, Betz R, Rogge K (2007) EU Emissions Trading – better job second time around? In: European Council for Energy-Efficient Economy (Paris): Proceedings of the 2007 eceee Summer Study. Saving energy – just do it! La Colle sur Loup, Côte d'Azur, France: ECEEE

TND Emnid (2004) Summary-Report zur Studie Emissionshandel – im Auftrag der deutschen BP AG

Weis H (2005) Marketing – Kompendium der praktischen Betriebswirtschaft. Olfert K (ed), Kiel

Prudence, profit and the perfect storm: climate change risk and fiduciary duty of directors[1]

Donna Lorenz

Maunsell Australia Pty Ltd
Level 9, 8 Exhibition Street
Melbourne Victoria 3000 Australia
donna.lorenz@maunsell.com

Abstract

A 'Perfect Storm' is named for its unique converging forces that individually would each create a 'strong storm' but together feed perfectly into each other to create the ultimate storm. Many warn such are the combining forces of climate change. Likely to happen concurrently and change constantly, the converging risks of climate change will take many forms. This paper reviews the potential business risks of climate change considered both highly likely and earliest to occur, and explores if Australian company directors have a fiduciary obligation to respond.

Businesses are already warned to expect exposure to a combination of physical changes, changing market conditions, new sources of competition and new government regulations due to climate change. These changes will likely affect different industries in different ways, different companies within industries in different ways and different regions and countries differently. This review found Australian directors have a fundamental fiduciary duty to manage business risks with the care and diligence comparable to a reasonable person in that position. They are also required to actively inform themselves while managing those material risks. It is clear that climate change represents a significant risk to many Australian businesses. As such, Australian directors have a fiduciary duty to assess the impacts of climate change on their business and, if sufficiently material, to disclose the exposure to the market and respond to mitigate the risk.

Keywords: Climate change, business risks, Australian perspective

---

[1] The research for this paper was conducted prior to 2005 and updated in 2008 for the purpose of this publication.

"Embedded in the challenge of climate change are both dangers and possibilities. Immense dangers for firms and investors who make bad choices, or no choices, about how to respond to the risks posed by climate change, and are then held accountable in the marketplace, the boardroom, or the courts; and immense possibilities for firms and investors to turn challenge into opportunity, acting prudently and creatively to help society reduce the risks that it faces from climate change and making money doing so."

John Holdren
John F. Kennedy School of Government, Harvard University

1 Introduction

There is growing if not definitive consensus climate change is the most threatening challenge society has ever faced. Unmitigated it will have enormous effects on all actors, elements and aspects of our world. Former White House Chief of Staff and former Director of the New York Stock Exchange, Leon Panetta, once described climate change as the *'the perfect storm in the making'* (INCR 2003). He argued *'powerful physical and financial forces are converging and have the potential to create financial havoc'*, damaging entire industries and creating huge unforeseen liabilities.

Just like an approaching storm, climate change is often seen as an environmental or social issue. Business has traditionally argued these areas are beyond the fiduciary responsibility of directors, which they say constrains them to act only to the *financial* benefit of the company. Pressures to act on social or environmental grounds have often prompted many dedicated economists to herald Milton Friedman's 1970 assertion *'[t]here is one and only one social responsibility of business – to use its resources and engage in activities designed to increase its profits so long as it stays within the rules of the game ... .'* (Friedman 1970). As the then Australian Federal Treasurer, Peter Costello, more recently echoed this ethos when he stated, *'a director's obligation is to their shareholders. ... At the end of the day directors are appointed by shareholders to act in the interests of shareholders. ... and the interests of shareholders consists of preserving and increasing value'* (*'Telstra Row Hits Shareprice'* 2005).

If Panetta's, Friedman's and Costello's arguments are all true, one might then consider *'what is a director's responsibility to shareholders in regard to the risks of climate change?'*.

In 2005 an Australian government review of the nation's risks and vulnerabilities to climate change warned Australia has much at risk. The study identified industry sectors then worth AUD$ 58.6 billion to the national economy and 8.2% of GDP as requiring 'attention' or 'urgent attention' (AGO 2005b). By both Friedman's and Costello's standards such material risks would seemingly mandate at least some, and perhaps many, directors have a responsibility to act on climate change to protect shareholder value.

This paper is an excerpt of a broader study of the fiduciary duties of Australian company directors and their current (as of late 2005) understanding and action on climate change risks. It explores climate change as a risk to Australian business and relevant Australian corporate law to determine if Australian directors' fiduciary responsibility requires them to consider and act on these risks.[2]

2 Methodology

The approach undertaken to assess the issues of evaluating climate change as a material risk to Australian business and directors' fiduciary obligations involved three techniques:

1. a review of industry, legal and scientific literature related to company director fiduciary duties and climate change risk in Australia,
2. anonymous, semi-structured interviews with relevant market stakeholders,
3. a comparative analysis of findings from literature review and interviews

Although the body of knowledge has ever increasing consensus and confidence, climate change research has a range of projected impacts and timeframes. This span of possibility often causes debate to centre on the disparity of potential impacts considered least likely to occur, and usually the most long term, rather than those already agreed as almost certain and soon. This study has therefore focused on risks widely considered by most Australian scientists and academics as the most probable and soon to occur.

Risk framework

The risk framework is a synthesis of (then) recent and widely accepted papers from government, scientific, environmental and investment sectors including:

a) the Australian government's 2005 *'Climate Change: Risk and Vulnerability'* report (CCRV Report) (AGO 2005b);
b) CSIRO's (Commonwealth Scientific and Industrial Research Organisation) 2003 *'Climate Change: An Australian Guide to the Science and Potential Impacts'*, (specifically focusing on risks deemed 'almost certain' or 'highly likely' to occur by 2030) (AGO 2003);
c) the Climate Change Action Network of Australia's (CANA) discussion paper *'Climate Risk: the financial risks that climate change presents to Australian companies'* (CANA 2003); and,
d) the 2005 AMP Capital Investors & Baker McKenzie paper for company analysts, *'Climate Change and Company Value'* (Woods and Wilder 2005).

---

[2] An industry sector review and interviews with directors of vulnerable industries were also undertaken as part of this broader research however these results are not presented here.

Stakeholder interviews

Relevant market stakeholders were identified in the literature review and included representatives of the insurance, banking and legal sectors, environmental NGOs, investor groups, government environmental and economic departments and a national industry association for company directors.

Interviews were semi-structured with similar questions asked of each participant. This mixture of standardised questions and narrative discussion resulted in unique interviews reflecting the priorities and actions of the interviewee. Comparison of results is therefore limited but the findings provide some insightful, if only indicative, understanding of the then current status of climate change discourse in the Australian market.

Question themes for stakeholders:

- Do you consider climate change a risk to business?
- Do you consider climate change an opportunity for business?
- Do you expect businesses to have a climate change strategy? Who manages it?
- Do you expect businesses to discuss climate change at board level? Should it be reported on?
- Do you consider managing material risks part of a director's fiduciary responsibility?

Relevant excerpts and quotes from research interviews are presented in text boxes as well as integrated with the text. The alias of the interview participant is noted underneath their quote

3 Fiduciary responsibility of company directors

It is a fundamental principle of Australian company law that the director, acting as a trustee of company assets, owes a fiduciary duty to the company (Tomasic et al. 1996, pp. 337). The core notion of the principle is 'that a person (the fiduciary) is subject to an obligation to do an act or thing for the benefit of another person' (the beneficiary).

Trust is a cornerstone of the concept. In 1967 Justice Mason[3] explained in the High Court of Australia, 'accepted fiduciary relationships are sometimes referred to as relationships of trust and confidence' (Tomasic et al. 1996). He further noted the critical element of these relationships is that the fiduciary undertakes or agrees to act for or on behalf of, or in the interest of, another person in exercising a power which affects the interests of the other person. Justice Dawson, hearing the same case, noted that also inherent in the relationship is vulnerability on the part of one party (the beneficiary or shareholder) requiring them to rely on the other party (the

---

[3] Hospital Products Ltd v United States Surgical Corporation (1984) 156 CLR 41 (Tomasic et al. 1996)

fiduciary) to act in good conscience in the 'protection of equity', namely theirs. That protection of equity requires managing risk and capitalizing on opportunity.

The prudent man acting reasonably

Acting in accordance with the precautions of a prudent man is broadly held as the performance benchmark in evaluating the actions of those in a fiduciary position (Web Finance, 1997). In 1995 Justices Clarke and Sheller[4] found directors should 'take reasonable steps' to position themselves appropriately to guide and manage the company and that ignorance or failing to ask was not a defence in negligence (Tomasic et al. 1996, pp. 390; Cassidy 2003, pp. 241; Baxt et al. 2000). They also adopted the 'ordinary prudent man' test or the 'reasonable man of ordinary prudence' test as being 'equally applicable to Australia' (Cassidy 2003, pp. 241).

While not all stakeholders interviewed agreed, legal and investment experts felt the 'reasonable test' for climate change had passed, suggesting directors now have an obligation to inform themselves, review the risks and, if necessary, act.

> "Climate change is everywhere. It's been on the cover of TIME a couple of times; it's been on the cover of National Geographic and Business Week; there's been a movie about it. The average person ... knows what you mean when you say 'climate change'".
>
> "Anyone that has properly had a look at [climate change] would [find it] pretty hard to justify not doing anything. ... There's a risk to directors and senior managers that don't understand the issue and are making decisions that might be affected by climate change. ... That's where we are."
>
> Climate Change Lawyer, Practice Leader
>
> "If you are a reasonable person, as a fiduciary, yes of course you [have a duty to look at climate change] ... Not taking this into consideration I think would be both unwise and perhaps not consistent with your fiduciary responsibilities."
>
> CEO, Superannuation Fund

Directors' duties

Underpinning fiduciary responsibility are directors' duties essentially 'designed to ensure they will use their powers with a reasonable degree of care for the company's best interests' (Cassidy 2003). Duties related to assessing and responding to material risks can be broadly categorised as diligence and disclosure. In carrying out these duties directors must act in 'good faith' and benefit the company as a whole (Tomasic et al. 1996, pp. 337). While increasingly debated, benefiting the company as a whole generally means protecting it as best as possible from material risks and positioning it to 'continually prosper' and maintain shareholder value (Armstrong 2000).

These duties are not vague suggestions, they are supported by common law, the Corporations Act 2001, the Corporate Law Economic Reform Program Act 1999

---

[4] Daniels v Anderson (1995) 16 ACSR 607 (Tomasic et al. 1996)

(Cth) (CLERP Act) (Cassidy 2003) and the Australian Stock Exchange (ASX) corporate governance guidelines.

Business judgment rule

That directors can today be held accountable for maintaining a duty of care and competence should not give rise to a lack of innovation or commercial risk taking. Indeed specifically to provide a 'safe harbour' for directors, a 1989 Senate Standing Committee report on the social and fiduciary duties and obligations of company directors recommended a 'business judgment rule' be introduced to Australian Company Law (Australian Senate 1989). Introduced into legislation by the *CLERP Act 1999* as s180(2) and (3) of the *Corporations Act*, this rule protects directors, with conditions, from the outfall of a bad business decision. If the questioned judgment was made abiding their obligation to inform themselves 'to the extent they reasonably believe appropriate', they exerted 'an active discretion', a reasonable amount of care, had no material interest in the decision and 'rationally' believed it was in the best interests of the company, the director is not liable (Baxt et al. 2000, pp. 250, Cassidy 2003, pp. 245).

The Business Judgment Rule only applies to the duty of care and diligence and is clearly defined as any decision 'to take or not to take action' (Baxt et al. 2000, pp. 250). It is important to note **it is not a failure to make a decision** (emphasis added), in which case Baxt et al. (2000) advise s180(2) would likely not be available.

While the Rule seems to further affirm a director's responsibility to make active and informed decisions, Cassidy (2003) argues it 'significantly weakens the standard required under the duty of care' and replaces 'reasonableness' with 'rationality'. She notes whether the director considers themselves properly informed is up to them and whether the decision is in the best interests of the company is largely based on what they believe. However, while Baxt et al. (2000) note the courts' usual hesitancy to 'substitute its opinion for management' they also note the requirement for directors to inform themselves about the subject matter is in line with recent decisions. They note the sleeping director 'has no place in corporate law' and the 'rather relaxed attitude' previously adopted 'will no longer be tolerated'.

In considering what would constitute being 'appropriately informed' on climate change under the business judgment rule one legal stakeholder advised *"everyone's learning path is probably going to be different"* (Climate Change Lawyer, Practice Leader).

Interview excerpt:
"The first thing to do is acknowledge there is something that they have to know about, then they have to find a way to inform themselves. For some people the best thing to do will be to … get their heads around the science (… perhaps IPCC website), then read a couple of analyst reports …, and then start a structured process of saying 'What does that mean for my business under different circumstances?'. They may come out and say 'I don't think we need to do anything' but it would be quite surprising if they did reach that conclusion."

Climate Change Lawyer, Practice Leader

4 Managing risk

The threshold for a risk triggering fiduciary obligations to act is based on whether that risk can be considered a potential material impact to the business (Tomasic et al. 1996). If directors are to fulfil their duties to inform themselves and act appropriately on climate change they must have a good understanding of the risks associated with it, the probability and timing of them occurring and the consequences for their business if they do. An evaluation of various risk assessment techniques is beyond the scope of this report, however some key factors to consider in assessing climate change risks include:

- Consensus vs. uncertainty
- Uncertainty vs. probability
- Short term vs. long term

Consensus vs. uncertainty

Those advising to avoid action on climate change often herald the positions of skeptics or the scientific uncertainty of the exact impacts as reason not to respond at all. Harvard Professor, Dr. John Holdren advises against this arguing 'all of the observed climate change phenomena are consistent with the predictions of climate science for greenhouse gas induced warming' and that no alternative 'culprit' yields this 'fingerprint match' (INCR 2003). He further notes that a credible skeptic would need to provide an alternative cause to the observed changes and a theory on how 'greenhouse gases are *not* having the effects that all current scientific knowledge indicates' and that no skeptic had ever achieved either.

George Monbiot, self-proclaimed skeptic of skeptics, says 'it is hard to convey just how selective you have to be to dismiss the evidence for climate change. You must climb over a mountain of evidence to pick up a crumb: a crumb which then disintegrates in your palm. You must ignore an entire canon of science, the statements of the world's most eminent scientific institutions, and thousands of papers published in the foremost scientific journals' (Monbiot 2005).

The scientific consensus on climate change comes from the international entity the Intergovernmental Panel on Climate Change (IPCC). The IPCC prepares regular, independent assessments on the 'state of knowledge on climate change' based on peer reviewed and published scientific and technical literature. The 1st, 2nd, 3rd and (now) 4th Assessments were prepared in 1991, 1995, 2001 and 2007 respectively. Each Assessment declares with increasing scientific assurance the significance of climate change risk and that it is almost undisputedly 'human-induced' through the release of greenhouse gases into the atmosphere (IPCC 2005). The 4th Assessment declared that it was now 'unequivocal' that the planet was warming, with eleven of the previous twelve years among the twelve warmest on record, and it was highly likely (more than 90%) due to human activities (IPCC 2007). While too extensive to detail here these Assessments represent the findings of a clear majority of the world's scientific community and are widely considered the benchmark reference for current status on climate change science. Thus they form the basis for many ongoing studies and reports including the Australian government's 'Climate Change Risk and Vulnerability' report (CCRV) (AGO 2005).

Uncertainty vs. probability

The Australian government's 2004 paper 'Economic Issues Relevant to Costing Climate Change Impacts' expressly noted 'decisions on climate change will rarely, if ever, be made under conditions of certainty' (AGO 2004). CSIRO notes probabilistic assessments of risk, accounting for uncertainties, are 'regarded as the way forward' (Pittock 2003, pp. 179). This suggests uncertainty does not justify inaction; indeed uncertainty is always a factor in risk assessment. In such assessments risk is defined as the probability of a projected climatic outcome multiplied by the consequence of that outcome (Risk°=°Probability°x°Consequence) (Pittock 2003, pp. 76; Jones 2005).

The CSIRO projects an 'almost certain' average temperature rise of 1°C by 2030; while under a low emissions scenario increases of 1°C to 2.5°C are expected by 2070, under a high emissions scenario, which the world is currently tracking to (Rahmstorf et al. 2007), an average increase of between 2.2°C and 5°C by 2070 is expected, with 'best estimates' at 3.4°C (CSIRO 2007).

There is also strong confidence in projections of the likely impacts associated with this warming. *Figure 1* shows a selection of CSIRO's projected environmental impacts for an up to 1-2°C temperature rise relevant to the industry sectors deemed vulnerable in the CCRV Report. Businesses should understand the consequences of these changes for them, their competitors, their buyers and suppliers.

Environmental impacts with up to a 1°C warming
- 10-40% shrinkage of snow-covered area, 18-60% decline in 60-day snow cover in Australian Alps
- Bleaching and damage to the Great Barrier Reef equivalent to 1998 and 2002 in up to 50% of years; 60% of reef is regularly bleached
- Decreases in water supply to ~10% in southern Australia
- Average crop yields increased in most regions but drought impact increased also
- Up to 20% increase in extreme daily rainfalls

Environmental impacts with up to a 1-2°C warming
- Up to 58-81% of the Great Barrier Reef is bleached every year, hard coral communities widely replaced with algal communities
- Up to 22% increase in storm surge heights in Cairns; flooded area doubles
- +2.5% to -25% change in runoff in the Murray Darling Basin

Fig. 1. Projected impacts associated with warming thresholds
Source: Adapted from Jones 2005; Pittock 2003

Short term vs. long term

Whether a risk is considered long term or short term generally determines the priority it holds for company directors, and short term usually has their focus. This short term view is perhaps justified if delayed action on the long term risk can still avert it, however, if a long term risk requires action today else is inevitable then prudence and diligence should provoke action on it today.

All stakeholders interviewed agreed climate change already presents current and increasing risks to business. The literature review also supported there are many valid risks and reasons for concern regarding climate change in this current and short term timeframe. Many also warned that the expected 'rate of change' may alter, meaning that current medium and long term risks can change to shorter term risks at any time and therefore require ongoing attention (Pittock 2003, 2005; Woods and Wilder 2005; CANA 2003; AGO 2005; Rahmstorf et al. 2007).

Political change can also quickly bring new conditions. During 2005 research interviews legal and government participants suggested a change in political leadership may see a national Emissions Trading Scheme introduced in Australia. Since then a federal election, held in 2007, which saw climate change high on the agenda, resulted in a change of government which, in their first week in office, ratified the Kyoto Protocol and brought forward commencement of an Emissions Trading Scheme to 2010 (Commonwealth of Australia 2008). Managing the obligations and liabilities of emissions trading has shifted from a 'maybe' for business in Australia to a 'reality' within two years.

One participant in 2005 said they had been waiting for the media to "wake up and do what the media does" for two years. He believed the media may not have understood the "full enormity of the situation" but warned that "when they do run with this" that's where "the biggest pressure" will come from and could "change the whole game". Since then media coverage of climate change has risen at an astronomical pace with special reports and front page exposés now common place; and clearly influential in putting climate change on the political agenda. This participant also advised Boards should not be arguing about whether climate change is taking place and whether they should respond to it but "the speed with which they have to do it". This required 'speed' has accelerated significantly since 2005.

5 Material risks of climate change to Australian industry

The former Australian government commissioned the 'Climate Change: Risk and Vulnerability Report' (CCRV Report) to identify sectors and regions 'that might have the highest priority for adaptation and planning' within a 30 to 50 year time frame (AGO 2005). The Report looks at the implications of the 'most likely' climatic change events the CSIRO projects and focuses on temperature change (including heatwaves and reductions in frosts), rainfall change, sea level rises and increases in cyclonic wind intensity. The industry sectors the Report deems to 'require attention' are ***Tourism*** and ***Forestry***; and those requiring 'urgent attention' are ***Energy Infrastructure, Buildings & Settlements***, and ***Agriculture***.

Broadening the risk framework

While providing a good review of the likely environmental impacts of climate change and the implications for industries and regions, the CCRV Report is perhaps misleading if positioned, as it claims, as a definitive 'framework for managing future climate risk' (AGO 2005). Both the literature review and interviews found market risks of climate change extend further than the projected environmental impacts, however the report focuses on the physical impacts of climate change and makes little or no mention of these broader risks.

This broader risk spectrum was well defined by the Climate Action Network of Australia (CANA), a non-profit, coalition of 30 Australian environmental, public health, social justice and research organisations (CANA 2003). The CANA paper outlined the financial risks climate change represented to Australian companies (CANA 2003). They sent the paper to all then company directors of the ASX200 "to remind them of their fiduciary duties to deal with financial risks to their companies" (Lawyer/Campaigner, Environmental NGO). A legal participant advised the paper effectively drew a 'line in the sand'; "They put them on notice. In five years time a director can't say 'well I didn't think it was relevant'" (Climate Change Lawyer).

In 2005 the Total Environment Centre released a 'Guide for Company Analysts: Climate Change and Company Value' (Woods and Wilder 2005). In this paper, authors AMP Capital Investors and Baker McKenzie advised investment analysts of the climate change-related regulatory risks they should now be considering in their market evaluations.

By also reviewing the risks detailed in these reports, this study aims to expand the environmental risk considerations reviewed in the CCRV Report into a broader framework. This paper proposes this expanded framework is more representative of the full scope of market issues and pressures many businesses will be exposed to due to climate change. Risk categories identified include:

- Operational risk
- Insurance risk
- Regulatory risk
- Shareholder risk
- Litigation risk
- Capital risk
- Competitive risk

These risk categories, their associated trends and participating stakeholder views are examined below.

Operational risk

Operational risk refers to impacts that may impede or prevent businesses from executing their normal business practices or purpose. In this context this is predominantly the environmental impacts of climate change where rising temperatures, drought, sea level rise, storm surge, flood and extreme weather events can impact a business's ability to operate or continue production (CANA 2003; AGO 2005b). A 2007 study found that of businesses shut down by extreme events, 25% usually did not reopen (City of Melbourne 2002).

Operational risks could also include water or fuel restrictions, as well as electricity blackouts should demand for energy exceed capacity or storms damage infrastructure. These business interruptions can have flow on affects to other businesses and communities relying on or supplying their products and services. Businesses should be aware of operational risks across their value chain and develop planned responses.

Insurance risk

Insurance companies are in the business of risk and are perhaps the 'canary in the coalmine' when it comes to climate change. The insurance participant agreed, "This is all about pulling risk off the community. Often, whereas individual sectors may not see their risks, I think we tend to see a lot of the effects first". Already some insurers and reinsurers are requiring boards of high risk companies provide their climate change strategy before they will provide or renew directors' and officers' insurance coverage (Mills et al. 2005). While not having a strategy may not necessarily result in a denial of cover the participant explains, "Certain companies probably are exposed currently to the possibility of getting sued and therefore [insurers] would have to take that into consideration on whether they would accept the risk and then what premium they would charge ... Climate change is a risk and we believe you should have a strategy".

Insurance Australia Group's (IAG) Chief Actuary, Tony Coleman, warns 'climate change threatens the insurance industry's ability to underwrite risk from weather events' (AGO 2005). This is a concern for both the industry itself and those to whom it provides coverage. The industry is experiencing increasing weather-related losses at an unprecedented pace and magnitude around the world (Woods and Wilder 2005; Mills et al. 2005). This places it under immense pressure to find the balance between commercial viability and coverage. The insurance participant explains, "In order to stay viable we have only two choices; to reduce cover or increase premium". They advised at this stage the industry is taking the approach to engage with business to increase awareness and try to reduce GHG emissions aiming to "actually reduce our exposure to risk". Areas where the research participant is focusing their engagements include building codes, infrastructure vulnerability, emergency procedures and engineering works. A US trend the interviewee advised may be introduced in Australia is the 'percentage deductible'. This is where the customer takes on more of the risk themselves by agreeing to a percentage of the cost rather than a fixed deductible, "It's one step away from excluding cover".

Research interviews identified many areas deemed vulnerable to climate change where insurance coverage is already either unavailable or considered unaffordable; including agricultural crop damage, wind damage coverage for forest plantations, and flood and storm surge coverage for many coastal or riverine homes or buildings. Currently flood and storm surge coverage is unavailable in Cairns, which is an area projected to have an up to 22% increase in storm surge height and a doubling in flood area (Pittock 2003; Jones 2005). If an extreme weather event wiped out a number of homes there the banks with mortgages on those properties may be exposed to increased bad debts. The insurance participant agreed, "That's the kind of thing presumably they [the banks] need to take account of". Uninsured costs are also a concern for governments whom are usually considered 'the insurer of last resort' and often expected to provide assistance in crisis circumstances.

Regulatory risk

Regulatory risk refers to policy, tax and legislative changes that may affect the business operations or dynamics. There are several forms of regulatory risk associated with climate change and a 'carbon constrained economy' with many legislated mechanisms already introduced in Australia and around the world. Perhaps the most significant on the global stage is the Kyoto Protocol which effectively introduced, among other mechanisms, a global Emissions Trading Scheme for participating countries (Woods and Wilder 2005). This mechanism currently runs through to 2012 with discussions on what shape it will take next already underway.

Australia long said no to a national Emissions Trading Scheme, however in 2007 the then federal government announced a national scheme would commence in 2011[5]. Such schemes effectively put a 'price on carbon' and a cap on emissions, translating to a risk for some and an opportunity for others. Companies that make savings in their emissions can create carbon credits tradable with high emitters needing more than their allocation (Woods and Wilder 2005). Although not then announced, many stakeholders interviewed were of the view that greenhouse gas regulation would be introduced; "Certainly there is a real prospect that we will see some form of greenhouse gas regulation within Australia between now and 2010" (Climate Change Lawyer, Practice Leader).

The potential risk of not factoring in a carbon price in investment decisions was exemplified by American Electric Power (AEP) who, in 2004, under shareholder pressure, was forced to look at "what a carbon price in the US might mean for their forward investment program" (Climate Change Lawyer, Practice Leader). "Between 2005 and 2010 there weren't really any problems but post 2010 they

---

[5] A November 2007 federal election resulted in a change of government which then ratified the Kyoto Protocol and brought the introduction of a national emissions trading scheme forward to 2010.

had about US$ 1.5 billion worth of investment that could effectively become stranded. ... It's a good indication of how it is a very serious sort of corporate governance issue and probably not as well recognised because it doesn't fit into some of those traditional boxes." Given a carbon price will soon be introduced in Australia, these concerns of stranded assets or incorrectly calculated investments are very real and relevant.

Regulatory considerations should also extend beyond borders. In a globalised economy it is not just state or national regulations that may present risks. In 2002 shares in Xstrata, BHP Billiton and Rio Tinto fell as much as 4% within 48 hours of Japan announcing it was considering a 'non-Kyoto' coal tax (Planet Ark/Reuters 2002). Xstrata took the 'biggest hit' and was particularly significant given the losses occurred very shortly after a public share offering for which their 350 page prospectus had only one sentence dedicated to climate change (Phillips Fox 2004). This sharp movement, although quickly recovered, indicates how sensitive the current status quo can be (Philips Fox 2004, Planet Ark/Reuters 2002).

Baker McKenzie partner and climate change specialist, Martijn Wilder, advises, 'For corporations, and particularly multi-nationals, the emergence of a diversity of climate related laws, in developed and developing countries, means a clear understanding is required as to the nature of legal liabilities that now exist or are likely to exist in the future and how best to position themselves' (in Pittock 2005, p. 237).

> "The regulatory risk is probably the one that would concern me the most in business."

Shareholder risk

Shareholder risk refers to impacts and influences existing shareholders may potentially have on the company based on a specific issue. In 2002 Innovest Chairman James Martin noted that as 'arguably the world's most pressing environmental issue' it is logical shareholders, being concerned about anything that could have 'material bearing' on a company's financial performance, should be concerned about climate change (CANA 2003). Shareholder resolutions based on climate change issues have been increasing in the US, with the number of filed resolutions tripling from 2001 to 2002 (CANA 2003; Baue 2002). The average support level for these resolutions also increased (CANA 2003). Thus far most resolutions have been aimed at the 'high emitting' sectors and demand disclosure around potential climate change risks (see above) however these actions could easily set the precedent for other vulnerable sectors (CANA 2003).

While shareholder action is growing in the US, interviews conveyed at the time of interviews there was no such parallel action in Australia. One banking participant stated "there is surprisingly little activism really on the shareholder side". This lack of shareholder interest was further indicated by the Australian Shareholder Association who, in declining an interview, advised, "We appreciate the importance of this issue, however at present it is not one that we are focusing on nor have the resources to cover". No participants advised they were experiencing shareholder pressures in the Australian market. This is particularly interesting considering the rising public concern of climate change in Australia (Climate Institute 2007), and may indicate future mobilisation of shareholders to agitate for climate change action. Research suggested the current predominant strategy in Australia to be engagement with business. This will presumably continue unless shareholders find that approach unsatisfactory. It is perhaps worth remembering momentum and action built quickly around the issue in the US indicating such a situation can change quickly (CANA 2003).

Litigation risk

Whether plaintiff or defendant, right or wrong, litigation is always a drain on resources and exposes a company to myriad risks. Climate change litigation is developing as a niche specialty within the legal community both in Australia and internationally (Kerr 2002; Grossman 2003). In 2002 the Australian Conservation Foundation (ACF) forecast the 'potential for climate change and its consequences to be the subject of litigation on a scale that could eclipse anything yet witnessed in any domestic or international jurisdiction' (Kerr 2002). Research participants however had varying views on how immediate or concerning this risk is in Australia. One climate change lawyer participant stated, "I think we're a long way from having a court anywhere saying X company is liable in damages for a breach in duty".

The first climate change litigation cases before the Australian courts have been administrative law actions mostly pressuring planning authorities to consider related greenhouse gases in development proposal decisions (Woods and Wilder 2005; Pittock 2005). Recent administrative decisions now mandate companies at least consider the greenhouse gas intensity and implications of their projects and greenhouse gas assessments are now required as part of most project works approvals under the Protocol for Environmental Managment (PEM) (Woods and Wilder 2005; EPA Victoria 2006). While such considerations have so far not changed a planning decision they have already played an important role in developing juris prudence and not winning should not necessarily be seen as failing. Indeed in 1994 when Greenpeace first brought the issue of greenhouse gas intensity before the Australian courts challenging the Redbank power station, the action was unsuccessful (Woods and Wilder 2005). However in 2003 the NSW Minister for Planning denied a second Redbank power station which had a higher GHG profile than any other power station in the region (Woods and Wilder 2005; Pittock 2005). These actions build awareness and set precedents. The participating Climate Change Lawyer suggests this legal foundation setting is where most of the legal development will take place at the moment, "… Our legal system is based on cases setting some precedent and, that being reliable, … build juris prudence into your administrative law decisions and use that to grow across to your tortious stuff".

Other legal participants acknowledged litigation has many objectives and outcomes. "A damages case is only one kind of case; cases can be brought seeking injunctions which require actions to be taken … the law can work in different ways", advises the interviewed Lawyer/Campaigner, Environmental NGO. The Climate Change Lawyer, Practice Leader confirms success may have different forms, "Increasingly environmental litigation is often brought with one eye to the media... cases that are very unlikely to succeed are brought anyway because of the profile that that litigation can bring with it".

> "Courts are being well used for legal agitation, profile raising and trying to call directors and companies to account; I think that's being done quite successfully. It's almost a way of agitating for change by using the judicial system but it's quite different to say ... a court will award damages ... because of their emissions of greenhouse gases."
>
> Climate Change Lawyer

A weakness of potential climate change litigation is the issue of whether a legal remedy has the capability to be a remedy for climate change. Legal participants agree that while Australia is the world's largest greenhouse gas emitter per capita (ACA 2004), as it contributes a relatively low amount in terms of overall global emissions it may make it difficult to prove 'causation' in tort cases; "... you might find a difficulty proving that connection because you might find [the courts] say that 'even if I ordered you to do something the problem would still be the same'" (Climate Change Lawyer, Practice Leader). However, the Lawyer/Campaigner participant advises this doesn't mean it won't happen; "The causation issue is a difficult one but it's not insurmountable ... I know work is being done at the moment to strengthen the links and also to make them specific enough to support a lawsuit".

If or when legal actions are initiated, identifying defendants is deemed relatively easy with the most likely candidates being those 'predominantly responsible' for GHG emissions (Kerr 2002; Grossman 2003). As scientific consensus attributes a clear majority of GHG emissions to the burning of fossil fuels the most likely defendants are considered to be a) the *producers* of fossil fuels, such as multinational coal and oil corporations and their industry associations; and b) the *users* of fossil fuels, such as multinational car manufacturers, electricity corporations, and aluminium corporations and their industry associations (Grossman 2003). "If you were trying to reach a threshold of contribution you would probably look more at the petroleum sector or the mining sector for targets", Lawyer/ Campaigner, Environmental NGO.

Many see the tobacco litigation of recent years as the 'blueprint' for climate change litigation (Baue 2003a). Baker McKenzie Chairman, James Cameron, warns the likelihood of successful litigation against a company would perhaps be greatly increased 'if they were deemed to have acted culpably by, say, lobbying against greenhouse gas regulations' (CANA 2003; Baue 2003a). The Lawyer/Campaigner, Environmental NGO adds, "... it is probable that companies will be sued in the reasonably foreseeable future and there will be a target selection process and companies that have been the most regressive are obviously more likely to be made the subject of lawsuits".

Increasingly these climate change cases are not isolated or unusual. In 2004 the World Wildlife Fund's Climate Justice Programme noted some ten legal actions worldwide aimed at combating climate change. Cases vary widely and have been brought before courts under public law, civil law, public international law and, in the case of the Inuit people, human rights law (WWF 2004a). Peter Roderick, Director of the Climate Justice Program, notes that 'not so long ago, the courts were seen as having little to say about global warming ... but since 2001, and the finding that human activities have caused most of the warming... there has been greater willingness on the part of environmentalists to take the issue to the judges, and on the part of the judges to realise its importance' (WWF 2004b). Roderick notes he is heartened and hopeful to see Australian and US judges 'going where their political leaders refuse to tread' (WWF 2004b).

Perhaps important in weighting this risk however is the fact that even the legal participants suggest litigation is not the first step in addressing climate change, "Litigation is always the last resort and it's not the most constructive way of achieving things in society", Lawyer/Campaigner, Environmental NGO.

> "... certainly if [actions] do happen they'll be in the context of people losing big piles of money when things change quickly ... . Who ever it is that took responsibility for the documentation, they're the ones that will be responsible."
>
> Climate Change Lawyer, Practice Leader

Capital risk

Capital risk refers to any negative influence a company's climate change risk profile may have on attracting or keeping capital investment. Attracting capital investment has always been a function of a risk/return ratio. Banks and institutional investors are increasingly recognising climate change as a significant business risk over the medium to long term, and are therefore developing means to factor it into investment evaluations (CANA 2003; Woods and Wilder 2005; Innovest 2202). Martijn Wilder of law firm Baker McKenzie advises, 'It is... critical for corporations to be aware of the way in which global capital – especially within the investment and insurance industries – is reassessing investments for carbon risk and the growing reluctance of shareholders to tolerate non-performance on greenhouse matters' (in Pittock 2005, p. 237).

This was supported by both the banking and superannuation research participants, "financial literacy is a major part of the sustainability program. ... Climate change issues are now being factored into the portfolio" (Climate Change Consultant, Australian retail bank). While they admit "there is difficulty in working it out", in lending considerations the banks are starting to look at climate risks and asking "will it affect the ability of that client to pay the money back or will it affect the asset value?". The bank advises, as part of its climate change strategy, it aims to "be very clear on [their] logic in terms of what [they're] funding" and is "looking at methodologies to look at the carbon intensity of [its] portfolio" (Climate Change Consultant, Australian retail bank).

Like the insurance companies though, banks and superannuation funds at this stage are engaging with companies to reduce risks associated with climate change. Seemingly to varying levels of success: "we're still trying to get super funds in a dialogue with the companies they invest in ... we are still a little way away [from] getting the buy in of the vast majority of super funds, but it's started"(CEO, Superannuation Fund). Super funds are "universal investors" explained the superannuation participant, "we will be permanent investors of the larger companies around the world and if you are a permanent investor then it is important to take a leading role as the owner or part owner of those companies".

Perhaps the most significant recent development in climate change and capital is the Carbon Disclosure Project (CDP), a global coalition of institutional investors informing investors and companies of the potential impact of climate change on company value (Innovest 2005). In 2007 CDP5 represented 315 institutional investors with assets of US$ 41 trillion (Innovest 2007). The CDP sends a questionnaire to the Financial Times' Global 500 companies requesting information related to the company's carbon performance and strategies. Since its inception in 2000, the Project has garnered increasing profile with market participants and increasing responses from questioned companies. In September 2005 an Australia/New Zealand CDP was launched asking the ASX100 and NZ50 to disclose their views and performance around carbon intensity and strategy. The Australian institutional investor coalition already consists of members worth over A$ 345 billion in assets and is in discussions with others to join (IGCC 2007). Only 50% of Australian and New Zealand companies responded to the CDP5 questionnaire, lower than the global average of 58% (IGCC 2007). This response did however include 68% of the ASX100. 97% of responding companies acknowledge climate change as a risk or opportunity to their business.

While perhaps in its relative early stages, indicators suggest carbon management and climate change positioning will increasingly be a key differentiator in a company's investment risk profile (Woods and Wilder 2005; Innovest 2002; Cogan 2004) Investment analysts are becoming increasingly sophisticated in their understanding of these risks.

Competitive risk

Competitive risk is perhaps the most natural and inherent risk of business. If competitors move first and distinguish themselves to the appeal of the market, a company can be left in a 'catch up' position or, at worst, irrelevant.

With rising community awareness and concern for climate change comes the opportunity to provide an appealing alternative, as well as avoid being the focal point for disconcerted consumers. Should any company provide a product or service that fits an anticipated low greenhouse gas intensity market appeal, they may be able to capture significant competitor advantage. Woods and Wilder (2005) note there is already evidence some companies are pursuing such a strategy. They also quote a recent UK report 'Brand Value Risk from Climate Change' which argues 'climate change could become a mainstream issue by 2010' and therefore has reputational implications for companies not seen to be addressing it (Woods and Wilder 2005).

A legal participant concurred, "There is a kind of growing appetite for products and services that are seen to do something about it" (Climate Change Lawyer, Practice Leader). He further notes the opportunity for innovative companies to make a full paradigm shift in their positioning. He points to Toyota as a prime example of this whom he says are using their *Prius* hybrid technology to "set the agenda for automobile transport in the future". Indeed he advised Ford has given up its own efforts to develop similar technology choosing to license it from Toyota instead. This has the potential to position Toyota "to become more like Microsoft" with their engines in many lines of cars. "There are opportunities for companies that can invest in technology to carve out a space for themselves where they're kind of untouchable … That's a hugely, bigger opportunity than any kind of benefit from carbon trading" (Climate Change Lawyer, Practice Leader).

> "If you can become the dominant provider of the next generation of car technology, or pick any kind of product you like, then that's the kind of thing that makes dynasties."
>
> Climate Change Lawyer, Practice Leader

Competitive forces can be both national and international. For example, while the Australian snow resorts are investing in snow technology and non-winter appeal to respond to already reduced snow fall, a 1998 NSW survey concluded much of that market would be lost to New Zealand and other holiday sports (Pittock 2003). For agriculture this transborder competition could come from newly arable land or capable production in developing economies where they may be able to provide product of equal quality at less cost.

> "One of the biggest imponderables though is in terms of what it will do for the global markets. There are whole areas that might be opened up to cultivation, particularly Siberia for grains, which have hitherto been covered in permafrost. … Russia could produce … enough to flood the world with exports."
>
> Strategic Policy Manager, State Government

The coal industry, Australia's largest export (ACA 2004), is not impervious to changes in global competition either. One legal participant advised his understanding that BHP Billiton in Europe was at the time selling coal with emissions allowances "stapled to it" (Climate Change Lawyer, Practice Leader). He noted,"you're not just getting the resource any more, you're getting the right to emit the carbon dioxide that will inevitably come out when you burn it". While that product offering was based in Europe he advised if their customers came to like it and other coal-hungry and Kyoto-ratified nations also found it appealing to have Kyoto-compliant coal, such as Japan, without a similar offering Australian coal may have found itself locked out of the Kyoto-participating market. He advised the big risk for Australian companies is that the Europeans will be way ahead. "When we do change all of [our] competitors will be ahead of [us] and we'll end up buying technology and know-how from the people that did it first" (Climate Change Lawyer, Practice Leader).

> "You only need a couple of things like that to happen and suddenly the business logic of the Kyoto Protocol all flips around. It goes from being an apparent disadvantage to an apparent advantage for some of the businesses that were the ones that were perhaps … speaking out strongest against it."
>
> Climate Change Lawyer, Practice Leader

6 Conclusion

Australian company directors have a fiduciary and statutory duty to respond to, and disclose, material risks to their company. The more significant the risk or risks to an industry or company the more amplified a director's duty of care and diligence. For Australia's identified vulnerable industries, and perhaps others, it is clear that climate change already represents a significant and multifaceted material risk. These risks are current and critical and extend beyond the environmental impacts of the Government's *Climate Change: Risk and Vulnerability* Report. As with the 'Perfect Storm', should any or all of these risks converge and occur concurrently potential damages increase exponentially. As such, it can be determined Australian corporate directors do indeed have a fiduciary duty to assess and, if necessary, respond to and disclose to the market, the risks climate change represents to their corporate charge.

Whether future courts will consider climate change an issue directors or officers should have prudently and reasonably acted upon will depend on the specifics of the company and the damages it has subsequently incurred. However, as with tobacco and asbestos litigation it is possible today's actions will be under future scrutiny. Certainly if action or inaction today precedes future losses due to climate change impacts, and shareholders or stakeholders can persuasively argue it was their lack of care or diligence that lead to it, directors could find themselves defending their behaviour in court. While viable, such proceedings may not occur for some time and inherently suggest remedies need to await, rather then preempt, damages. Wisdom may therefore question if we can afford to await legal rectification or rely only on potential legal pressure for proactivity. Perhaps the most effective protection from the perfect storm of climate change is not post-event litigation for failing fiduciary obligations, but shareholder diligence and consumer consciousness demanding preemptive leadership and preparedness from business, market and government leaders.

References

Armstrong P (2000) Corporate Governance in Commonwealth Countries, Commonwealth Association for Corporate Governance, (Accessed 7 June, 2005)

ASX (2002) ASX Corporate Governance Council, Minutes of 2nd Meeting, 19 September, 2002, Australian Stock Exchange, Sydney Australia. (Accessed 7 June, 2005 http://www.asx.com.au/about/pdf/CorporateGovernanceCouncil190902.pdf)

ASX (2003a) ASX Corporate Governance Council, Principles of Good Corporate Governance and Best Practice Recommendations, March 2003, Australian Stock Exchange, Sydney Australia

ACA (Australian Coal Association) (2004) Coal 21: Reducing Greenhouse Gas Emissions Arising from the Use of Coal in Electricity Generation. A National Plan of Action for Australia, March 2004, Australian Coal Association, Canberra, Australia

AGO (Australian Greenhouse Office) (2003) Climate Change: An Australian Guide to the Science and Potential Impacts. Prepared by CSIRO for the Commonwealth of Australia, Canberra, Australia, ISBN 1 9208040 12 5

AGO (Australian Greenhouse Office) (2004) Economic Issues Relevant to Costing Climate Change Impacts. Prepared by Marsden Jacob Associates, Commonwealth of Australia, Canberra, Australia

AGO (Australian Greenhouse Office) (2005) Climate change: Risk and Vulnerability Report. Prepared by Allen Consulting Group, Commonwealth of Australia, Canberra, Australia

Australian Senate (1989) Company Directors' Duties: Report on the Social and Fiduciary Duties and Obligations of Company Directors, by the Senate Standing Committee on Legal and Constitutional Affairs. November, 1989. Australian Government Publishing Service, Canberra

ASIC (Australian Securities and Investment Commission) (2004) Continuous Disclosure Obligations: Infringement Notices. An ASIC Guide. May 2004. (Accessed November 6, 2005 at http://www.asic.gov.au/asic/pdflib.nsf/LookupByFileName/infringement_notice_guidelines.pdf/$file/infringement_notice_guidelines.pdf)

Baue W (2002) CERES Calls on Institutional Investors and Corporate Directors to Address Global Climate Change Risk. SRI World Group, May 01, 2002, (Accessed 19 April, 2005, http://www.socialfunds.com/news/print.cgi?sfArticleId=833)

Baue W (2003a) Climate Change Litigation Could Affect Companies' Market Value. SRI World Group, July 23, 2003, (Accessed 19 April, 2005, http://www.socialfunds.com/news/print.cgi?sfArticleId=1180)

Baue W (2003b) Institutional Investors Send Wall Street Wake-Up Call to Address Climate Risk. SRI World Group, November 25, 2003, (Accessed 19 April, 2005, http://www.socialfunds.com/news/print.cgi?sfArticleId=1276)

Baue W (2003c) UN Institutional Investor Summit Considers Opportunities of Addressing Climate Change. SRI World Group, November 26, 2003, (Accessed 19 April, 2005, http://www.socialfunds.com/news/print.cgi?sfArticleId=1277)

Baue W (2004) Three more companies address climate risk, So shareowners withdraw resolutions. SRI World Group, May 4, 2004, (Accessed 19 April, 2005, http://www.socialfunds.com/news/print.cgi?sfArticleId=1412)

Baxt R, Renard I, Simkiss R, Webster J (Baxt et al.) (2000) "CLERP" Explained. The Corporate Law Economic Reform Program Act 1999, in association with Ian Ramsay, CCH Australia Limited, Sydney, Australia

Bell C (2006) A New Manhattan Project. The Diplomat, February/March 2006, Diplomat Media, Sydney, Australia

Bracks S (2005) Bracks Releases Climate Change Action Update, Media Release. Office of the Premier, Victorian State Government, Melbourne, Australia, April 6, 2005 (Accessed at http://www.dpc.vic.gov.au/domino/Web_Notes/newmedia.nsf/0/9482236bf820903fca256fdc00077ca6?OpenDocument on February 19, 2006)

Cassidy J (2003) Concise Corporations Law. 4th Edition, The Federation Press, Leichhardt, NSW

CANA (Climate Action Network Australia) (2003) Climate Risk: the financial risks that climate change presents to Australian companies. July 2003 (Accessed 19 April, 2005, http://www.cana.net.au/documents/legal/discussion_paper.doc)

City of Melbourne (2002) Mind Your Business. An Emergency Management Planning Guide for Business in the City of Melbourne, Melbourne, Australia

Climate Institute (2007). Climate of the Nation. Sydney, Australia: The Climate Institute Australia

Cogan DG (2003) Corporate Governance and Climate Change: Making the Connection. A CERES Sustainable Governance Project Report, by Investor Responsibility Research Center, Washington DC, USA

Cogan DG (2004) Investor Guide to Climate Risk, Action Plan and Resource for Plan Sponsors, Fund Managers and Corporations. CERES Report, July 2004

Commonwealth of Australia (2002) Continuous Disclosure: The Financial Reporting Framework, Treasury, Commonwealth of Australia, Parkes, Australia

Commonwealth of Australia (2008) First 100 Days. Achievements of the Rudd Government, Office of the Prime Minister, Commonwealth of Australia, Parkes, Australia

Friedman M (1970) The Social Responsibility of Business Is To Increase Its Profits. New York Times Magazine, September 13, 1970. (Accessed at http://www.umich.edu/~thecore/doc/Friedman.pdf)

Goodman S, Kron J, Little T (2003) The Environmental Fiduciary. The Case for Incorporating Environmental Factors into Investment Management Policies. The Rose Foundation for Communities and the Environment, Oakland, USA

Grossman DA (2003) Warming Up to a Not-So-Radical idea: Tort-based Climate Change Litigation, vol 28 (1): Columbia Journal of Environmental Law

IGCC (Investor Group on Climate Change) (2005) (Accessed at http://www.igcc.org.au on February 14, 2006)

INCR (Investor Network on Climate Risk) (2003) Institutional Investor Summit on Climate Risk. Final Report, November 21, 2003, UN Headquarters, New York City, USA

Innovest (Innovest Strategic Value Advisors) (2002) Value at Risk: Climate Change and the Future of Governance. CERES Sustainable Governance Report, April 2002

Innovest (Innovest Strategic Value Advisors) (2005) Carbon Disclosure Project 2005. Brochure, September 2005

Innovest (Innovest Strategic Value Advisors) (2007) Carbon Disclosure Project Report 2007, (Accessed at http://www.cdproject.net/cdp5reports.asp on March 31, 2008)

IPCC (Intergovernmental Panel on Climate Change) (2005) About IPCC, (Accessed from http://www.ipcc.ch/about/about.htm on January 17, 2006)

IPCC (Intergovernmental Panel on Climate Change) (2007) Fourth Assessment Report (AR4). Climate change 2007: Summary for Policymakers, (Accessed from http://www.ipcc.ch/pdf/assessment-report/ar4/syr/ar4_syr_spm.pdf on February 24, 2008)

Jones RN (2005) Recent Science – What is Dangerous Climate Change? Presentation, CSIRO Marine and Atmospheric Research (Accessed October 30, 2005 http://www.cana.net.au/documents/2005conf/CANA05_dangerous_Jones.pdf)

Kerr M (2002) Tort-based Climate Change Litigation in Australia. Discussion Paper, Australian Conservation Foundation, March 2002, Melbourne, Australia (Accessed at http://acfonline.org.au/uploads/res_climate_change_litigation.pdf on August 28, 2005)

Mills E, Roth R, Lecomte E (2005) Availability and Affordability of Insurance Under Climate Change: A growing Challenge for the US, Ceres, USA

Monbiot G (2005) Junk Science. The Guardian, May 10, 2005 (Accessed at http://education.guardian.co.uk/higher/sciences/story/0,,1480375,00.html on September 10, 2005)

Phillips Fox (2004) The Day After Tomorrow. Lessons for Business, Climate Change Update 01, July 2004, Phillips Fox, Melbourne, Australia

Pittock AB (ed) (2003) Climate Change: An Australian Guide to the Science and Potential Impacts. Australian Greenhouse Office, Paragon Printers, Australasia

Pittock AB (2005) Climate Change: Turning Up the Heat. CSIRO Publishing, Collingwood, Australia

Planet Ark, Reuters (2002) Japan Considers Coal Tax in Greenhouse Push, September 6, 2002, Reuters News (Accessed at http://www.planetark.org/dailynewsstory.cfm/newsid/17634/story.htm on November 3, 2005)

Planet Ark, Reuters (2004) AEP, Cinergy in shareholder pact on emissions curbs, February 23, 2004, Reuters News (Accessed at http://www.planetark.org/dailynewsstory.cfm/newsid/23916/newsDate/23-Feb-2004/story.htm on November 3, 2005)

Rahmstorf et al. (2007) Recent climate change observations compared to projections. Science, February 1, 2007 (Science DOI: 10.1126/science.1136843)

Telstra Row Hits Shareprice (2005) 7.30 Report. ABC Television, September 5, 2005 (Transcript accessed at: http://www.abc.net.au/7.30/content/2005/s1453739.htm on February 13, 2006)

Tougher Regime for Big Polluters (2006) The Age, February 19, 2006, Melbourne, Australia (Accessed at http://www.theage.com.au/news/national/tougher-regime-on-big-polluters/2006/02/18/1140151851231.html on February 19, 2006)

Tomasic R, Jackson J, Woellner (1996) Corporations Law, Principles, Policy and Process. 3rd Edition, Butterworths/Reed International, Australia

Web Finance Inc (1997) Prudent Man Rule, Investor Words (Accessed 9 June, 2005 http://www.investorwords.com/3927/Prudent_Man_Rule.html)

Woods and Wilder (2005) Climate Change and Company Value. A Guide for Market Analysts, Total Environment Centre, November 2005 (Accessed at http://www.igcc.org.au/ProdImages/Climate_Risk_and_Company_Value.pdf on January 29, 2006)

WWF (World Wildlife Fund) (2004a) Climate Change Litigation. Summary of legal actions as of December 2004. (Accessed at http://www.climatelaw.org/cases/elaw/litigation.summary.doc on September 11, 2005)

WWF (World Wildlife Fund) (2004b) Australia Can't Ignore its Greenhouse Gas Emissions. Climate Change News. (Accessed at http://www.panda.org/about_wwf/what_we_do/climate_change/news/news.cfm?uNewsID=16330 on September 11, 2005, published on November 8, 2005)

Business risks and opportunities from climate change in large developing countries – a case study focusing on China

Xianli Zhu, Xiangyang Wu

Research Centre for Sustainable Development
Chinese Academy of Social Sciences
Jiangguomeiwai Dajie 5, Beijing 100732, China
xianl_@yahoo.com

Abstract

Most developing countries are of high vulnerability and lack the necessary institutional arrangements, financial instruments, expertise, and experience for risk management related to such complicated and expansive issues as climate change. The business risks and opportunities associated with climate change in large developing countries are different from those in the developed world. This paper will conduct a case study on China about the climate change-related business risks and opportunities.

Keywords: Business risk, opportunities, climate change, case study, China

1 Introduction

China is the most populous country in the world and a developing country in the sense of rapid economic development and social transition. It has a coal-dominated energy mix and among its fossil fuel reserves, 94% is coal; petroleum and natural gas together represent a share of 4% (Dadi Zhou 2003). As a result, China's production and consumption of coal is the highest in the world. The last few years of rapid energy demand and consumption increase have seen it surpass the United States to become the leading greenhouse gas (GHG) emitter on earth. Due to resource availability and technology constraints, in the next few decades, China will have to continue to rely on coal as its main source of energy. Such factors lead to relatively high GHG intensity in China and emission reduction difficulties.

In light of the rapidly developing economy and the transition from a highly centralized planned economy toward integration into the global market after WTO entry, China's laws, regulations, and policies are experiencing rapid changes. Especially the regulations on environmental protection and pollution control, and some measures, such as emission trading of SO_2, are still being implemented on a trial basis. One typical example is that the Law of Pollution Control was updated twice in the last 10 years, once in 1995 and again in 2000.

China is facing the double pressure of development and emission reduction. Under the current energy mix and technology constraints, China's economic development will lead to higher emissions and an increase in climate change risks. However, China lacks the technology, expertise, and capacity for disaster control, making it highly vulnerable to climate change.

2 Climate-change related business risks

2.1 Sources of climate change risks

High vulnerability to climate change impact

It has become common knowledge that the poor are likely to be hit hardest by climate change, and that capacity to respond to climate change is lowest in developing countries and among the poorest people in those countries. As a developing country, China is highly vulnerable to the negative impact of climate change.

Table 1. Key observed climate trends and variability in China

Change in Temperature	Warming during past 50 years, more pronounced in winter than summer, rate of increase more pronounced in minimum than in maximum temperature
Change in precipitation	Annual rainfall has decreased in past decade in North-east and North China, has increased in Western China, Changejiang River and along south-east coast
Extreme weather events	Significantly longer heat-wave duration, the frequency of occurrence of more intense rainfall events has increased, increasing frequency and intensity of droughts, and higher frequency and intensity of tropical cyclones originating in the Pacific

Source: IPCC 2007

Agriculture, the sector highly sensitive to climate change, still contributes 11.8% of the country's GDP and 62% of employment (China National Statistics Bureau, 2007), much higher than the share in developed countries. China is a country of water shortage and uneven water distribution around the year and among different regions. The per capita water resource possession is only one fourth of the world average. The distribution of precipitation among different regions is highly imbalanced, declining from the southeast coastal areas to the northwest inland. According to the Inter-governmental Panel on Climate Change, in China, a 30 cm sea-level rise would inundate 81,348 km^2 of coastal lowland (IPCC 2007). As indicated in Table 1, although only having less than one-tenth of the country's land, eastern China is home to more than one-third of the country's population and contributes more than half of the Chinese economy. Sea level rise and sea water intrusion due to climate change would post a major threat to the future of this region.

Table 2. The role of China's Eastern region in Chinese economy – 2005

	National total	East China	
		Absolute	% in National total
Area (million km^2)	9.6	0.9	9.5
Population (million)	1308.0	464.0	36.0
GDP (trillion RMB)	18.3	111.0	55.6

Source: China Statistical Yearbook 2006, China National Statistical Bureau, 2006

More than half of the Chinese territory is arid and semiarid land. The further decrease in precipitation, higher runoff, and evaporation will deteriorate the water shortage in China and impede economic and social development.

Highly diversified climate, typology, economic, and social conditions

China is a big country with highly diversified climate conditions. From south to north, the country covers the following zones: equatorial, tropical, subtropical, temperate, medium temperate and frigid temperate. Based on precipitation, the country could be divided into 4 zones from southeast to northwest: humid, semi-humid, semi-dry, and dry zones (www.china.org.cn). Combined with the complicated typology, the weather of different parts of the country at the same latitude is considerably diverse. Often some parts are suffering from floods, while other parts are facing severe droughts.

Furthermore, the economic development level and social conditions of different parts of the country are also considerably diverse. Generally, the eastern coastal areas are much more economically advanced and of higher population density; the western regions are less developed and have lower population density. This makes the form and degree of climate-related disasters and risks extremely varied in different regions. The lack of detailed data and the country's rapid economic development and social transition make risk assessment a big challenge.

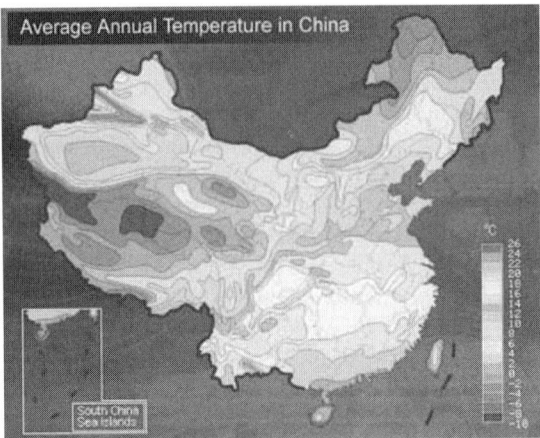

Fig. 1. Temperature in different parts of China
Source: www.china.org.cn

Due to all these factors, it is extremely difficult for an enterprise to assess the climate change related risks it is facing in China, making it difficult for enterprises to take the necessary business decisions.

Uncertainties about the future of the international climate change regime

Climate change is a global issue and definitely requires multilateral cooperation in order to address it. However, the Kyoto Protocol only defines the greenhouse emission obligations of industrialized countries during the 2008–12 period. The negotiations on post–2012 international climate change regime formally started at the end of 2005.

However, due to the disagreements among the major emitters, it is highly uncertain whether and when China will undertake some form of emission reduction commitments and how and when such emission commitments will influence different economic sectors and enterprises.

Lack of transparent, clear, and long-term legal and policy framework

In China, one source of climate-related risks is the lack of a transparent, clear, and long-term legal and policy framework. Business decisions need to be based on stable laws and regulations, as a project or enterprise often lasts for decades. However, in China, due to rapid development, laws, regulations, and policies are also in quick transition and change. For example, in 2001, there was no automobile emission standard, by 2004, the EU II standards were widely introduced, which will be replaced by the EU IV standards by 2008. By the end of 2006, the country's Constitution issued in 1982 had experienced 4 periods of revision. Between 1978 and 2005, the State Council submitted hundreds of bills to the National People's Congress, and more than 650 administrative regulations and laws have been formulated and are in force. In addition to frequent changes in government plans, policies, regulations, many policies lack transparency, contain vague stipulations, and are ineffectively enforced, bringing about diverse risks to related industries and enterprises (China Democratic Politics Construction White Paper, 2005).

Low capacity of mitigation and adaptation

Given the tremendous threat of climate change, the reactions include mitigation and adaptation. With its high vulnerability to the impact of climate change, China urgently needs to take a proactive attitude toward climate change adaptation. However, its mitigation and adaptation capacity is low due to lack of expertise, technology, resources and effective institutional arrangements. The vast population, diverse climate, topographic, hydrologic and resource conditions in different parts of China all require great flexibility in climate change adaptation. However, local communities, especially those poor ones located in remote and ecologically fragile regions, are most in need of adaptation but lack the means for adaptation.

Enterprises always operate with a certain economic, social, and cultural background, and the national and local circumstances inevitably exert influence on business operations. The high vulnerability and low adaptation capacity represent high business risks for enterprises.

2.2 Climate change perception of different economic actors

Although the overall trend of climate change in China is in line with international scientific reports and widely recognized by the public, the awareness among the public about the causes of climate change is not so high. However, as greenhouse gas emission is increasingly recognized as a kind of pollution, taking energy saving measures and being "green" will improve corporate image, and sharpen corporate competitive edge.

Central government: The Chinese central government, while admitting the seriousness of the climate change threat, still categorizes climate change as a low priority within the domestic policy hierarchy. Therefore, although many climate-friendly policies and measures are implemented in China, they are more often than not for energy considerations, instead of for protection of the global climate system.

Local governments: in the pursuit for local economic development, local governments tend to neglect or impede the enforcement of environmental protection and pollution control laws and regulations. Therefore, climate change, a global problem and with much lower policy priority in China than local environmental pollution, tends to be neglected in the administrative activities of local government agencies.

Enterprises: the business risks and opportunities associated with climate change are little recognized among enterprises, although many enterprises have been taking measures to increase their energy efficiency, which is more for economic considerations than for climate change control.

2.3 The specific forms of future possible climate impacts

The trend of climate change in China corresponds to the general trend of global warming. Several studies on the extreme climate events show that the extreme cold events are likely to decrease, while the extreme hot temperature events are likely to increase, and the droughts and floods are likely to be more severe and at higher frequency.

Climate change impact on natural systems

According to IPCC (IPCC 2007), the major impact of global warming on China will involve a large northward shift in the subtropical crop areas. Large increases in surface runoff, leading to soil erosion and degradation, frequently floods in the south, and spring droughts in the north ultimately will affect productivity. Permafrost in northeast China is expected to disappear if temperatures increase by 2°C or more. The northern part of China would be most vulnerable to the hydrological impact of climate change; future population growth and economic development here may seriously exacerbate the existing water shortage. The vegetation zones or climate zones would move to high latitudes or westwards, and there would be corresponding changes for scope, acreage, and demarcation lines of vegetation zones.

Deltaic coasts in China would face severe problems from sea-level rise. China is a country with long coastlines and the coastal areas are also the regions with the most advanced economy in China. Sea level increase will pose a high degree of losses and costs.

Agriculture and food security

The adverse impact of climate change would increase the costs of future agricultural production. Simulation indicates that potential food production would decrease by 10% due to climate change and extreme climate events during 2030–2050 under the present cropping system, present crop varieties, and present management levels. There would be an overall decreasing trend in the yield of China's main crops: wheat, rice, and maize. Currently, China's 1.3 billion population is still growing at a rate of around 10 million per year, and reductions in major grain yield will curtail the prospects of food security in China.

Human health

An increase in the frequency and duration of severe heat waves and humid conditions during the summer will increase the risk of mortality, principally in older age groups and urban poor populations of China. Heat-stress related chronic health ailments are also likely for physiological functions, metabolic processes, and immune systems. Adverse health impact also results from a build-up of high concentrations of air pollutants such as NO_2, ozone, and air-borne particulates in large urban areas. Moreover, the occurrence of vector-borne infectious diseases will rise. This will increase public medical-care costs, health-costs for families and enterprise losses in the form of productivity.

Human settlements, energy, and industry

Changes in climate patterns and worsening water shortages will affect human settlement, especially during rapid urbanization. At the current speed, it is estimated that in the next 20 years, more than 200 million rural residents will move into towns and cities. The availability of water is always a major factor influencing the establishment and development of cities and towns.

Extreme weather events will damage infrastructure facilities and disrupt energy supply. In all countries, the GHG mitigation burden is first shared by the energy sector and energy-intensive industries. These industries are faced with the double burdens of GHG emission mitigation and adaptation costs.

Insurance and other financial services

Climate change will increase the uncertainties in risk assessment, leading to greater pressure on the financial sector and mistakes in risk assessment, causing increases in risk remedy, slowing down the expansion of foreign financial services to China and growth of the Chinese financial sector.

In China, state-owned banks have a much higher non-performing loan ratio than foreign banks. China's WTO entry and the opening up of the financial sector bring about ever-fiercer competition to the Chinese banking sector. Climate change risks and impacts will complicate the assets management of banks and cause additional risk exposure to Chinese banks.

On the futures market, climate change impacts will influence the production and supply of agricultural products, which in turn, will lead to fluctuations in the prices of agricultural future products, including soy bean, cotton, corn etc. Moreover, the futures price of crude oil is also highly sensitive to climate.

For the insurance sector, the impact of climate change may be most severe and direct. The object of insurance coverage is generally low incidence events, but due to climate change, the frequency of "natural disasters" increases, which in turn leads to an increase in insurance pay-outs.

2.4 The availability of risk management tools

Since 1978, the Chinese insurance sector has maintained an average annual growth rate of 30%, far exceeding the country's GDP growth. The opening up of the Chinese insurance sector has made some progress. So far 41 foreign insurance companies have set up 97 operations, and 131 foreign institutions have established 191 representative offices in China (Guoliang Wang 2005).

However, the Chinese insurance sector is still in the fledging phase and is unable to meet the requirements of national economy and social development. Specifically, the insurance sector is small in size, with a tiny share in GDP. By the end of 2004, insurance penetration was around US$ 40 per capita and insurance revenue was about 3.4% of the country's GDP, still much lower than world average levels (Guoliang Wang 2005).

Statistics from the China International Decade for Natural Disaster Reduction (IDNDR) indicate that in recent years over 200 million people, annually for the most part, have been stricken by natural disasters, and that direct economic losses from natural disasters exceed RMB 200 billion, but in China, insurance pay-outs are about 1% of the disaster losses; in contrast, the ratio is 20% in Europe.

2.5 Corporate climate risk strategies

Companies and institutional investors alike face growing financial and legal risk that climate change will adversely affect the value of the assets for which they have fiduciary responsibility. To address the risk of climate change, company directors and institutional investors are tackling climate change risks in the following ways:

- Localize climate change risk management strategy: manage climate change risks based on local climate, typology, economic, and social circumstances
- Get sufficient expertise to make informed and responsible decisions regarding climate change.
- Make comprehensive assessments and regular updates of the company's current and probable risk exposure to the financial and competitive consequences of climate change.
- Develop, announce, and implement explicit strategies on climate change and integrate them into the company's overall business strategy.
- Link executive payment to the company's performance on climate change objectives.
- Develop and follow best practice standards for disclosing the level of climate change exposure to investors and other external and internal stakeholders.
- Incorporate climate change considerations into the overall investment strategy.
- Explore the commercial potential of new, "climate-friendly" investment products.

Climate change is a new challenge for the whole world. To manage climate risks, enterprises need to keep close track of the relevant research on climate change trends, impacts, risks, relevant government policies, as well as good practices adopted by enterprises of the same region and same industry.

3 Climate-change related business opportunities

3.1 Large potential market for energy efficiency and renewable management technologies

Under pressure from energy shortages, the public's environmental concerns, and international pressure regarding cross-border pollution from CO_2 and acid rain, the Chinese government is granting a high priority to energy-mix optimization and improvements in energy efficiency. In 2004, the Chinese government issued the 2020 Energy Conservation Program which stipulates that by 2010, China's energy intensity (based on fixed prices) shall decline 16% and that by 2020, the energy intensity level shall decline 43% from the 2002 level. In Feb. 2005, China issued a Renewable Energy Promotion Law. Furthermore, the 2020 Renewable Energy Development Program under review also set targets involving an increase in the share of commercial renewable energy in China's energy mix from the current approximately 8% to 10% by 2010, and to 15% by 2020. To realize the targets, some regulations and laws, policy support systems, supervisory and administrative systems, as well as technology service systems will be established and enforced, creating a huge market for energy efficiency and renewable development technologies and facilities.

3.2 International partnerships

Due to China's crucial role in the international co-efforts against climate change, in addition to efforts under the UNFCCC, some technology partnerships have been established for climate change, which will also provide attractive business opportunities in relevant fields.

The Asia-Pacific Partnership. In July 2005, six nations - the US, Australia, India, China, South Korea, and Japan formed the new "Asia Pacific Partnership for Clean Development and Climate (APP-CDC)". The 'vision' of this pact is to develop and implement new technologies that will allow the economies of these nations to grow, while mitigating the environmental degradation that has always accompanied such rapid economic growth. Central to this economic strategy are new technologies that will allow the exploitation of coal with relatively low emissions of pollutants, including greenhouse gases such as CO_2 (US Department of State 2005).

Partnership with the EU. Two months later, a similar technology cooperation partnership was formed between China and the EU. During the 8th China-EU Summit held in Beijing in September 2005, a Joint Declaration on Climate Change between China and the EU was issued in order to tackle the serious challenges of climate change through practical and results-oriented cooperation. This partnership will strengthen cooperation and dialogue on climate change, including clean energy, and promote sustainable development. It will include cooperation on the development, deployment and transfer of low carbon technology, including advanced near-zero-emissions coal technology through carbon capture and storage (EU External Relations, 2005).

3.3 CDM project opportunities

The World Bank estimates that China contains 50% of CDM project implementation potential in the world (World Bank, 2004). In the last two years, there have been significant improvements in the Chinese government's attitudes and policies on CDM. Initially, China took a prudent attitude toward CDM implementation. The domestic regulatory framework on CDM project implementation was not established until June 2004. Since then, with Kyoto Protocol breakthroughs taking effect and the demonstration effects of massive CDM implementation in other developing countries, the Chinese government agencies have sped up efforts of promoting CDM implementation in China. In Oct. 2005, the Chinese government replaced the Interim Measures on CDM Project Operation with a permanent one. Besides, the Chinese government has sped up the approval of CDM projects. As of 27 Aug 2007, 737 CDM projects from China have entered the international CDM pipeline, accounting for 31% of the world total. These projects will have accounted for 53.4% of the total expected CERs from all existing projects by the end of 2012 (UNEP Risoe Centre 2007).

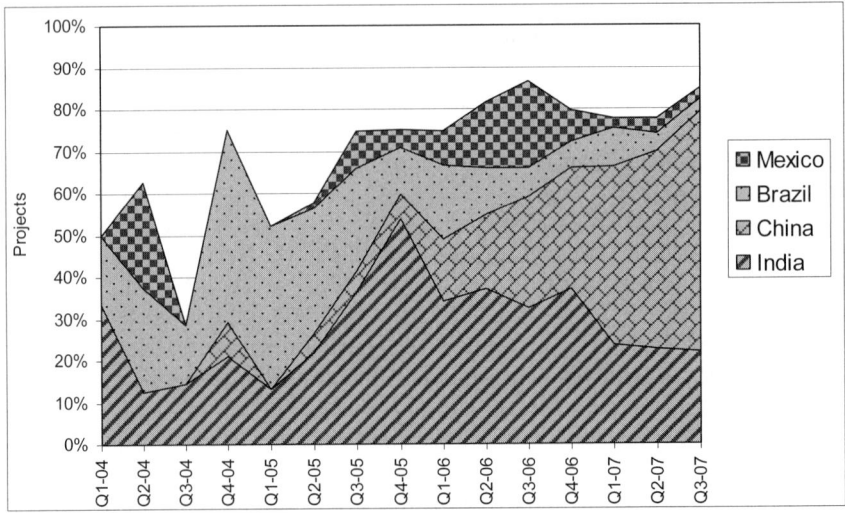

Fig. 2. CDM Projects for 4 major Hosting Countries – India, China, Brazil, and Mexico
Source: UNEP Risø CDM Pipeline dated 27–08–2007

China's national rules on CDM are unique in several aspects: 1) enterprise eligibility requirements. In most developing countries, there is no restriction on the participation of foreign-funded enterprises in CDM implementation. However, China stipulates that only Chinese-funded enterprises and Chinese holding enterprises (with 51% share of the equity) are eligible to implement CDM projects in China. 2) Government levies on the CER sales revenue from CDM projects: 65% for HFC and PFC emission reduction projects; 30% for N_2O emission reduction projects; and 2% for energy efficiency, renewable, and methane capture and utilization projects as well as forestation projects. 3) Moreover, CER sales price is one of the aspects subject to government approval, meaning that the government can reject a project simply on the basis that it finds the CER sales price 'unreasonably' low.

3.4 Business opportunities for insurance and other risk management services

With ever-increasing exposure to climate-change risk and constantly increasing costs of climate change impact, the demand is huge for insurance products to cover such risks from sectors sensitive to climate change risks such as farming, fishery, tourism and transportation, and this creates enormous business potential for insurance and other risk management services.

The Chinese government has set very ambitious targets for energy efficiency improvement and renewable energy development. According to the 2020 Energy Conservation Plan, China plans to reduce the energy intensity of its GDP by 50% during the 2003–2020 period, creating a huge market for energy efficiency improvement services and investments (NDRC 2005).

Besides, the country has committed itself to increasing its share of renewable energy in its total primary energy supply from the current 7% to 20% by 2020 (NDRC 2007). A series of preferential policies and measures, including government subsidies for bio-gas and higher grid-access tariffs for electricity from renewable sources, have been issued to encourage private investment in renewable energy. Based on the existing status and development trends of various technologies, the Chinese government estimates that to realize the target, 2 trillion RMB (around 200 billion Euros) of investment will be needed by the year 2020, creating a huge market for various renewable energy technology providers and equipment manufacturers (China Economic Daily, 5 Sept 2007).

4 Conclusions

Climate change presents both business risks and opportunities to the business sector. It is increasingly creating a wide range of impacts, which pose large business risks for certain sectors. In large developing countries like China, the risks are especially high due to the high vulnerability and lack of effective risk management expertise and tools.

However, as the most attractive forest direct investment (FDI) destination in the world, China also provides an excellent arena for some climate-change risk related business activities. The huge demand for energy efficiency improvement and renewable energy development technologies, the large potential and increasing government support for CDM project implementation, as well as the market niche for insurance products and other financial buffers against climate-change risks all provide lucrative opportunities for prepared enterprises.

References

China (2004) China Initial National Communication on Climate Change. China Plan Press, Beijing, pp 12–13
China Economic Daily (2007) 'China Will Enhance Investment in Renewable Energy'. 5 Sept 2007
China State Statistics Bureau (2007) 2006 China Statistics Communiqué. (in Chinese), http://www.stats.gov.cn
EU Commission (2005) Press Release, MEMO/05/298, EU and China Partnership on Climate Change, Brussels
IPCC (2007) Vulnerability, Impacts and Adaptation. Chapter 10 Asia, pp 472–475
NDRC (China National Development and Reform Commission) (2007) Outlines of China 2020 Energy Conservation Programme. www.ndrc.gov.cn
NDRC (China National Development and Reform Commission) (2007) Outlines of China 2020 Renewable Energy Development Programme, www.ndrc.gov.cn
UNDP (2005) Human Development Report, p 227
UNEP Risø Centre, CDM Pipeline dated 27–08–2007
US Department of State (2005) Vision Statement for a New Asia-Pacific Partnership on Clean Development and Climate
Wang GL (2005) Promote Commercial Insurance and Create a Harmony Society. (in Chinese), Shanghai Securities News, 25 Oct. 2005
World Bank (2004) Clean Development in China, p 108